Food Forensics

Stable Isotopes as a
Guide to Authenticity and Origin

Food Forensics
Stable Isotopes as a
Guide to Authenticity and Origin

Editors

James Francis Carter
Queensland Health Forensic and Scientific Services
Queensland Government
Coopers Plains, Qld, Australia

and

Lesley Ann Chesson
IsoForensics, Inc.
Salt Lake City, Utah, USA

CRC Press
Taylor & Francis Group
Boca Raton London New York

CRC Press is an imprint of the
Taylor & Francis Group, an **informa** business
A SCIENCE PUBLISHERS BOOK

CRC Press
Taylor & Francis Group
6000 Broken Sound Parkway NW, Suite 300
Boca Raton, FL 33487-2742

First issued in paperback 2021

© 2017 by Taylor & Francis Group, LLC
CRC Press is an imprint of Taylor & Francis Group, an Informa business

No claim to original U.S. Government works

Version Date: 20170302

ISBN-13: 978-0-367-78208-5 (pbk)
ISBN-13: 978-1-4987-4172-9 (hbk)

Library of Congress Cataloging-in-Publication Data

Names: Carter, James Francis, editor. | Chesson, Lesley Ann, editor.
Title: Food forensics : stable isotopes as a guide to authenticity and origin / editors, James Francis Carter, Queensland Health Forensic and Scientific Services, Queensland Government, Coopers Plains, Qld., Australia, and Lesley Ann Chesson, IsoForensics, Inc., Salt Lake City, Utah, USA.
Description: Boca Raton, FL : CRC Press, 2017. | "A Science Publishers book."
| Includes bibliographical references and index.
Identifiers: LCCN 2017004913| ISBN 9781498741729 (hardback) |
ISBN 9781498741736 (e-book)
Subjects: LCSH: Food--Analysis. | Food--Quality. | Stable isotopes. | Trace elements. | Food adulteration and inspection.
Classification: LCC TX545 .F554 2017 | DDC 664/.07--dc23
LC record available at https://lccn.loc.gov/2017004913

Visit the Taylor & Francis Web site at
http://www.taylorandfrancis.com

and the CRC Press Web site at
http://www.crcpress.com

Foreword

The horse meat scandal of 2013 refocused global attention on food authenticity and origin and the harm that food fraud inflicts. Consumers are duped into spending money that some can ill afford, trust is lost and business reputations are damaged. Food safety is also compromised, for example the deaths that have arisen from counterfeit alcohol products. The subsequent Review by Professor Elliott*, in which I was privileged to play a part, described how 'cutting corners' elides through food fraud into '*food crime*', an organised activity by groups which knowingly set out to deceive, and/or injure, those purchasing food. The UK Government accepted the Elliott recommendations, including setting up National Food Crime Units within the Food Standards Agency and Food Standards Scotland. These Units' 2016 baseline threat assessment was reassuring in that it found that organized crime groups have not made substantial in-roads into UK food and drink in the way they have in other countries. However, more worryingly, a small number of UK food businesses are believed to have links to organized crime.

Within the systems approach based on eight pillars of food integrity advocated by the Elliott Review laboratory services is a key component. Stable isotope analysis was one of the techniques Professor Elliott and I considered. We were anxious that the promise of this demanding technique, reaching maturity in its mainstream criminal forensic applications, should not be compromised by lack of forensic probity in its application to food.

Hence *Food forensics: stable isotopes as a guide to authenticity and origin* edited by Jim Carter and Lesley Chesson is timely and valuable. This volume, the first to focus exclusively on food authentication and geo-location by stable isotope analysis and allied techniques, provides a sound basis to their scientific practice with forensic rigor. The international contributors are all acknowledged experts in their fields.

The topics covered appropriately commence with an introduction to stable isotopes, reference materials, traceability, scientific good practice and reporting. Since they are essential to the proper forensic application of any technique it is pleasing to see chapters on sampling and sample preparation of foodstuffs and other forensic exhibits as well as quality assurance and control of data, data analysis and interpretation. The latter, drawing on international criminal forensic standards includes common statistical approaches for exploring food isotopic data alongside examples of investigating food authenticity and origin.

Well-established theories to explain variations in the natural abundances of hydrogen, carbon, and oxygen isotopes in plants are dealt with in a dedicated chapter providing the food scientist with sufficient information to predict variations in the isotopes of these elements in plant-based foods. Interpretations of variations in nitrogen, sulfur, and strontium isotope ratios of plant and plant-based foods are also discussed. There are separate chapters on flesh foods, fruits and vegetables, alcoholic beverages, dairy products, and edible oils. The chapter on food webs, the detailed ecological history of an organism and its relationship with other organisms on different time scales, shows how stable isotope analysis answers questions such as where was the animal living and what was it eating when its tissue was formed? While some techniques may be challenged in distinguishing conventional food from food described as 'organic' emerging compound-specific isotope techniques, in combination with other multi-marker strategies and multivariate statistical tests, are more promising, as described in a chapter dedicated to organic food. A final chapter deals with miscellaneous topics such as bottled water, carbonated soft drinks, caffeine, vanilla/vanillin, essential oils, sweeteners, honey, maple syrup and other food products as well as isotope effects during food preparation.

The passion of the editors and authors, who form a distinguished scientific community, contributes to the breadth of coverage of this book, the depth of each chapter, and its readability. Thus the book will serve as an informative reference text for new users as well as a reminder of best practice for seasoned professionals. I congratulate the editors and contributors for successfully compiling this excellent volume that describes the state of the art of a valuable scientific asset for the detection and prevention of food fraud and food crime.

Dr. Michael Walker MChemA, FIFST, FRSC

Former Public Analyst, now Referee Analyst in the Laboratory of the Government Chemist, UK and Subject Matter Expert to the Elliott Review.

* Elliot C. 2014. Elliot review into the integrity and assurance of food supply networks—Final report—A national food crime prevention framework. Edited by Food and Rural Affairs Department for Environment and Food Standards Agency: HM Government.

Michael Walker

Preface

Food forensics is a science that seeks to determine the authenticity and origins of (typically premium) foodstuffs using techniques more commonly applied in high profile criminal forensics. The science of food forensics can be traced back many centuries and has always used cutting-edge technology to try and remain "one step ahead" of the knowledgeable fraudster. Because sophisticated fraudsters can often, convincingly mimic the physical and chemical compositions of targeted foods, a current tool of choice in food forensics is stable isotope analysis.

The invisible "signature" or "fingerprint" inherent in the stable isotope ratios of bio-elements (hydrogen, carbon, nitrogen, oxygen, and sulfur) is often combined with elemental concentration data and the isotope ratios of trace radiogenic elements, such as strontium and lead. The complexity of this isotopic "signature" is impossible to detect without access to highly sophisticated analytical instrumentation and is extremely difficult to counterfeit without a detailed knowledge of numerous physical, chemical, and biological processes that can affect isotope abundances in foodstuffs. Food forensics is a science that draws on many, diverse disciplines, including plant and animal biology, hydrology, analytical chemistry, and statistics.

An understanding of plant and animal nutrition and metabolism is important to understanding and interpreting the ways in which stable isotopes are assimilated by plants and the animals consuming these plants. Through this understanding it is possible to interpret the isotopic "signature" and to infer the cultivation practices and origins of plants and the diets and origins of animals. This volume presents current, widely-accepted theories on the processes that impact food isotopic "signatures" and the interpretations of food source an investigator may be able to draw based on those theories.

A cornerstone of forensic science is that both raw analytical data, and any interpretation based on these data, will withstand scrutiny in a court of law. An understanding of analytical chemistry is also essential to generate "forensic" data, which must include an understanding of the need for traceability, quality control, and quality assurance. The analysis of large, complex datasets requires sophisticated statistical tools and when interpretations based on such data are to be presented to a court of law, it is essential that statistical treatments are both appropriate and valid. It is often desirable that large, complex datasets can be presented in a simple graphical form that can be presented to a layperson serving on a jury. As described throughout this volume, stable isotope analysis, along with allied techniques, supplies quantitative data that is well-suited for forensic interpretation, in both investigative and court settings.

This book draws together many diverse branches of science, and many internationally renowned scientists, to present *food forensics*: stable isotope analysis (and allied techniques) as a guide to the origin and authenticity of the most common (and not-so-common) foodstuffs.

James F. Carter
Lesley A. Chesson

Acknowledgements

On reflection, I spent too long writing the introduction to this volume—thinking it would be the chapter that most people would read, but writing this acknowledgement proved more difficult, principally because it marks the end of a long but ultimately very rewarding process. I am now faced with the questions of who to acknowledge, for what, and in what order.

There is no question that the first person I must thank is my beloved wife, Gabrielle, for her endless patience and encouragement (especially through some recent bad times) and for allowing me to spend most evenings of 2016 writing and editing this volume. I must also thank Caroline Vouvoulis for being the most wonderful step-daughter and for all the trust she puts in me.

Next (and again without question) I must thank Lesley for the boundless energy she expended on this project. Clearly this project sparked our unlikely mutual passions for food chemistry and stable isotopes and what began as a vague idea of "putting a book together with some friends" turned into something real. I must also thank Lesley for teaching me American English (of which "liters" was the most painful lesson) and for our regular exchanges of emails that kept me sane although some might argue otherwise.

I am, of course, indebted to all the contributors to this volume and I am honestly surprised that they responded to my initial e-mails, let alone spent time and effort to share their knowledge. Writing/editing this volume was a wonderful interaction between the various authors—everyone worked when their input was needed, everyone respected other's input, and no-one pushed themself to the front. Thanks to all of you.

I would very much like to acknowledge the UK Association of Public Analysts (APA) for the great insight I gained from attending their training sessions and summer schools. I don't think this volume would exist without the inspiration and knowledge I took away from those courses. It is wholly appropriate that the preface to this volume is written by one of the mentors for the APA.

Appropriately as a footnote, I would like to thank the Information Research Services (Library) at Queensland Health Forensic and Scientific Services for helping me cite some of the more unusual works on food authentication.

Dr. Jim Carter
(December 2016)

First, I need to extend a huge "thank you" to Jim Carter for the invitation to join him on this editing adventure. Having never edited a volume before, I was very glad to muddle through the process with Jim, whose logical and pragmatic approach and dry sense of humor about the missteps and setbacks we encountered helped the project go more smoothly than I could have ever imagined. I'd do it again, Jim… but perhaps not right away!

I was thrilled when I first learned of this volume, because it encompassed many of the things I'm passionate about—food and forensics and isotopes. Since finishing my master's degree in 2009, there had been relatively few opportunities for me to sit and think critically about these topics all at once. I enjoyed the chance to delve into the published literature and contribute to some chapters myself.

However, the truth is that this volume would simply not have been possible without the other chapter contributors. Several I knew personally and had worked with in the past; others I knew by professional reputation only. I feel privileged to have had the opportunity to work with them all. You would be hard-pressed to find a more experienced and better-qualified group of professionals to discuss the applications of stable isotope analysis to food forensics. It was an absolute honor to have them contribute works to this volume.

I am extremely appreciative of IsoForensics, Inc. and my colleagues/friends at the company. Thank you for your patience, support, and listening ears during the publication process. It meant a lot to me.

On the topic of "listening ears," I am profoundly grateful to my husband (Mike) and my parents (Boon and Mary) for their unflagging encouragement and enthusiasm—not just in the 3+ years it took to publish this volume, but for all the phases of my professional and personal life. Thanks for being there, always.

Finally, I want to thank the readers—I hope you will find this volume as interesting and valuable as I do. I look forward to hearing your thoughts. Thanks in advance for your positive response and feedback.

Lesley A. Chesson
IsoForensics, Inc.
(November 2016)

Contents

Isotope Ratio Measurements for Food Forensics

James F. Carter

1.1 Introduction

1.1.1 A bit of introduction...

"To such perfection of ingenuity has the system of counterfeiting and adulterating various commodities of life arrived in this country, that spurious articles are every where to be found in the market, made up so skilfully, as to elude the discrimination of the most experienced judges."

Frederick Accum

Although the words above were written in 1822, in the first edition of Accum's *Treatise on Adulterations of Food and Culinary Poisons* (Accum 1820), they provide a fitting opening to a modern text on food fraud. Indeed, even two centuries ago Accum went on to record *"... it is evident that this practice has been carried on for a long time"*. Today, bogus goods are purchased (both willingly and unknowingly) on a daily basis, ranging from fripperies such as designer jeans and DVDs to items that may, very likely, make a difference between life or death—pharmaceuticals, aircraft components, etc. Although the true extent of this market can never be known, various sources across the Internet estimate that somewhere in the region of 10% of all traded goods are dishonestly presented for sale in some way (UK Government Scientific Advisor 2015).

Forensic and Scientific Services, Health Support Queensland, 39 Kessels Road, Coopers Plains QLD 4108, AUSTRALIA.
Email: Jim.Carter@health.qld.gov.au

What Accum described as *nefarious practice* can be summarized as:

> *Counterfeiting*—the practice of manufacturing goods, typically of inferior quality, and selling them as a premium product without authorization.
>
> *Substitution*—the practice of presenting an inferior product as a premium product.
>
> *Adulteration*—the practice of diluting a premium product with an inferior product or with substances intended to mimic the product.

These terms are frequently used inter-changeably and may well be used as such in the following chapters. Often viewed as simply supplying a harmless "black" or "grey" market in which the none-too-fussy shopper can obtain seemingly "luxury goods" at "knock-down prices," the sale and purchase of counterfeit, substitute, or adulterated goods, especially foodstuffs, compromise both manufacturers and consumers through poor brand experience, loss of brand reputation, financial loss, and danger to health. The criminal nature of such practices gives rise to legislation intended for the general protection for purchasers of food:

> *If a person sells to the purchaser's prejudice any food which is not*
> *(a) of the nature, or (b) of the substance, or (c) of the quality,*
> *of the food demanded by the purchaser, he is guilty of an offence.*

The text above forms part of the UK Food Act 1984 but almost all countries of the world have similar legislation. In most cases legislation distinguishes between offences involving deception and those deemed to be more serious, such as the intentional or neglectful preparation and/or sale of injurious foods.

The manufacturers and retailers of many products, not just foodstuffs, now adopt a range of easily recognized devices to deter both potential counterfeiters and persons intent on deliberate and harmful contamination:

> *Sophisticated products*—products which are themselves difficult to copy.
>
> *Sophisticated packaging*—packaging with fancy designs incorporating devices such as holographic images and metallic strips.
>
> *Tamper evident packaging*—packaging which cannot be opened without leaving some obvious record.

Although such devices are primarily intended to reassure customers, the use of blatant product sophistication and garish designs can also serve to advertise a product—a practice far more common in designer goods such as sports shoes (trainers or sneakers) than foodstuffs. Like any new technology, anti-counterfeiting measures will be expensive to implement but difficult to copy. Unfortunately, as technology becomes readily available, the price inevitably falls and counterfeiters are able to adopt the same techniques. In an effort to stay at least one step ahead of ever-sophisticated counterfeiters, food scientists are looking to intrinsic and difficult-to-copy properties of foods for authentication.

Many global organisations, national governments, and manufacturing associations have defined the physical and chemical properties of foodstuffs, by means of legislation, standards, and codes-of-practice. Although the primary function of these regulations is to ensure that food is safe and wholesome through defined composition, they can also provide a means to guarantee the authenticity of the product including the Country of Origin (CoO). Two properties of special interest for foodstuffs are the stable isotopic and elemental compositions of natural and manufactured products, which are controlled by a large number of parameters:

> *geography and climate*—will affect rainfall and temperature and hence the localized hydrogen and oxygen isotopic composition of water,
> *underlying geology* and *agricultural practices*—will affect nutrient availability from soil (nitrogen, micro- and macro-elements) to both plants and animals, and
> *manufacturing processes*—may introduce or remove components of a food and/or change the physical or chemical nature of some or all of these components.

The combined effects of these, and many other possible processes, gives rise to a *signature, fingerprint, footprint* or *profile* for a food product, which can be highly characteristic and which has been compared to the specificity of DNA analysis (Meier-Augenstein 2006). Stable isotope analysis, often combined with trace elemental profiling and chemical composition measurements, is at the forefront of *food forensics*, a discipline that seeks to use modern analytical techniques and sophisticated statistical tools to establish the authenticity and origin of foods.

1.1.2 A bit of history...

In 1913, Frederick Soddy, working at the University of Glasgow, discovered that the radioactive decay of elements such as thorium and uranium produced substances that exhibited chemical properties identical to those of known elements, such as lead. To account for these findings Soddy used the word *isotope* (meaning "same place") to explain what appeared to be different chemical elements occupying the *same place* in the periodic table. The origin of the word *isotope* is often attributed to a family friend of Soddy, Margaret Todd.

During the same rich period of experimental physics, both J. J. Thompson and Francis Aston, working at the Cavendish laboratories in Cambridge, observed the existence of naturally occurring isotopes of neon, using instruments which were soon to be known as *mass spectrometers*.

The combined importance of isotopes and mass spectrometers to modern science was acknowledged by Thompson's Nobel Prize for physics in 1906, Soddy's Prize for chemistry in 1921 and Aston's Prize for chemistry in 1922.

1.1.3 A bit of theory...

Atomic nuclei (*nuclides*) are characterized by:

> *atomic number Z*—the number of positively charged protons and,
>
> *atomic mass number A*—the total number of nucleons (protons + neutrons).

The information needed to describe an isotope is easily (and universally) summarized by the combination of the chemical symbol (equivalent to the atomic number) and the atomic mass number in the form; ^{11}B, ^{12}C, ^{20}Ne, ^{56}Fe, ^{208}Pb, etc. Every known element exists as a number of isotopes, both naturally occurring and man-made atoms of the same element with varying numbers of neutrons such as ^{10}B, ^{13}C, ^{22}Ne, ^{54}Fe, ^{206}Pb, etc. Importantly, changes in the nucleus have a subtle effect on the chemistry of the orbiting electrons, which, in part, explains why analysis of isotopic composition can be a powerful tool to study the origin and fate of many materials— sometimes termed *isotope chemistry.*

Isotopes are classed as either *stable isotopes* or *radio-nuclides* (note that a greater or lesser number of neutrons does not necessarily make a nucleus more or less stable). The distinction is that stable isotopes endure over geological time scales, whereas radio-nuclides have characteristic half-lives $(t_{1/2})$ and spontaneously transform into stable isotopes through the emission of radioactive particles. The stable products of radioactive decay, such as ^{87}Sr, ^{206}Pb, and ^{208}Pb, are often described as *radiogenic* isotopes and were the original subject of Soddy's experiments.

To demonstrate some of these ideas, consider carbon, which exists as 15 known isotopic forms but predominantly as two stable isotopes: ^{12}C and ^{13}C, with relative abundances of approximately 98.9 and 1.1%. The ratio of these isotopes has not changed since the Earth was formed, although localized variations occur as a result of numerous physical, chemical, and biological processes. To record a stable isotopic *fingerprint* simply requires a very precise measurement of the $^{13}C/^{12}C$ ratio of a sample. In contrast, the next most abundant form of carbon, ^{14}C, is continuously formed in the Earth's upper atmosphere in very small amounts— $^{14}C/(^{12}C + ^{13}C)$ approximately 10^{-12}. All three forms of carbon are assimilated into plants, animals, and minerals to become part of the carbon *isotopic signature.* The ratio of stable isotopes $^{13}C/^{12}C$ is largely controlled by plant metabolism and/or animal diet and is frequently used to interpret these conditions. The radioactive form of carbon decays with a predictable half-life ($t_{1/2}$ = 5,730 yrs) and the residual ^{14}C activity provides a useful means by which to estimate the age of materials over an approximate 1,000 to 10,000 years (from present).

As an aside: the 1950s and 1960s witnessed a significant increase in the concentration of atmospheric ^{14}C as a result of nuclear weapons testing—the *bomb-pulse*. The residual activity from these tests can be used to date biological materials post the mid-1950s with a reported ± 2-year resolution. Both conventional *radiocarbon* and *bomb-pulse* carbon dating have found forensic applications (Tuniz et al. 2004); for example, *bomb-pulse* carbon dating (Asenstorfer et al. 2011) and ^{137}Cs decay (Hubert et al. 2009) have been applied to authenticate vintage wines.

1.2 Reporting and reference materials

1.2.1 Reporting isotopic composition

Isotopic compositions can be expressed as either absolute or relative measurements, appropriate to particular applications. The following text strives to conform to the recommendations of the Commission on Isotopic Abundances and Atomic Weights (CIAAW) of the International Union of Pure and Applied Chemistry (IUPAC). For a comprehensive explanation of this sometimes bewildering nomenclature the reader is referred to Ty Coplen's comprehensive text on the subject (Coplen 2011).

When isotopically labelled materials are used, for example as tracers in metabolic studies, absolute values are determined; the *isotopic* or *radiochemical purity* is the fraction of the label present in a specified chemical form, the position within a molecule, or even the chiral form of the compound. *Radionuclidic purity* is the fraction of the total radioactivity present as a specified radionuclide and conveys nothing about the chemical form. These macroscopic measurements are expressed as the *atom fraction* (x), the amount (n) of a specified isotope divided by the total amount of atoms of the element within the sample. For an element with two major isotopes, such as carbon, the atom fraction in a sample is expressed by Equation 1.1.

$$x(^{13}C) = \left(\frac{n(^{13}C)}{n(^{12}C) + n(^{13}C)} \right)$$

Equation 1.1

Atom fraction is a dimensionless quantity but mmol/mol, μmol/mol and similar units are acceptable to most peer reviewed journals. Use of the term *atom percent* (atom%) is no longer recommended.

The small variations in the natural abundances of stable isotopes of the bio-elements (H, C, N, O, and S), which are of interest in food forensics, are almost universally expressed as relative measurements. The important reason for this is that the precision with which differences can be measured will always better that the accuracy with which absolute values can be measured (Hayes 2001). The *isotope ratio* (R) of a sample is the ratio of the number of atoms of one (heavier) isotope to the number of atoms of another (lighter) isotope of the same chemical element, Equation 1.2:

$$R(^{13}C/^{12}C) = \frac{N^{13}C}{N^{12}C}$$

Equation 1.2

In Equations 1.1 and 1.2, the symbols n and N are used respectively to denote the amount of a given isotope on macroscopic and microscopic scales.

The widely accepted form of reporting isotopic composition (shown in Equation 1.3) is known as *delta-notation* and dates from 1948 when the journal *Science* reported a lecture by Harold Urey in which he expressed the oxygen isotopic composition of various marine carbonates relative to the shell of a sea snail (Urey 1948). Relative isotope-ratio differences, expressed as *delta*-values, δ-values or simply δ can be the difference between any two substances, but are almost universally reported relative to an international reference material (RM). For example, the δ-value for the carbon isotopic composition of a sample (*samp*) is defined by:

$$\delta^{13}C_{RM} = \left(\frac{R(^{13}C/^{12}C)_{samp} - R(^{13}C/^{12}C)_{RM}}{R(^{13}C/^{12}C)_{RM}} \right)$$

Equation 1.3

The first published form of Equation 1.3 appeared discretely in a 1950 paper, which focused on electronic circuits for a new design of stable isotope ratio mass spectrometer (McKinney et al. 1950). The variations in isotopic abundances for most commonly studied chemical elements are typically small, and values are reported in part per hundred (% or percent), part per thousand (‰ or per mil) or part per million (ppm or permeg). It is no longer considered appropriate to include these fractions in Equation 1.3. Because of Urey's role in the genesis of δ-notation it has been proposed that the SI units of relative isotopic ratio be named the *Urey* (Brand and Coplen 2012), with the more common per mil replaced by milli-Urey (mUr).

In food forensics, and many other fields, it is often important to study the differences between the isotopic compositions of the various components of a system, for example between the carbon isotopic composition of a plant and the soil in which it grows. This can be conveniently expressed in Δ-notation. For the example above;

$$\Delta^{13}C_{plant/soil} = \delta^{13}C_{plant} - \delta^{13}C_{soil}$$

Equation 1.4

Note that the delta symbol, both upper and lower case, is italicized. Also, since δ-values are difference measurements, authors are encouraged to prefix values greater than zero with the + symbol.

1.2.2 Traceability

The *take home message* from the previous section is that stable isotope abundance measurements are reported relative to reference materials (RMs) and for that reason the correct selection and use of RMs is the cornerstone of measurements that are both repeatable (within a laboratory) and reproducible (between laboratories).

   All measurements reported in SI units should be traceable to a unique reference, which is accepted throughout the world. Most SI unit are now based on quantum phenomenon, although some are still based on artefacts such as the International Prototype of the Kilogram (IPK), a platinum/iridium cylinder, which defines at present the SI unit of mass. In the case of stable isotope abundance measurements, the *primary reference materials* fall very much under the heading of *artefact*. These materials define the zero point on the respective δ-scale—mostly!

   The concept of traceability depends on a chain of measurements linked back to the appropriate international primary RM (artefact or quantum phenomenon) through a series of calibrations—i.e., comparisons between two materials in the chain. Provided the uncertainties of the comparisons are known, results obtained through calibration against one of these RMs will itself be traceable to the agreed reference. Table 1.1 lists the internationally agreed reporting scales and primary RMs for the five major elements in the biosphere. The RMs used for stable isotope ratio measurements can be confusing and the materials shown in Table 1.1 deserve some explanation.

Table 1.1 The primary reference materials for reporting isotope ratio measurements.

Element	Ratio	Reporting scale	Artefact	Current RM
Hydrogen	$^2H/^1H$	Vienna Standard Mean Ocean Water (VSMOW)	SMOW/VSMOW water	VSMOW2 water, SLAP2 water
Carbon	$^{13}C/^{12}C$	Vienna PeeDee Belemnite (VPDB)	PDB PeeDee Belemnite	NBS-19 calcium carbonate, LSVEC lithium carbonate
Nitrogen	$^{15}N/^{14}N$	Atmospheric nitrogen (N_{2AIR})	Air	Air
Oxygen	$^{18}O/^{16}O$	Vienna Standard Mean Ocean Water (VSMOW)	SMOW/VSMOW water	VSMOW2 water, SLAP2 water
Sulfur	$^{34}S/^{32}S$	Vienna Canyon Diablo Troilite (VCDT)	CDT Canyon Diablo Troilite	IAEA-S-1 silver sulfide

1.2.3 VSMOW

VSMOW was a physical sample of water prepared by the International Atomic Energy Agency (IAEA), which defined the zero point of the δ-scales for both hydrogen and oxygen measurements with no associated uncertainty. VSMOW replaced an earlier RM, SMOW, which was both a physical sample of water and a defined isotopic composition. The prefix "V" was added to avoid further confusion. The original supply of VSMOW has now been exhausted and since 2009 has been replaced by VSMOW2, also with a value of 0.0‰. Because VSMOW2 has been calibrated against the original VSMOW, there is uncertainty associated with the composition, albeit very small ($\delta^2H \pm 0.3‰ / \delta^{18}O \pm 0.02‰$) and this uncertainty is likely to become smaller with subsequent use. The hydrogen and oxygen δ-scales have been further anchored with a second sample of water, Standard Light Antarctic Precipitation (SLAP2), such that $\delta^2H_{VSMOW-SLAP}$ of SLAP2 = −427.5‰ and $\delta^{18}O_{VSMOW-SLAP}$ of SLAP2 = −55.50‰, with the uncertainties given above.

1.2.4 VPDB

Like VSMOW, VPDB began as a physical material (PDB), a sample of calcium carbonate (limestone) derived from a belemnite fossil from the Pee Dee region of South Carolina, USA (hence PDB). This material was very quickly exhausted and was replaced by a defined value (Coplen 1994). The carbon isotopic composition of a material with $\delta^{13}C = 0‰$ relative to VPDB is:

$^{12}C = 0.988\ 944(28)$

$^{13}C = 0.011\ 056(28)$

The values in brackets indicate the uncertainty in the final decimal places despite this being a defined value. Recognizing the limited practical value of this definition, the IAEA also defined the scale relative to another calcium carbonate material NBS-19 such that: $\delta^{13}C_{VPDB}$ of NBS-19 = +1.95‰ exactly.

1.2.5 N_{2AIR}

In many ways atmospheric nitrogen is an ideal RM for nitrogen isotopic composition as it is freely available and shows negligible global variation (Mariotti 1983; Mariotti 1984). The main disadvantage of atmospheric nitrogen as a RM is the difficulty in comparing a gaseous material with nitrogen evolved from solid samples. For convenience several solid reference materials (mostly ammonium and nitrate salts) have been prepared with δ-values ranging from approximately +180 to −30‰ *vs.* N_{2AIR}. Unfortunately, the assigned δ-values for all of these materials have relatively large uncertainties, typically ± 0.2‰.

1.2.6 VCDT

Canyon Diablo Troilite (CDT) was developed from meteorite fragments from the Barringer meteor crater (Canyon Diablo, Arizona, USA), which contained a number of minerals including troilite, a form of iron sulfide. For any number of reasons choosing a Martian meteorite as the primary standard for terrestrial sulfur isotope abundance was not ideal but ultimately it was a lack of homogeneity ($>0.4‰$) that lead to the replacement of this material with a defined value (VCDT) by assigning IAEA-S-1 silver sulfide reference material $\delta^{34}S_{VCDT} = -0.3‰$ exactly (Krouse and Coplen 1997).

1.2.7 Practical reference materials

In practice a laboratory will rarely use primary RMs to perform routine stable isotope ratio measurements. The possible exceptions are hydrogen and oxygen stable isotope measurements of water samples for which VSMOW2 and SLAP2 are readily available and relatively inexpensive. Other primary material may be incompatible with the sample introduction system or simply unavailable. It is, however, vital that a laboratory obtains a range of primary and/or secondary RMs as the starting point for inter-laboratory comparability. Table 1.2 shows a typical hierarchy of RMs that should be available in a stable isotope laboratory.

Primary and secondary RMs will principally be used to calibrate a range of in-house RMs, which should be a close physical and chemical match for

Table 1.2 Typical hierarchy of reference material used for stable isotope ratio measurements.

RM Type	Description	Example	Comments	Uncertainty (1 standard deviation)
Primary	Physical materials which define the international δ-scales	NBS-19 IAEA-S-1	Limited supply—0.5 g available for purchase every 3 years; Impractical for day-to-day use	None
Secondary	Physical materials with internationally agreed δ-values and uncertainty	USGS-40 L-glutamic acid, IAEA-N-1 ammonium nitrate, IAEA-CH-7 polyethylene	Limited supply; Limited range of materials and δ-values	$<0.05‰$ $\delta^{13}C$ $\sim0.2‰$ $\delta^{15}N$ $\sim2‰$ $\delta^{2}H$
In-house	Materials calibrated against Primary and/or Secondary RMs		Used on a daily basis for QA and QC; Chemically and physically similar to typical laboratory samples	Ideally $<0.15‰$ $\delta^{13}C$, $\delta^{15}N$ and $\delta^{18}O$ $<2.0‰$ $\delta^{2}H$
Working gas	Gas from high pressure cylinders		Calibrated daily against in-house RMs	

samples that are routinely analyzed at the facility. These in-house RMs will be used on a day-to-day basis for calibration and quality control (QC) and Quality Assurance (QA) (Carter and Fry 2013a). Individual samples are typically measured relative to laboratory working gases delivered from high pressure cylinders. The isotopic composition of these gases will change over time and must be calibrated against in-house RMs, ideally on a daily basis.

This chain of calibration will ensure that laboratory measurements are traceable to the primary RMs and, hence, to the international isotope reporting scales. Each stage of calibration will necessarily introduce additional uncertainty into the final result but it is possible to take steps to minimize the overall uncertainty:

- select RMs bracketing the range of possible δ-values expected;
- select RMs characterized by low uncertainty; and
- analyze RMs multiple times.

The laboratory must also periodically analyze primary and/or secondary RMs as a form of in-house proficiency test to ensure that the calibrations obtained from in-house materials are both accurate and repeatable.

1.3 Isotope ratio measurements

1.3.1 Data normalization

The aim of traceability is to ensure inter-laboratory comparability. To achieve this measured data must be *normalized* (sometimes described as *stretched* and *shifted*) to the appropriate international δ-scale. All of the numerical values that follow are taken from the CIAAW website (http://ciaaw.org, accessed December 2015).

1.3.2 Hydrogen measurements

The process of normalization for hydrogen isotopic measurements is illustrated in Figure 1.1 based on the 1993 recommendation of the CIAAW (Coplen 1994):

> "δ^2H values of all hydrogen-bearing materials be measured and expressed relative to VSMOW reference water on a scale normalized by assigning consensus values of −428‰ to SLAP reference water, and authors should clearly state so in their reports."

As noted above the supplies of both VSMOW and SLAP have been exhausted and replaced by two closely matched water samples christened VSMOW2 and SLAP2. These materials are ideal for the measurement of water samples as the isotopic compositions are widely spaced and encompass virtually all natural waters, with the possible exception of

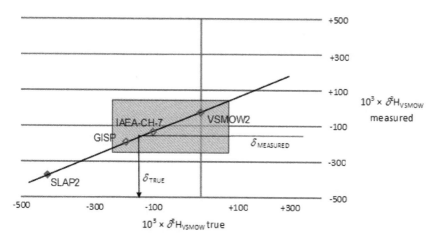

Figure 1.1 Normalization of δ^2H data to the internationally accepted reference materials VSMOW2 and SLAP2. GISP water or IAEA-CH-7 polyethylene can be used as a third calibration point and/or act as QC material. The shaded area shows the typical δ^2H composition of organic materials.

some tropical waters. An additional reference water, Greenland Ice Sheet Precipitation (GISP) has an isotopic composition conveniently between VSMOW2 and SLAP2 and can serve as a third calibration point and/or a QC material. A simple plot of the consensus *vs.* measured isotopic composition (Figure 1.1) then provides the means to normalize measured data to the international hydrogen δ-scale.

Measurements of hydrogen isotopic composition have become increasingly important in determining CoO and although VSMOW2 and SLAP2 are ideal RMs for the measurement of water samples, they are less than ideal for the measurement of solid organic materials. First, liquid samples are not readily compatible with the auto-samplers and inlet systems used to analyze solid materials. Second, the isotopic compositions of the reference waters do not extend to the more enriched values of some organic materials. The shaded area in Figure 1.1 shows the typical range of δ^2H compositions of organic materials.

Going some way to solve this problem, the United States Geological Survey (USGS) now supplies water RMs sealed in silver capsules (typically 0.15 or 0.25 µL) to facilitate introduction via a normal auto-sampler (Qi et al. 2010). A number of solid organic materials are also available (e.g., polyethylene foil, oil, and hair) with δ^2H values ranging from approximately −120 to −50‰. These materials provide options for additional calibration points and/or QC materials. The latter point is important to ensure that the inlet behaves the same towards silver encapsulated water and organic materials. Any off-set from the calibration line in Figure 1.1

will suggest that the inlet system does not behave the same towards water and organic samples.

RM suppliers are constantly working to develop new and replacement RMs to meet the increasing range of applications of stable isotope analysis techniques. The reader is strongly recommended to visit the websites of IAEA and USGS to check the availability of reference materials and the currently accepted δ-values and uncertainties.

1.3.3 Oxygen measurements

The normalization of oxygen measurements is very much akin to hydrogen with a CIAAW recommendation (Coplen 1994):

"$\delta^{18}O$ values of all oxygen-bearing materials be expressed relative to VSMOW such that $\delta^{18}O_{SLAP/VSMOW} = -55.5\%$."

The normalization principles shown in Figure 1.1 also apply to $\delta^{18}O$; both liquid and silver encapsulated SLAP2, GISP, and VSMOW2 provide the primary anchors with a range of solid materials available for calibration and QC. A number of solid RMs have significantly enriched isotopic compositions, which provide a convenient means to extend the normalization to positive $\delta^{18}O$ values—e.g., benzoic acids IAEA-601 (+23.1‰) and IAEA-602 (+71.3‰).

Although it is unlikely to be relevant to the field of food forensics, it should be noted that separate reporting scales exist for carbonates ($\delta^{18}O_{VPDB}$) and gaseous oxygen ($\delta^{18}O_{AIR-O2}$). The relationships between these scales are well established but not permanently defined and so, for simplicity, the isotopic compositions of oxygen bearing materials should be reported relative to VSMOW.

1.3.4 Carbon measurements

As noted above, the VPDB scale has been defined relative to NBS-19 limestone. To improve both the accuracy and precision of measurements, in 2005 the CIAAW recommended that (Coplen et al. 2006):

"$\delta^{13}C$ values of all carbon-bearing materials be measured and expressed relative to the VPDB on a scale normalized by assigning consensus values of –46.6‰ to LSVEC lithium carbonate and +1.95‰ to NBS 19 calcium carbonate, and authors should clearly state so in their reports."

The principle of normalization for carbon isotopic compositions is, therefore, the same as applied to hydrogen and oxygen measurements. When the $\delta^{13}C_{VPDB}$ scale was redefined by reference to NBS-19 and LSVEC, both the assigned values and associated uncertainties of many of the secondary reference materials were changed (Coplen et al. 2006). In general, this was a good thing, as uncertainties of many RMs improved; however, when comparing measurements reported prior to 2005, it may be necessary to re-scale the data.

In some respects NBS-19 and LSVEC are ideal RMs for the normalization of carbon isotopic ratios as they have widely different isotopic compositions, which span the likely composition range for organic compounds (Figure 1.2). Unfortunately, these RMs also come with a number of limitations. First, these materials are not readily available—supplies were always restricted, for conservation, but both NBS-19 and LSVEC are currently *out-of-stock* and no replacements are currently available. Second, the VPDB scale is defined by the carbon dioxide liberated from NBS-19 limestone by treatment with phosphoric acid. In contrast, the most common IRMS sample preparation/inlet device is an Elemental Analyzer (EA) in which a sample is combusted in the presence of oxygen; trying to convert $[CO_3]^{2-}$ to CO_2 in an oxidizing environment is not a favorable reaction.

To overcome the availability and/or suitability for the primary RMs, a laboratory has a number of options:

- normalize measured data using secondary RMs [e.g., USGS-40 ($\delta^{13}C$ = −26.39‰) and USGS-41 ($\delta^{13}C$ = +37.6‰)]
- prepare in-house RMs to closely match the isotopic composition of the primary RMs (Carter and Fry 2013b).

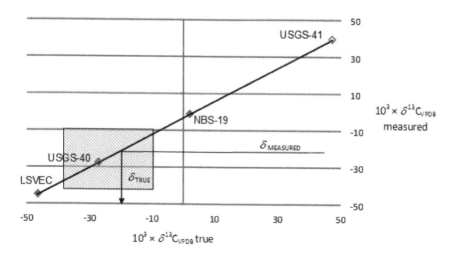

Figure 1.2 Normalization of $\delta^{13}C$ data to the internationally accepted reference materials NBS-19 and LSVEC. USGS-40 and USGS-41 provide alternative calibration points and/or act as QC material. The shaded area shows the typical $\delta^{13}C$ composition of organic materials.

The plus point of both these options is replacing carbonates with organic compounds, which can be a close matrix match for typical laboratory samples and will be far more readily converted to CO_2 during combustion in an EA. As a negative point, the overall uncertainty budget for measurements must increase due to the additional links in the chain of calibration; however, with careful selection of RMs, this increase in uncertainty can be kept to a minimum.

1.3.5 Nitrogen and sulfur measurements

Both the $\delta^{15}N_{AIR}$ and $\delta^{34}S_{VCDT}$ scales are still defined relative to a single primary RM in a similar manner to the original definitions of VPDB:

a material with $\delta^{15}N = 0‰$ relative to N_{AIR}

$^{14}N = 0.996\ 337(4)$
$^{15}N = 0.003\ 663(4)$

IAEA-S-1 silver sulfide = $-0.3‰$ exactly.

No doubt, at some time, these scales will be formally defined by reference to two primary RMs. In the interim, a number of secondary RMs exist that meet some of the basic requirements for RMs including widely spaced δ-values, bracketing likely sample values. Some examples are given in Table 1.3.

Table 1.3 Secondary reference materials suitable for the normalization of nitrogen and sulfur stable isotope measurements.

Material	Nature	Value (‰)	Unit	SD
USGS25	ammonium sulfate	−30.3	$\delta^{15}N_{AIR}$	0.4
IAEA-N-1	ammonium sulfate	+0.4	$\delta^{15}N_{AIR}$	0.2
IAEA-N-2	ammonium sulfate	+20.3	$\delta^{15}N_{AIR}$	0.2
IAEA-S-2	silver sulfide	+22.6	$\delta^{34}S_{VCDT}$	0.16
IAEA-S-3	silver sulfide	−32.5	$\delta^{34}S_{VCDT}$	0.16

1.4 Good practice

1.4.1 Good practice in practice

The Forensic Isotope Ratio Mass Spectrometry (FIRMS) network, in conjunction with the UK National Measurement facility, has produced an excellent Good Practice Guide (GPG) to stable isotope measurements. This is available to download for free (forensic-isotopes.org/assets/IRMS%20 Guide%20Finalv3.1_Web.pdf) and the reader is strongly encouraged to obtain a copy and to absorb its contents.

Always remember that the need for good practice and quality control in food forensics is not limited to stable isotope measurements and must extend to all physical and chemical measurements that are applied. It is not the intention of this chapter to repeat the advice of the FIRMS' GPG but simply provide a summary of what is necessary for a laboratory to obtain data that are:

- repeatable over a long time period within the same laboratory
- comparable to data from other laboratories

Typically, food forensics is a comparative technique and the strength of evidence that can be presented is very much dependent on the size (and quality) of any background database. Both of the requirements above must be met if data are to be reliable in building reference datasets (or *databases*).

The best way to achieve reliable data can be summarized in two points:

- the Principle of Identical Treatment (PIT)
- traceability to the international δ-scales

The PIT simply requires that samples and reference materials are treated in an identical manner and should be applied to all physical and chemical treatments of the sample. Also (and most importantly) the inlet of the instrument (typically an EA and open-split arrangement) must behave in an identical manner to both samples and standards. In general this requires that the elements in a sample are converted quantitatively to CO_2 and N_2 (for measurement of carbon and nitrogen isotope abundances, respectively), H_2 and CO (for measurement of hydrogen and oxygen isotope abundances, respectively), or SO_2 (for measurement of sulfur isotope abundances) and that these gases are transported to the ion source of the instrument without isotopic fractionation (or at least with consistent fractionation).

Once samples have been prepared and measured according to the PIT it is important to ensure that identical treatment is applied to all subsequent processing of the data; sometimes described as the Principle of Identical Correction (PIC) (Carter and Fry 2013a).

1.4.2 On-board (hidden) corrections

Instrument software has become increasingly sophisticated and allows an analyst to apply a number of corrections that were previously performed in external spreadsheets or Laboratory Information Management Systems (LIMS). Most modern IRMS instrument software will automatically apply a number of corrections to the measured data before presenting results to the analyst. Many of these are discussed in the FIRMS' GPG, but in summary, data will be "normalized" to a single calibration point (often a working gas) (Paul et al. 2007) and, for certain gases, a correction is applied for $^{16}O/^{17}O/^{18}O$ ratios (Brand et al. 2010). The software may give an analyst various options for these corrections but so long as they are applied evenhandedly to samples and reference materials, results should be consistent.

1.4.3 Optional corrections

In addition to instrument-based corrections and the essential normalization to international δ-scales, a number of other corrections are sometimes applied to isotope ratio measurement results. When establishing analytical protocols a laboratory must determine whether these corrections are necessary or useful. Applying too many "corrections" to data can introduce as many problems as are solved. As a heuristic technique, a laboratory should only apply corrections that have a significant and positive effect or they simply become *fudge-factors*!

1.4.4 Blank correction

The purpose of a *blank correction* is to subtract the contribution of any gases evolved during a blank measurement (i.e., an empty tin or silver capsule) and for a number of reasons this correction is best avoided. *Blank correction* only becomes important when sample size is limited or the abundance of a certain element in the matrix is very low. Typically in food forensics sample sizes are large, given that only a few hundred micrograms of any element is required for an analysis. Homogeneity is much more of a problem as discussed in Chapter 2.

   In addition, the apparent isotopic composition of small peaks can vary wildly with varying integration parameters and determining the true isotopic composition of an instrument blank is not as simple as measuring the response from a few empty capsules and subtracting the value. The true isotopic composition of the blank must be determined by analyzing a series of dilutions for two reference materials with different isotopic compositions (Carter and Fry 2013a).

   A much more robust practice is to establish the acceptable size of a blank analysis; if a blank is less than 1% of a normal sample peak height, it can have little effect on the measured isotopic composition. Once suitable acceptance criteria have been established, it is important to measure the size of the blank before any sample measurements. A high blank is indicative of contamination—capsules, forceps, preparation area, etc. When a high blank is identified the important action is to identify and address the cause and not simply correct for it.

1.4.5 Size correction

Modern IRMS instruments have detectors and amplifiers with excellent linearity and very wide dynamic range. Unfortunately, because the ion source of the instrument is tightly enclosed, to maximize sensitivity, the apparent isotopic composition of a sample may vary dependent on the amount of gas entering the ion source (i.e., the peak size) as shown in Table 1.4. The linearity of an IRMS instrument can be measured on a daily basis either by introducing samples or working gas pulses of varying size as illustrated in Figure 1.3. When validating a measurement protocol it is important to measure linearity by both means and, hopefully, obtain the same correction factor.

The linearity is simply the slope of the data (δ-value *vs.* amplitude); in this example the slope is 0.02‰ per 1,000 mV. This may seem a very small

Table 1.4 An example of linearity measurement data of an IRMS instrument.

Signal amplitude (mV)	δ-value (‰)
1078	−0.109
4797	0
6537	+0.040
11181	+0.102
15185	+0.189
22273	+0.344

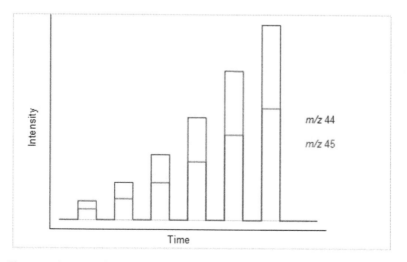

Figure 1.3 An example of linearity measurement data display of an IRMS instrument.

correction but, if sample peaks vary from 5,000 to 15,000 mV the correction would amount to 0.2‰, which is larger than typical experimental error for $\delta^{13}C$ determination. If the linearity correction is small it can simply be ignored but, like blank correction, it is important to *measure and monitor* the linearity. A gradual trend or sharp change in linearity may point to a problem with the IRMS instrument or inlet systems.

1.4.6 Drift correction

Despite the high stability of modern instrument electronics, the δ-scale of an IRMS will inevitably *drift* over time most often due to changes in the isotopic composition of background gases within the ion source. The phenomenon is often most pronounced when measuring hydrogen isotopes as this gas can diffuse against the flow of the vacuum pumps back into the ion source. Drifts in the δ-scale are readily identified through the regular analysis of in-house QC materials. To be useful, such a material must be analyzed every 5–10 samples depending on the number of replicates for each sample.

A number of publications describe how to adopt an effective QC regime (e.g., Brand 2009; Coplen and Qi 2009) typically based on control charts or Shewhart charts with warning (the accepted value ± 2 standard deviations) and action limits (the accepted value ± 3 standard deviations).

A good analytical protocol (and a good analyst) will define appropriate actions once the defined control limit for QC is breached:

- stop the analytical sequence,
- re-test the QC material to ensure that the breach was not an outlier, and/or
- re-test all analytical samples following the last successful QC.

The trade-off in this process is that if you run a QC material every 5 samples you will analyze more QCs but you risk only 5 sample results; if you run a QC material every 10 samples you analyze fewer QCs but you risk 10 sample results.

An alternative approach is *drift correction*, which uses the QC results as a continuously variable offset for the analytical data. This approach relies on two assumptions:

- the instrument drift has an equal effect on QC material and samples, and
- the instrument drift is linear between QC material measurements.

If these assumptions are considered to be correct, the process is easily implemented via instrument software or external spreadsheets. When adopting this approach it is sensible however, to run a second QC material between drift-correction RMs to ensure that the drift correction is working effectively, otherwise this becomes a *fudge-factor*.

1.4.7 Memory correction

For certain measurements, notably high temperature reduction over glassy carbon, the measured isotopic composition of a sample can be affected by the composition of the previous samples. For example, six replicate measurements of SLAP2 (δ^2H = −427.5‰) followed by six replicate measurements of VSMOW2 (δ^2H = 0.0‰) might well show a pattern:

| −428.0 | −428.1 | −427.8 | −428.0 | −428.3 | −428.1 |
| −12.2 | −4.0 | −1.9 | −0.4 | −0.3 | −0.3 |

In this example, the δ-values measured for VSMOW2 exhibit a memory effect from the SLAP2 previously analyzed, which contains far less ^2H relative to VSMOW2.

Researchers have presented algorithms that correct the measured δ-value of a sample by subtracting some proportion of the measured value of up to three previous samples (e.g., Gröning 2011). These algorithms are typically applied to water samples in which the peak size is relatively constant; for solid samples in which the peak size can be variable, an algorithm would also need to account for the size of preceding samples.

A simpler and more robust solution to overcome memory effect is to ignore the first few replicate analyses. The disadvantage of this approach is that it is wasteful of analytical time especially for the δ^2H or $\delta^{18}O$ analysis of solid samples for which the silver residue must be emptied every one hundred to two hundred samples.

1.4.8 Summing up

In order for stable isotope ratio measurements to be useful as a tool in food forensics they must be both repeatable (within-laboratory) and traceable to international reporting scales (between-laboratory). To achieve results that are fit-for-purpose the following steps are recommended (Carter and Fry 2013a):

1. obtain RMs with internationally agreed δ-values (spanning the likely range of δ-values for samples),
2. prepare matrix matched in-house RMs (calibrate these against the international RMs and calculate the overall uncertainty),
3. use RMs to determine which corrections are necessary or desirable (linearity, blank, drift, memory),
4. use RMs to demonstrate adherence to the Principle of Identical Treatment of sample preparation and inlet systems (focus on the difference, Δ, between RMs),
5. prepare in-house QC materials (chemically similar to routine samples and with typical δ-values),

6. develop an analytical protocol that includes the routine analysis of QC materials (with every batch of samples!),
7. ensure that sample and QC data are processed and corrected in an identical manner (by both instrument and external software),
8. monitor the δ-values of QC materials and act on any deviations or trends, and
9. periodically analyze international RMs as an in-house proficiency test (act on any significant deviations or trends).

1.5 Nomenclature

When reporting stable isotope ratio measurements it is important to conform to a standard nomenclature:

- δ and Δ should be presented in italic font
- δ-values should be preceded by a + or − symbol

the subscript text can have a subtle but important meaning:

Example

δ^2H	A general form of reporting isotopic difference
δ^2H_{VSMOW}	The data are traceable to the international reporting scale
$\delta^2H_{VSMOW-SLAP}$	The data are traceable to the international reporting scale through two point normalization using the accepted isotopic compositions of the specified RMs.

δ-values are typically presented as ± the standard uncertainty at one sigma level, which can be derived from the standard deviation of replicate measurements or from the complete uncertainty budget, which should be defined. When large tables of data are presented, it can be acceptable simply to report the median uncertainty for measurements of a given element.

References

Accum, F. 1820. *There is death in the pot: A treatise on adulterations of food and culinary poisons*, London: Longman.

Annual Report of the Government Scientific Advisor. 2015. Forensic science and beyond: Authenticity, Provenance and Assurance. Evidence and case studies. London: Government Office of Science.

Asenstorfer, R. E., G. P. Jones, G. Laurence and U. Zoppi. 2011. Authentication of red wine vintage using bomb-pulse [14]C. *ACS Symposium Series (Progress in Authentication of Food and Wine)* 1081: 89–99.

Brand, W. A. 2009. Maintaining high precision of isotope ratio analysis over extended periods of time. *Isotopes in Environmental and Health Studies* 45: 135–149.

Brand, W. A., S. S. Assonov and T. B. Coplen. 2010. Correction for the [17]O interference in $\delta^{13}C$ measurements when analyzing CO_2 with stable isotope mass spectrometry (IUPAC Technical Report). *Pure Applied Chemistry* 82: 1719–1733.

Brand, W. A. and T. B. Coplen. 2012. Stable isotope deltas: tiny, yet robust signatures in nature. *Isotopes in Environmental and Health Studies* 48: 1–17.

Carter, J. F. and B. Fry. 2013a. Ensuring the reliability of stable isotope ratio data—beyond the principle of identical treatment. *Analytical and Bioanalytical Chemistry* 405: 2799–2814.

Carter, J. F. and B. Fry. 2013b. "Do it yourself" reference materials for δ^{13}C determinations by isotope ratio mass spectrometry. *Analytical and Bioanalytical Chemistry* 405: 4959–4962.

Coplen, T. B. 1994. Reporting of stable hydrogen, carbon, and oxygen isotopic abundances. *Pure and Applied Chemistry* 66: 273–276.

Coplen, T. B., W. A. Brand, M. Gehre, M. Gröning, H. A. J. Meijer, B. Toman and R. M. Verkouteren. 2006. New guidelines for δ^{13}C measurements. *Analytical Chemistry* 78: 2439–2441.

Coplen, T. B. and H. Qi. 2009. Quality assurance and quality control in light stable isotope laboratories: A case study of Rio Grande, Texas, water samples. *Isotopes in Environmental and Health Studies* 45: 126–134.

Coplen, T. B. 2011. Guidelines and recommended terms for expression of stable-isotope-ratio and gas-ratio measurement results. *Rapid Communications in Mass Spectrometry* 25: 2538–2560.

Gröning, M. 2011. Improved water δ^2H and δ^{18}O calibration and calculation of measurement uncertainty using a simple software tool. *Rapid Communications in Mass Spectrometry* 25: 2711–2720.

Hayes, J. M. 2001. Fractionation of carbon and hydrogen isotopes in biosynthetic processes. *Reviews in Mineralogy and Geochemistry* 43: 255–277.

Hubert, P., F. Perrot, J. Gaye, B. Medina and M. S. Pravikoff. 2009. Radioactivity measurements applied to the dating and authentication of old wines. *Comptes Rendus Physique* 10: 622–629.

Krouse, H. R. and T. B. Coplen. 1997. Reporting of relative sulfur isotope-ratio data. *Pure and Applied Chemistry* 69: 293–295.

McKinney, C. R., J. M. McCrea, S. Epstein, H. A. Allen and H. C. Urey. 1950. Improvements in mass spectrometers for the measurement of small differences in isotope abundance ratios. *Review of Scientific Instruments* 21: 724–730.

Mariotti, A. 1983. Atmospheric nitrogen is a reliable standard for natural N-15 abundance measurements. *Nature* 303: 285–287.

Mariotti, A. 1984. Natural N-15 abundance measurements and atmospheric nitrogen standard calibration. *Nature* 311: 251–252.

Meier-Augenstein, W. 2006. Stable isotope fingerprinting—Chemical element 'DNA',. *In*: T. Thompson and S. Black (eds.). *Forensic human identification: An introduction.* Boca Raton, Florida, USA: Taylor & Francis Group.

Paul, D., G. Skrzypek and I. Fórizs. 2007. Normalization of measured stable isotopic compositions to isotope reference scales—a review. *Rapid Communications in Mass Spectrometry* 21: 3006–3014.

Qi, H., M. Gröning, T. B. Coplen, B. Buck, S. J. Mroczkowski, W. A. Brand, H. Geilmann and M. Gehre. 2010. Novel silver-tubing method for quantitative introduction of water into high-temperature conversion systems for stable hydrogen and oxygen isotopic measurements. *Rapid Communications in Mass Spectrometry* 24: 1821–1827.

Tuniz, C., U. Zoppi and M. A. C. Hotchkis. 2004. Sherlock Holmes counts the atoms. *Nuclear Instruments and Methods in Physics Research B* 213: 469–475.

Urey, H. C. 1948. Oxygen isotopes in nature and in the laboratory. *Science* 108: 489–496.

CHAPTER 2

Sampling, Sample Preparation and Analysis

James F. Carter[1,*] *and Lesley A. Chesson*[2]

2.1 Introduction

This chapter does not set out to provide a comprehensive overview of sampling strategies. Many excellent texts exist on the requirements for the sampling of forensic exhibits (Aitken 1999; Aitken and Taroni 2004; Gy 2004a–c; Petersen et al. 2005) and foodstuffs in particular (Esbensen 2015; Paoletti and Esbensen 2015). Individual chapters may also provide overviews of the sampling requirements for specific sample types. This chapter is instead intended to provide some general guidelines on sampling foodstuffs as well as other forensic exhibits and preparing them for isotopic and elemental analysis.

2.2 Sampling

2.2.1 Why do we sample?

There are many practical and theoretical reasons why we analyze a sample of a population:

- To study the whole population is not possible
- To study only one or two cases is not representative
- To save time and money

The first point above is axiomatic: the entire population of any foodstuff will typically be far too large to analyze in its entirety and, therefore only a sample of the population can be tested.

[1] Forensic and Scientific Services, Health Support Queensland, 39 Kessels Road, Coopers Plains QLD 4108, AUSTRALIA.
[2] IsoForensics, Inc., 421 Wakara Way, Suite 100, Salt Lake City, UT 84108, USA.
  Email: lesley@isoforensics.com
* Corresponding author: Jim.Carter@health.qld.gov.au
With contributions from Ujang Tinggi.

Although modern manufacturing practices strive for highly consistent products, foodstuffs can demonstrate significant heterogeneity, both within-samples and between-samples. Before collecting samples the analyst must consider what kind of sampling and sample preparation can make the analytical samples not only homogeneous but, more importantly, representative? The first stage in developing a reliable sampling strategy or methodology is to demonstrate within-batch homogeneity. The second stage in developing a useful methodology is to demonstrate between-batch heterogeneity.

2.2.2 How and what do we sample?

In the field of food forensics it will always be difficult to obtain truly authentic samples. When a researcher goes to extreme trouble to obtain authentic samples—e.g., juice pressed from a single cultivar of apples for a single orchard—such samples may not be typical of those available to retail customers (i.e., real-world samples). A researcher must, therefore, choose a sample type and sampling strategy that address a specific question. Often, the most pragmatic means to obtain a suite of samples is simply to source reliable wholesale or retail outlets. The results from such a survey must, however, always be viewed with slight cynicism as nothing may be what it claims.

Individual samples from a given manufacturing batch will often appear (at the very least) visually homogeneous whereas samples from different brands may appear markedly different. Even if samples appear visually identical, it is not reasonable to assume that all individuals share the same chemical and isotopic characteristics; this must be demonstrated empirically. Typically, five to 20 individual samples must be tested to establish whether a specific characteristic is homogenous across the population. Once a degree of homogeneity is established, and without evidence to the contrary, it is reasonable to extend this assumption to the entire population.

Possibly the most difficult task is to determine which components of a sample are to be analyzed. It is usually possible to simply grind samples to a fine powder and analyze the resulting powdered material as being representative. Although this approach is simple and robust, it has two significant drawbacks. First, most foodstuffs are complex matrices comprised of varying amounts of:

- water
- protein
- fat
- carbohydrate (both complex carbohydrates and simple sugars)
- trace organic compounds (alkaloids, vitamins, colors, etc.)
- inorganic compounds and trace elements

The possibility to sub-divide a sample is almost endless and the key question to be addressed is, "Where does the important information lie?" Is it contained in the whole sample? Is it contained in the fat fraction? Is it contained in an individual fatty acid? By combining all components into a single composite sample we potentially lose meaningful and characteristic information.

Second, and perhaps most important, not all samples will contain the same components in the same proportions. For example, comparing the isotopic composition of very lean meat with very fatty meat would be a meaningless and misleading exercise. Before using complex and costly analytical techniques, it is essential to ensure that we are comparing *like-with-like* and literally not comparing apples with oranges. Determining which fraction or components to examine is a critical step in developing a robust sampling strategy and analytical protocol. If the fraction is too "broad" the isotopic signal will be blurred or obscured by other information; if the fraction is "narrow" it may require significant time and effort to isolate and analyze.

To speed the development of reference data sets, it is common to analyze composite samples by physically combining material from several individuals. Typically, a composite sample should be made from five to 20 individual samples. In practice, the information gained from analyzing a composite sample is equivalent to the median information that would be gained from analyzing individual samples. What is lost is knowledge of the range of values within the population; what is gained is a greater throughput of samples and a larger data set in a shorter time.

2.2.3 How big a sample?

Having decided on a sample type and component(s) of interest, it is necessary to collect the samples that will form the background *database* against which comparisons can be made. Although the word *database* can have significant meaning in certain fields of application, in the context of food forensics it is simply used to describe a collection of data derived from sampling of the general background or population.

As a heuristic approach, a database should contain measurements from at very least 20 individual samples and ideally 20 individual samples from each discrete grouping (e.g., field, feeding regime, etc.). Also, the larger the database, the greater the strength of evidence that can be drawn from comparisons against it (Carter et al. 2014).

Before we consider how big our database needs to be, we should first consider the following:

- What is it we are trying to estimate?
- How precise do we want the estimate to be?
- What are we going to do with the estimate once we have it?

As a useful starting point, some legislation provides guidelines on sampling specific foods—e.g., The Feeding Stuffs (Sampling and Analysis) Regulations 1999 (UK) and the Australia and New Zealand Food Standards Code (ANZFSC). Although these regulations were never intended for stable isotopic analysis they can provide a useful starting point for developing a sampling strategy and can be easily defended in legal settings.

2.3 Practical sampling and sample preparation

2.3.1 Sampling scallops—a practical example

Figure 2.1 illustrates the preparation of scallops to form a single composite sample for isotopic and trace element analyses. The object of the following steps was to prepare a representative sample that was homogeneous to the size typically used for isotopic analysis, *ca.* 2 mg of sample for each element.

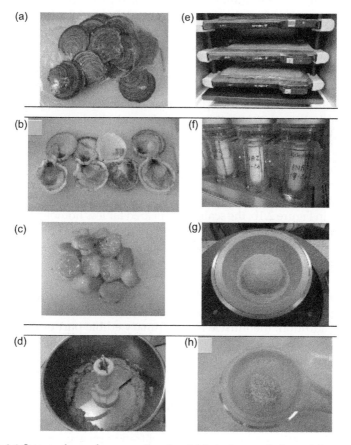

Figure 2.1 Stages of sample preparation for stable isotope and elemental concentration analyses of scallops.

Each composite sample was prepared from approximately 1 kg, or twelve individual scallops, purchased from local retail fishmongers (Queensland, Australia). This survey targeted Australian scallops intended for human consumption and should only be considered to be representative of this specific sub-group (unless proved otherwise).

Figure 2.1a shows a sample of twelve scallops, which were of a comparable size with visually similar shell coloration and patterns. Physical differences between individuals must always be considered; large scallops may be inherently (isotopically and elementally) different to small scallops and scallops with light shells may be (isotopically and elementally) different to scallops with dark shells. As noted above, it is misleading to compare apples with oranges unless we have explicitly demonstrated that this is a valid comparison.

The shells of all the individuals were intact and closed, which gave some confidence in the authenticity of the samples; specimens sold in half-shells were often found to be shark meat! When analyzing foodstuffs it is always good practice to check the integrity of the packaging and note any identification marks such as batch numbers or *best before* dates.

2.3.2 Sample preparation—a practical example

To prepare the scallops for isotopic and trace element analyses, we could simply grind all twelve individuals (shells and all), but common sense (and previous experience) tells us that this will *obscure* or *blur* the isotopic signature. It is also very likely that the incorporation and subsequent turn-over of isotopes and trace elements from diet and the environment into the scallop are very much slower in the shell than in the muscle, and governed by different factors.

As seen in Figure 2.1b, the scallops appeared far less homogeneous once they were opened, with different development of the sexual organs and some containing sand. If the inner content of the shells had been combined, the resulting sample would combine many isotopic signals—and a lot of sand!

One consistent feature of all the scallops was the adductor muscle, shown in Figure 2.1c, and this was chosen to be the basis of the sampling strategy. In addition, the muscle is the primary edible part of the scallop, which can be sold separately—often cleaned and presented on a shell (but not necessarily the original shell). By adopting this sampling strategy, we did not need to test the whole scallops, just the *meat* that might be sold frozen or form part of a meal. In this example, the muscle tissue was isolated and rinsed. This process typically does not need to be exhaustive as a small inclusion of other tissues is unlikely to make a significant contribution to the final isotope ratio and elemental concentration measurements.

The combined muscle tissue from the twelve individual scallops was then coarsely ground using an industrial food processor, as seen in Figure 2.1d. Although the bowl and blades of the processor were made of steel, this does not introduce significant metal contamination to the sample as the contact time is only a few seconds and the surface area is relatively small. Subsequent handling stages in which contact times were long and the surface area was large were performed using metal-free materials.

The coarsely ground combined muscle tissue sample was than packed into a polyethylene bag, frozen, and freeze dried (Figure 2.1e). The weight loss on drying was recorded so that (if required) the trace element compositions could be corrected to the original weight of the sample. In a previous study, seafood purchased frozen was found to contain significantly more water than that purchased fresh (Carter et al. 2015). In this study the trace element data were reported on sample dry weight as it appeared water had been added during processing—the packaging of one sample claimed "contains 10% added seawater for freshness"!

Figure 2.1f shows the removal of fat from the freeze-dried sample by refluxing with hexane. The fat was removed primarily to improve the grinding characteristics, as samples that contain a high proportion of sugar or fat tend to coalesce rather than grind to a fine powder. In addition, the fat content of seafood is highly polyunsaturated and readily oxidizes, causing samples to degrade rapidly. Once the fat is removed, the dried, de-fatted meat samples are durable at room temperature for long periods. Finally, studies have found that the lipid content of seafood tissue has a significantly different isotopic composition (especially for δ^2H values) to the flesh (protein). By removing the lipid component, the isotopic signature should be more representative of the muscle.

Once the sample was fat-free and completely dry of residual solvent, it was ground to a fine, visibly homogeneous powder (Figure 2.1g). As noted previously, stable isotopic analysis required approximately 2 mg of material and any inhomogeneity would be apparent in poor replicate analyses. In this example, the sample was ground using a zirconium oxide ball mill. This grinding medium has two advantages: it is extremely hard (hence producing a very fine powder) and it does not introduce metallic contamination (with the possible exception of zirconium). For some intrinsically flexible materials, such as a prawn shells, it is necessary to cool the sample with liquid nitrogen in order to make it brittle before grinding.

Finally, as show in Figure 2.1h, the ground sample was passed through a fine, non-metallic sieve to remove traces of connective tissue that did not grind without the addition of liquid nitrogen. In this example, the original 1 kg sample of whole scallops was reduced to approximately 3 g of fine homogeneous power of scallop adductor muscle.

Using the preparation method described here it would also be possible to recover and analyze the shell, moisture (water), and fat from the scallops.

Without prior knowledge, derived from earlier research, it would be prudent to analyze each of these components to determine where the important isotopic and trace element concentration differences lie.

2.4 Preparing samples for stable isotope ratio analysis

Solid samples intended for stable isotopic analysis are enclosed in tin or silver capsules as shown in Figure 2.2a. Tin capsules are used for ^{13}C, ^{15}N, and ^{34}S analysis and silver capsules are used for ^{2}H and ^{18}O analysis. The primary function of both types of capsule is physically to contain the finely powdered sample. Tin capsules have an additional role in the combustion

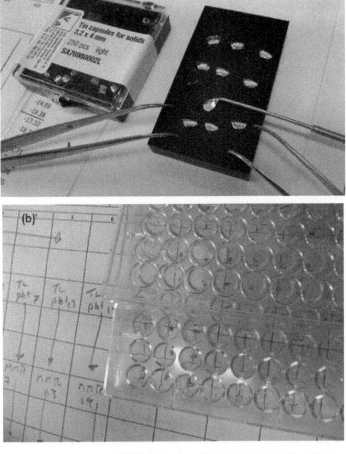

Figure 2.2 (a) Samples intended for IRMS analysis measured into tin capsules prior to crimping and (b) crimped capsules stored in a 96-well plate prior to analysis (note completed tracking template in the background).

process as tin readily reacts with oxygen, raising the localized temperature to approximately 1700°C promoting the combustion of the enclosed sample.

As tin capsules have a natural oxide coating, silver capsules must be used for ^{18}O analysis or the oxide coating would contribute to the oxygen formed from the sample. Silver capsules are significantly more expensive than tin capsules and it might seem sensible to use tin capsules for ^2H analysis. Unfortunately, the high temperature reduction (Schuetze/ Unterzaucher reaction) used for hydrogen isotope analysis is complex and empirical evidence shows that it is adversely affected by the presence of tin, causing poor repeatability—so silver it is!

Many laboratories routinely weigh samples into capsules using a six- or seven-place microbalance, which is a very time consuming process but such mass precision is necessary if elemental compositions are to be calculated. For isotopic analysis, samples only need to be weighed during method validation—i.e., to demonstrate that a specific sample type is quantitatively converted to H_2, CO_2, N_2, CO, or SO_2 during elemental analysis. The validation process will define the minimum and maximum range of peak heights and determine the need for linearity correction (see Chapter 1), which negates the need to weigh samples precisely. It is, however, good practice periodically to check conversion efficiency by analyzing some precisely weighed samples.

Once powdered sample material is weighed into capsules, the capsules are crimped into tight balls to prevent sample loss and to exclude atmospheric gases. Capsules must be crimped on a clean, hard ceramic, glass, or metal surface using tweezers (forceps) and both the tweezers and surface should be cleaned thoroughly after each sample to prevent cross-contamination between samples.

At this stage the samples are anonymous grey (tin) or silver balls and 96-well plates provide a convenient means to identify individual samples and track them on their journey to the mass spectrometer, as shown in Figure 2.2b. Unlike samples destined for analysis by techniques such as GC or HPLC, it is not possible to label the samples directly and Figure 2.3 shows a simple template (partially visible in Figure 2.2b) that can be used to record the identity of samples stored in a 96-well plate. It is also important that each plate is uniquely identified as a busy laboratory is likely to have numerous 96-well plates with samples ready for analysis at any time. Although printed versions of this template are convenient to use at the laboratory bench, electronic versions can also be developed, which can reduce transcription errors and automate subsequent data corrections.

It is good practice to wash 96-well plates with water several times after each use, to remove particles or sticky residues that may have escaped from the previous occupants. Because 96-well plates are relatively inexpensive, some laboratories simply discard used plates; this is generally

Tray #:

Prepared by:

Water Equilibration Date:

Project/Case:

Date:

Drying Date:

Run date:

		1	2	3	4	5	6	7	8	9	10	11	12
A	identifier 1												
	identifier 2												
	weight												
B	identifier 1												
	identifier 2												
	weight												
C	identifier 1												
	identifier 2												
	weight												
D	identifier 1												
	identifier 2												
	weight												
E	identifier 1												
	identifier 2												
	weight												
F	identifier 1												
	identifier 2												
	weight												
G	identifier 1												
	identifier 2												
	weight												
H	identifier 1												
	identifier 2												
	weight												

Figure 2.3 An example of a spreadsheet template for tracking samples weighed into 96-well plates for IRMS analysis.

an unnecessary precaution unless samples are highly isotopically labelled or in some way toxic. It is good practice for a laboratory to have dedicated 96-well plates for standards, blanks, and QC materials. Putting the same samples in the same positions of the same 96-well plates negates any need for cleaning but, more importantly, reduces the possibility of cross-contamination.

Homogeneous liquid samples with a high water content and low viscosity can be injected directly for analysis in the same manner as water samples. This works well for δ^2H and $\delta^{18}O$ measurements, but is less reliable for $\delta^{13}C$ measurements for a number of reasons. First, encapsulated solid samples drop into the combustion zone of the reactor whereas liquid injections may stick to upper (cooler) wall of the reactor and combustion may be slow or incomplete. Second, if the liquid contains dissolved CO_2 (e.g., beers, ciders, or sparkling wines), the measured $\delta^{13}C$ value will represent a combination of the dissolved CO_2 and any solid material, typically sugars. To obtain a reliable $\delta^{13}C$ measurement of a particular component, it is better practice to dry the sample to a sticky residue and analyze it as a solid sample. Chapters 8 and 9 outline methods and applications for measuring the isotopic composition of the dissolved CO_2 found in drinks like beers, ciders, and sparkling wines.

Liquids with low viscosity may flow up the walls of capsules, contaminating the outside of the capsule and potentially the preparation surface and the auto-sampler carousel. One solution is to add a small amount of an absorbent material such as Chromosorb™ diatomaceous earth to the capsule. The disadvantage of this approach is that it increases the accumulation of ash in the reactor and it is vital that the adsorbent material does not contain the analyte element of interest. Viscous liquid samples may be enclosed in tin or silver capsules and analyzed in the same manner as solid samples. Very sticky samples such as honey are easier to handle if diluted with a solvent that does not contain the analyte element of interest; water is a suitable solvent for $\delta^{13}C$ analysis of honey.

2.4.1 Preparing samples for δ^2H measurements

It should be apparent that the factors discussed next will be less important when analyzing liquid samples, which already comprise a high proportion of water, and more important when analyzing hygroscopic materials.

It must be stated that this subject is currently controversial and what is presented here is the best understanding that the authors have of the subject at the time of writing.

- What is certain is that the measured δ^2H composition of a sample is affected by contact with water, both from the sample matrix and from atmospheric moisture.

- What is uncertain is whether moisture attaches to a sample by means of absorption, adsorption, and/or chemical exchange.
- What is certain is that the measured δ^2H composition of a sample should be corrected to account for extraneous water.
- What is both uncertain and controversial is the means by which this correction is made.

The two forms of hydrogen are referred to as intrinsic (the hydrogen permanent within a sample) and extrinsic (the hydrogen present due to interactions with external water sources).

The measured δ^2H composition of a sample (*meas*) is simply the arithmetic combination of the intrinsic (*int*) and extrinsic (*ext*) components, Equation 2.1,

$$\delta^2H_{meas} = \delta^2H_{int} \times (1 - f_{ext}) + \delta^2H_{ext} \times f_{ext}$$
Equation 2.1

where f_{ext} is the fraction of extrinsic hydrogen. Strictly, the isotope ratio data should be combined as isotopic ratios (R) rather than as δ-values (see Chapter 1), as the δ-scale is not linear. However, over a small range the following equations provide a close approximation.

Some of the sample preparation methods described below simply aim to produce consistent values for δ^2H_{meas}, whereas others seek to determine δ^2H_{int} through an estimate of f_{ext}. Table 2.1 shows the range of values determined for f_{ext} for a number of biopolymers that have been the subject of stable isotope studies. Equilibration temperatures ranged from ambient to 130°C with the more extreme temperatures generally exhibiting larger f_{ext} values possibly due to break down of the biopolymer structure.

Equilibration with laboratory air—Undoubtedly, the simplest solution to any problem is to do nothing and this approach is certainly a close approximation to that solution; samples are simply weighed into silver capsules and left to equilibrate with laboratory air for several days to several weeks. This process assumes that the isotopic composition of the moisture

Table 2.1 The fraction of extrinsic hydrogen (f_{ext}) in common biopolymers; based on Chesson et al. 2009.

Material	Example	f_{ext}	α
cellulose	cotton	0.16–0.30	1.08–1.30
chitin	exoskeleton	0.15–0.16	na
collagen	bone	0.17–0.23	1.08
keratin	hair	0.06–0.22	1.00–1.10
protein	meat	0.13–0.20	1.06–1.10

na = data not available

in laboratory air is relatively constant and that long equilibration with an infinite volume of air will give consistent results. Note that if samples are to be equilibrated with water vapor or dried, the silver capsules must be left open, or partially crimped, to allow the free movement of water vapor.

Drying—This option includes: drying at elevated temperatures, drying under vacuum, and drying over desiccants together with every possible combination with drying times varying from several days to several weeks. It seems unlikely that any drying process will completely remove tightly bound extrinsic water and even if this was achieved, a sample would quickly attract atmospheric moisture once removed from the drying chamber. It should also be noted that elevated temperatures may change the isotopic composition of a sample through loss of volatile components.

Both *equilibration with laboratory air* and *drying* can produce very consistent within-laboratory results and both methods (or combinations of methods) are appropriate for comparative studies. However, the ultimate aim is to determine $\delta^2 H_{int}$, a parameter which should be characteristic only of a sample and, therefore, readily comparable between laboratories.

Equilibration with water vapor—This popular method aims to determine the $\delta^2 H_{int}$ composition of a sample by equilibration with vapor from one, two, or three waters of known isotopic composition. Samples are placed, in open capsules, in a chamber together with a reservoir of water with a well characterized isotopic composition. This process assumes that over sufficient time the extrinsic water in a sample will completely exchange with the water vapor in the chamber. Unfortunately, the water vapor does not have the same isotopic composition as the water and another factor is introduced, α—the ratio of the isotopic composition of the water vapor to that of the water. Table 2.1 shows the range of α values at equilibration temperatures ranging from ambient to 130°C.

$$\alpha = \delta^2 H_{ext} / \delta^2 H_{water} \qquad\qquad \text{Equation 2.2}$$

Equation 2.2 assumes that after equilibration the isotopic composition of the extrinsic water ($\delta^2 H_{ext}$) in a sample will be the same as that of the water vapor ($\delta^2 H_{water} \times \alpha$). The experiment, therefore, becomes a simultaneous equation with:

2 knowns $\delta^2 H_{meas}$ $\delta^2 H_{water}$

3 unknowns $\delta^2 H_{int}$ f_{ext} α.

In order to solve this problem, sub-samples must be equilibrated against two or more waters, ideally with a wide range of δ-values and the assumption that both f_{ext} and α are constant for all experiments (Chesson et al. 2009). Following equilibration, f_{ext} can then be calculated from Equation 2.3 and Equation 2.1 is easily rearranged to calculate $\delta^2 H_{int}$.

$$f_{ext} = \frac{\delta^2 H_{meas1} - \delta^2 H_{meas2}}{\delta^2 H_{water1} - \delta^2 H_{water2}}$$

Equation 2.3

Following equilibration with water vapor, samples must be carefully dried to remove any condensation from the capsules; various approaches to drying, described above, are recommended by different authors (e.g., Bowen et al. 2005; Coplen and Qi 2012).

Once samples have been prepared through various stages of equilibration and/or drying they must be isolated from atmospheric moisture and it is important, therefore, to utilize a *zero-blank auto-sampler* for hydrogen isotopic analysis. In a conventional auto-sampler typically only one position of the auto-sampler carousel is purged with helium, whereas a *zero-blank* design maintains the entire carousel in an inert atmosphere.

A number of Inter-Laboratory Comparisons organized by the Forensic Isotope Ratio Mass Spectrometry (FIRMS) network (Carter et al. 2009; Carter and Fry 2013) have highlighted that $\delta^2 H$ data are far less comparable between laboratories than other stable isotope measurements. All of the methods described above can produce very consistent within-laboratory results but questions still remain as to whether they truly measure $\delta^2 H_{int}$ and to what extent results are comparable between laboratories.

2.4.2 Preparing samples for elemental analysis

In general, the techniques described above will produce homogeneous samples suitable for both isotope ratio and elemental concentration analysis. Samples intended solely for elemental analysis do not need to be homogenized to the same extent because larger samples are analyzed—typically several hundreds of milligrams compared to one or two milligrams for isotopic composition. What is important for elemental concentration analysis is that contamination with and/or losses of elements are avoided during all stages of sample preparation and handling. Sources of contamination may include laboratory equipment such as hands/gloves, sampling tools (homogenizer blades, knives, etc.), containers, glassware, reagents, and even air-born dust particles. To avoid this, more conventional laboratory ware may need to be replaced with, for example, poly(tetrafluoroethylene) chopping boards, zirconium oxide bladed knives, and platinum dishes.

Prior to instrumental analysis, the organic matrix of a sample must be decomposed, typically by *wet digestion* or *dry ashing*. Wet digestion decomposes samples through the use of mixtures of strong oxidizing acids such as nitric acid, sulfuric acid, and/or perchloric acid, combined with heat—either using a simple hotplate or a sophisticated microwave digestion system. For dry ashing, samples are heated in a muffle furnace (using resistance or microwave heating), and the resultant ash is then dissolved in dilute acid (Tinggi et al. 1992, 1997; Locatelli 2008). Samples are typically ashed at temperatures below 550°C to avoid the loss of volatile elements such as Group I metals, As, Cd, and Hg.

Once digested, and in solution, a wide range of techniques are available for the determination of metal concentrations, but modern laboratories rely mostly on inductively coupled plasma atomic emission spectrometry (ICP-AES) for major nutritional elements (Group I and II metals, S and P) and ICP-mass spectrometry (ICP-MS) for the rest of the periodic table—especially for elements present at trace and ultra-trace concentrations.

Matrix matched standard reference materials (SRMs) are available for many food types, for quality control and assurance. Suppliers include the National Institute of Standards and Technology (NIST), the International Atomic Energy and Agency (IAEA), and the Institute for Reference Materials and Measurements (IRMM). When SRMs are not available, which is often the case for some trace and ultra-trace elements, in-house reference materials (RMs) and *spike recovery* are used, particularly for routine analysis.

2.4.3 Preparing samples for isotope ratio analysis of radiogenic elements

The techniques used to prepare samples for element concentration analysis are generally the same as those used to prepare samples for isotope ratio analysis of radiogenic elements, such as $^{87}Sr/^{86}Sr$ or $^{208}Pb/^{204}Pb$, $^{207}Pb/^{204}Pb$, and $^{206}Pb/^{204}Pb$. Briefly, a solid sample is first decomposed using wet digestion or dry ashing. Element concentration is next measured before the analyte of interest (Sr, Pb) is prepared for isotope ratio analysis via multi-collector ICP-MS (MC-ICP-MS) or thermal ionization mass spectrometry (TIMS).

To prepare samples for $^{87}Sr/^{86}Sr$ analysis, strontium is isolated by ion-exchange chromatography from the sample solution—digest, dissolved ash, or liquid sample (e.g., bottled water). Element isolation requires the use of specialized ion-exchange resins, which bind a particular element until a specific eluent is added. The most commonly used resins for isolating Sr are produced by Eichrom®, which typically consist of crown ethers immobilized on an insert matrix. Sr is loaded onto the resin from a nitric acid solution, such as the digest or dissolved ash generated for element concentration analysis. Water samples must be acidified with nitric acid prior to Sr isolation. Sr is then eluted using ultrapure water.

Ion-exchange chromatography can also be used to isolate the lead fraction prior to isotope ratio analysis. However, it's possible to measure lead isotope ratios—e.g., $^{208}Pb/^{204}Pb$, $^{207}Pb/^{204}Pb$, and $^{206}Pb/^{204}Pb$—directly from a sample solution via MC-ICP-MS. In these cases, a purified tracer, or *spike*, can be added to correct for mass bias (see below).

Because Sr and Pb are present in foodstuffs at trace or ultra-trace concentrations, as compared to the bio-elements (H, C, N, O, S), sample preparation prior to isotope ratio analysis should take place in a *clean laboratory* environment. This can be especially important for Pb since it is a common environmental contaminant. All reagents used for sample preparations—acids, water, etc.—must be ultrapure.

2.5 Isotope ratio analysis—bio-elements (H, C, N, O, and S)

It will be apparent, in the following chapters, that the vast majority of samples analyzed in the name of food forensics are solid or liquid samples measured via Elemental Analyzer (EA)-IRMS or Thermal Conversion EA (TC/EA)-IRMS. Both instrumental setups operate with a continuous flow of helium and comprise an auto-sampler, one or two furnaces, and a gas chromatography (GC) column—but there the similarities end. At the time of writing, very few researchers have employed coupled chromatographic techniques, such as GC-IRMS or LC-IRMS, for food authentication. For this reason, it seems beyond the scope of the current edition to discuss these techniques other than to recommend the guidelines in Chapter 1 of this volume and the FIRMS' *Good Practice Guide to Isotope Ratio Mass Spectrometry* (Carter and Barwick 2011). Note that the requirements for good practice must apply to each compound in a chromatogram (to be reported). Many of the published applications use chromatographic separation to isolate a single component (or limited number of compounds) from a mixture for which it is possible to bracket each compound with matrix matched standards with widely spaced isotopic composition. However, this requirement becomes very demanding for a complex sample!

2.5.1 The Elemental Analyzer (EA)

Figure 2.4a shows a schematic layout of the most common configuration used for elemental analyzers, in which samples are combusted in an oxidizing environment. A volume of high purity oxygen is introduced at the same time (or slightly following) the sample, causing the tin capsule, in which the sample is enclosed, to oxidize rapidly; this creates a localized temperature exceeding 1700°C. The evolved gases are carried through a heated bed of catalyst (typically chromium oxide) to ensure complete oxidation and then through a second catalyst (typically silver cobaltous oxide) to remove halide and sulfur compounds. Next, the combustion gases pass through a bed of high purity (electrolytic) copper to remove excess oxygen, which might otherwise damage the filament of the mass spectrometer. The final stage is water removal, which can be achieved using chemical traps, cryogenic traps, or drying membranes (e.g., Nafion™); the most common method is a simple chemical trap filled with magnesium perchlorate (Anhydrone™). At this stage the carrier stream should only contain N_2 and CO_2 gases, which are separated using an isothermal packed GC column. If only nitrogen isotope ratios are to be measured a further

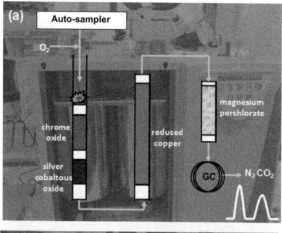

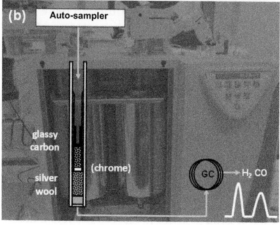

Figure 2.4 Components of a typical (a) Elemental Analyzer (EA) for carbon and nitrogen analysis and (b) Thermal Conversion Elemental Analyzer (TC/EA) for hydrogen and oxygen analysis.

CO_2 absorbing trap (e.g., Carbosorb™) can be added before the GC column, which will greatly speed analysis.

Because the EA operates by oxidation there can (naturally) be difficulties in converting elements already in high oxidation states, such as nitrates (NO_3^-), to the desired gases (CO_2 and N_2). A number of solutions have been proposed to this problem but recent research (Lott et al. 2015) has proposed a simple solution to this problem by switching off the oxygen flow during sample introduction—whereby samples thermally decompose rather than combust.

2.5.2 The Thermal Conversion EA (TC/EA)

Superficially, the TC/EA is a much simpler device than the EA. The original design comprised a single reactor, partially packed with glassy carbon operated at 1300 to 1450°C, shown schematically in Figure 2.4b. In complete contrast to the chemistry of the EA, the TC/EA system provides a strongly reducing environment in which the hydrogen and oxygen present in water or organic materials are converted to H_2 and CO via the Schuetze/Unterzaucher reaction (Santrock and Hayes 1987), shown in Equation 2.4.

$$C_xH_{2y}O_z + nC \rightarrow zCO + yH_2 + (n+x-z)C$$
$$H_2O + nC \rightarrow CO + H_2 + (n-1)C$$

Equation 2.4

Akin to the EA, a plug of silver wool in the lower part of the TC/EA reactor serves to remove halide and sulfur compounds. In common with an EA, the H_2 and CO gases evolved from samples are separated using an isocratic packed GC column. Although there is no necessity to place chemical traps between the TC/EA reactor and GC column, most laboratories include traps with varying combinations of water, CO_2, and VOC absorbing materials to prevent contamination of the GC column.

Although N_2 is also formed by the Schuetze/Unterzaucher reaction, it is generally considered to be not representative of the isotopic composition of the sample (Kornexl et al. 1999) due to side reactions that form compounds such as HCN (hydrogen cyanide). For the same reason, δ^2H measurements of nitrogen containing compounds can be incorrect and recent research recommends the addition of elemental chromium or combinations of glassy carbon, chromium and manganese to the reactor (Armbruster et al. 2006; Gehre et al. 2015; Nair et al. 2015; Gehre et al. 2017). It is not possible to make $\delta^{18}O$ measurement with chromium present in the reactor as this reacts with oxygen and it is good practice to reserve reactors that have contained chromium only for isotopic measurements of hydrogen.

A problem in $\delta^{18}O$ analysis is the need to eliminate the N_2 evolved from N-rich samples, such as proteins. Although N_2 and CO are separated by the GC column, nitrogen reacts with residual oxygen at the MS filament to form $[^{14}N^{16}O]^+$ which remains long after gaseous N_2 has left the ion source. This gas is detected as m/z 30 together with $[^{12}C^{18}O]^+$ and can cause large (and random) fluctuations in measured δ-values. Various approaches have been suggested to eliminate N_2 from the ion source, typically diverting and/or diluting the TC/EA eluent as the N_2 peak is evolved (Hagopian and Jahren 2012; Qi et al. 2011; Hunsinger and Stern 2012). It is also recommended to increase the length of the GC column, to increase the separation of N_2 and CO and provide a wider time-window in which to divert or dilute the nitrogen peak. Materials that contain a small proportion of nitrogen, such as plant tissues, can typically be analyzed without recourse to these measures—but when $\delta^{18}O$ measurements prove inconsistent, this is a likely cause.

2.5.3 Elementary good practice

The Principle of Identical Treatment (outlined in Chapter 1) requires that:

- samples and RMs are subject to the same chemical and physical processing,
- samples and RMs are introduced through the same instrument preparation/inlet system.

The practical implication of these requirements is that the instrument preparation/inlet devices must behave in an identical manner towards both samples and RMs, i.e.:

- the EA must convert both samples and RMs quantitatively to N_2 and CO_2,
- the TC/EA must convert both samples and RMs quantitatively to H_2 and CO.

In order to validate *identical treatment* for a new sample type, it is necessary to prepare and analyze a range of accurately weighed samples (Carter and Fry 2013). A plot of the peak size corresponding to the CO_2 (or other gas) evolved against the mass of carbon (or other element) should be linear and have a slope indistinguishable from compounds with excellent combustion efficiency, such as glucose. Laboratories analyzing samples via EA-IRMS might develop quality assurance (QA) standards for different sample types—vegetable oils, sugars, animal protein, etc.—to be routinely run alongside samples. For analysis using the TC/EA, closer matching of QA materials to the samples appears necessary.

2.5.4 Isotope Ratio Mass Spectrometry (IRMS)

A high precision IRMS instrument, for the determination of variations in isotope abundance, was first described by Alfred Neir in 1940 (De Laester and Kurz 2006) and the essential components of the instrument remain largely unchanged to the present day, with obvious improvements in electronics and vacuum technology. IRMS instruments are highly specialized for the analysis of $^2H/^1H$, $^{13}C/^{12}C$, $^{15}N/^{14}N$, $^{18}O/^{16}O$, and $^{34}S/^{32}S$, through the analysis of the gases H_2, CO_2, N_2, CO, and SO_2, respectively, using a high efficiency, tightly closed electron ionization (EI) source, a high transmission magnetic sector, and multiple Faraday cup collectors (Brand 2004). The use of multiple collectors, as opposed to the single electron multiplier commonly used in *organic* mass spectrometers, leads to reduced sensitivity but greatly enhanced precision, because the intensity of two (or three) ion beams are measured simultaneously, rather than sequentially. Another essential feature of IRMS instruments is an inlet system that admits gas through a capillary leak and allows for the rapid switching of sample and working gases.

Figure 2.5 illustrates a simple $\delta^{13}C$ measurement for a solid sample; measurements of other elements follow the same process. The first (rectangular) peak corresponds to an automated introduction of working gas—e.g., CO_2 from a high pressure cylinder—while the second (approximately Gaussian) peak corresponds to the CO_2 evolved from combustion of the sample. The IRMS instrument simultaneously records the intensity of m/z 44 $[^{12}CO_2]^+$ and m/z 45 $[^{13}CO_2]^+$, which the instrument software compiles into the traces shown (as a function of time).

The raw ratio data, from which δ-values are calculated, are simply the ratios of the integrated areas under the corresponding peaks from each ion: $[m/z\ 45]/[m/z\ 44]$. The δ-value of the CO_2 evolved from the sample is initially calculated by reference to the *working gas* (Equation 2.5) and in the example in Figure 2.5 the sample has a δ-value of $-11.16‰$ relative to the working gas. This δ-value must then be calibrated against two RMs with widely spaced isotopic compositions (see Chapter 1).

$$\delta = \frac{0.0109670 - 0.0110908}{0.0110908}$$
Equation 2.5

A full description of the set-up and operation of IRMS instruments can be found in the FIRMS' *Good Practice Guide to Isotope Ratio Mass Spectrometry*; please read this book in conjunction with that guide.

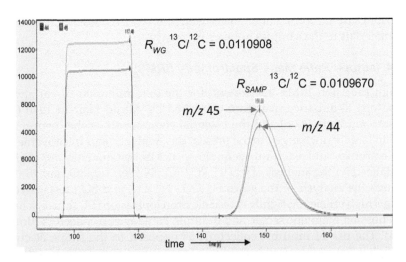

Figure 2.5 A typical display for the carbon isotopic analysis of a solid sample by EA-IRMS; R_{WG} and R_{SAMP} are the measured m/z 45/m/z 44 ratio of the working gas and sample gas, respectively.

Chapter 1 provides further information about *good practice* as related to data acquisition and processing.

2.5.5 *Isotope Ratio Infrared Spectroscopy (IRIS)*

Over the last 20 years, a new system of measuring isotope ratios has developed—isotope ratio infrared spectroscopy (IRIS). This laser-based spectroscopic technique measures the photo absorption of gas molecules containing different concentrations of isotopes. There are many manufacturers of IRIS instruments and each manufacturer uses a slightly different—and proprietary—method for measuring absorption and converting that measurement into isotopic ratios. At present, the IRIS technique is limited to the analysis of relatively few gases: H_2O, CH_4, N_2O, and CO_2. Despite this, there are advantages to IRIS as compared to IRMS— typically lower cost, in terms of both purchase price and ongoing operation, since IRIS instruments require no reference gases and helium is not needed as a carrier gas. IRIS instruments are also smaller than IRMS instruments, fitting on bench-tops, and are more rugged, making IRIS instruments field-deployable in settings were IRMS instruments are not.

Similar to IRMS instruments, IRIS instruments require samples to be in a gaseous form for analysis and some are equipped with EAs for conversion of solid samples (e.g., the Picarro® ^{13}C Combustion Module). Sample analysis via IRIS must adhere to many of the principles of good practice described previously in this chapter and in Chapter 1. IRIS analysis requires the use of RMs for data correction and normalization of measurements to the international isotope scales. This is especially important for samples—like alcoholic beverages—containing compounds that can generate spectral interferences in IRIS analysis, producing δ-values that deviate considerably from those measured for the same samples via IRMS. Multiple studies have investigated differences in the δ^2H and $\delta^{18}O$ values of alcohol-containing samples measured via IRIS *vs.* IRMS (e.g., Brand et al. 2009; Chesson et al. 2010). We note that at least one manufacturer seems to have found a way to effectively compensate for this interference and is currently marketing a "wine isotope analyzer" (Los Gatos Research).

2.6 Radiogenic elements (Sr and Pb)

In contrast to the bio-elements (H, C, N, O, and S) far fewer food studies have measured the isotope ratios of the radiogenic elements Sr and Pb. This is likely due to the overall low abundance of these elements in foodstuffs and the clean laboratory setup required for sample preparation prior to isotope analysis of radiogenic elements. In addition, relatively little is known about the biological, chemical, environmental, and geological processes impacting Sr and Pb isotopic variations in foodstuffs as compared to the bio-elements and so interpretation of Sr and Pb isotope abundances can be difficult without large reference databases. One exception to this is the use of $^{87}Sr/^{86}Sr$ in fish otoliths ("earstones") to investigate provenance.

2.6.1 Multi-collector ICP-MS (MC-ICP-MS)

The first commercially available multi-collector ICP-MS (MC-ICP-MS) instrument was introduced in 1992 with the aim of combining the versatile characteristics of an ICP source with the precision attainable with a multi-collector array. A magnetic sector mass analyzer is essential for this type of instrument because it is the spatial separation of the ion beams that allows the deployment of a multi-collector detector system for the simultaneous collection of isotopes of interest (Sr, Pb, Nd, U, etc.) (Douthitt 2007). For food analysis, MC-ICP-MS has largely replaced Thermal Ionization MS (TIMS), which requires time-consuming sample preparation to ensure high quality chemical separation of the analyte. Newer MC-ICP-MS instruments have the same precision as TIMS instruments and sample introduction is generally easier. However, TIMS instruments are still in use in many facilities and the technique is referenced occasionally in studies of animal-derived foodstuffs (e.g., Baroni et al. 2011; Rees et al. 2016).

Following isotope ratio analysis of radiogenic elements, data are corrected for instrumental *mass bias*. Both MC-IPC-MS and TIMS instruments are subject to mass bias, although analysis via MC-ICP-MS tends to be characterized by larger effects. While the causes of mass bias are not fully understood, they affect the movement of isotopes through the sampler and skimmer cones into the mass spectrometer, with heavier isotopes more likely to be transmitted. The effect of mass bias can be addressed by using either a *spike* or *internal standardization* method for correction. The *spike* method collects data from two separate sample analyses, one without a spike and a second in which tracers of known isotopic composition are added. The combination of analysis/tracer uniquely defines the mass bias corrected isotopic composition of an analyte element (Yang 2009). This correction method is suited for elements with four or more isotopes, some or all of which display natural variations in abundance (e.g., Pb). For correction via internal standardization, a correction factor is calculated using two isotopes of the analyte element of interest that are believed to be invariant in nature (Yang 2009)—for example, correction of $^{87}Sr/^{86}Sr$ data using a $^{86}Sr/^{88}Sr$ ratio that is assumed to be constant.

Measurement uncertainty is calculated from repeated analyses of reference materials. International reference materials available for Sr and Pb isotope analysis include SRM-987 (Strontium Carbonate Isotopic Standard) and SRM-981 (Common Lead Isotopic Standard), both available from NIST. These reference materials can be used to correct for instrumental drift during analysis, but measurement results are not normalized using the international reference materials in the same way δ-values of bio-elements are normalized. The isotope abundances of radiogenic elements are typically reported as simple ratios, without conversion to the δ-notation used for bio-elements (Chapter 1).

A specialized formed for MC-IPC-MS—laser ablation MC-ICP-MS—is used in an increasing range of forensic applications, for the "non-destructive" (at least on a macroscopic scale) testing of samples, such as (human) hair (Santamaria-Fernandez et al. 2009), drug packaging (Santamaria-Fernandez and Wolff 2010), glass (Orellana et al. 2013; Sjåstad et al. 2013), and paint (Orellana et al. 2013). A recent publication by Orellana et al. (2013) provides a comprehensive overview of LA-MC-ICP-MS specifically for the characterization of forensic evidence. To date, there have been relatively few publications using LA-MC-ICP-MS on foodstuffs. The examples are limited to fish otoliths and the measurement of $^{87}Sr/^{86}Sr$ along annual growth rings to understand fish movements through time (Brennan et al. 2015a,b). Additional details are provided in Chapter 6.

References

Aitken, C. G. G. 1999. Sampling—How big a sample? *Journal of Forensic Sciences* 44: 750–760.

Aitken, C. G. G. and F. Taroni. 2004. *Statistics and the Evaluation of Evidence for Forensic Scientists.* 2nd ed. Statistics in Practice. Chichester: Wiley.

Armbruster, W., K. Lehnert and W. Vetter. 2006. Establishing a chromium-reactor design for measuring δ²H values of solid polyhalogenated compounds using direct elemental analysis and stable isotope ratio mass spectrometry. *Analytical and Bioanalytical Chemistry* 384: 237–243.

Baroni, M. V., N. S. Podio, R. G. Badini, M. Inga, H. A. Ostera, M. Cagnoni, E. Gallegos, E. Gautier, P. Peral-García, J. Hoogewerff and D. A. Wunderlin. 2011. How much do soil and water contribute to the composition of meat? A case study: Meat from three areas of Argentina. *Journal of Agricultural and Food Chemistry* 59: 11117–11128.

Bowen, G. J., D. A. Winter, H. J. Spero, R. A. Zierenberg, M. D. Reeder, T. E. Cerling and J. R. Ehleringer. 2005. Stable hydrogen and oxygen isotope ratios of bottled waters of the world. *Rapid Communications in Mass Spectrometry* 19: 3442–3450.

Brand, W. A. 2004. Mass spectrometer hardware for analyzing stable isotope ratios. In: P. A. de Groot (ed.). *Handbook of Stable Isotope Analytical Techniques, Volume-I.* Amsterdam: Elsevier B.V.

Brand, W. A., H. Geilmann, E. R. Crosson and C. W. Rella. 2009. Cavity ring-down spectroscopy versus high-temperature conversion isotope ratio mass spectrometry; a case study on δ²H and δ¹⁸O of pure water samples and alcohol/water mixtures. *Rapid Communications in Mass Spectrometry* 23: 1879–1884.

Brennan, S. R., D. P. Fernandez, C. E. Zimmerman, T. E. Cerling, R. J. Brown and M. J. Wooller. 2015a. Strontium isotopes in otoliths of a non-migratory fish (slimy sculpin): Implications for provenance studies. *Geochimica et Cosmochimica Acta* 149: 32–45.

Brennan, S. R., C. E. Zimmerman, D. P. Fernandez, T. E. Cerling, M. V. McPhee and M. J. Wooller. 2015b. Strontium isotopes delineate fine-scale natal origins and migration histories of Pacific salmon. *Science Advances* 1: e1400124.

Carter, J. F., J. C. Hill, S. Doyle and C. Lock. 2009. Results of four inter-laboratory comparisons provided by the Forensic Isotope Ratio Mass Spectrometry (FIRMS) network. *Science and Justice* 49: 127–137.

Carter, J. F. and V. Barwick (eds.). 2011. Good practice guide for isotope ratio mass spectrometry. FIRMS.

Carter, J. F. and B. Fry. 2013. Ensuring the reliability of stable isotope ratio data—beyond the principle of identical treatment. *Analytical and Bioanalytical Chemistry* 405: 2799–2814.

Carter, J. F., S. Doyle, B.- L. Phasumane and N. NicDaeid. 2014. The role of isotope ratio mass spectrometry as a tool for the comparison of physical evidence. *Science and Justice* 54: 327–335.

Carter, J. F., U. Tinggi, X. Yang and B. Fry. 2015. Stable isotope and trace metal compositions of Australian prawns as a guide to authenticity and wholesomeness. *Food Chemistry* 170: 241–248.

Chesson, L. A., G. J. Bowen and J. R. Ehleringer. 2010. Analysis of the hydrogen and oxygen stable isotope ratios of beverage waters without prior water extraction using isotope ratio infrared spectroscopy. *Rapid Communications in Mass Spectrometry* 24: 3205–3213.

Chesson, L. A., D. W. Podlesak, T. E. Cerling and J. R. Ehleringer. 2009. Evaluating uncertainty in the calculation of non-exchangeable hydrogen fractions within organic materials. *Rapid Communications in Mass Spectrometry* 23: 1275–1280.

Coplen, T. B. and H. Qi. 2012. USGS42 and USGS43: Human-hair stable hydrogen and oxygen isotopic reference materials and analytical methods for forensic science and implications for published measurement results. *Forensic Science International* 214: 135–141.

De Laester, J. and M. D. Kurz. 2006. Alfred Nier and the sector field mass spectrometer. *Journal of Mass Spectrometry* 41: 846–854.

Esbensen, K. H. 2015. Materials properties: Heterogeneity and appropriate sampling modes. *Journal of AOAC International* 98: 269–274.

Gehre, M., J. Renpenning, T. Gilevska, H. Qi, T. B. Coplen, H. A. J. Meijer, W. A. Brand and A. Schimmelmann. 2015. On-line hydrogen-isotope measurements of organic samples using elemental chromium—an extension for high temperature elemental-analyzer techniques. *Analytical Chemistry* 87: 5198–5205.

Gehre, M., J. Renpenning, H. Geilmann, H. Qi, T. B. Coplen, S. Kümmel, N. Ivdra, W. A. Brand and A. Schimmelmann. 2017. Optimization of on-line hydrogen stable isotope-ratio measurements of halogen- and sulfur-bearing organic compounds using elemental analyzer-chromium/high-temperature conversion-isotope-ratio mass spectrometry (EA-Cr/HTC-IRMS). *Rapid Communications in Mass Spectrometry* 31: 475–484.

Gy, P. 2004a. Sampling of discrete materials—a new introduction to the theory of sampling: 1. Qualitative approach. *Chemometrics and Intelligent Laboratory Systems* 74: 7–24.

Gy, P. 2004b. Sampling of discrete materials: II. Qualitative approach—sampling of zero-dimensional objects. *Chemometrics and Intelligent Laboratory Systems* 74: 25–38.

Gy, P. 2004c. Sampling of discrete materials: III. Quantitative approach—sampling of one-dimensional objects. *Chemometrics and Intelligent Laboratory Systems* 74: 39–47.

Hagopian, W. M. and A. H. Jahren. 2012. Elimination of nitrogen interference during online oxygen isotope analysis of nitrogen-doped organics using the "NiCat" nickel reduction system. *Rapid Communications in Mass Spectrometry* 26: 1776–1782.

Hunsinger, G. B. and L. A. Stern. 2012. Improved accuracy in high-temperature conversion elemental analyzer $\delta^{18}O$ measurements of nitrogen-rich organics. *Rapid Communications in Mass Spectrometry* 26: 554–562.

Kornexl, B. E., M. Gehre, R. Höfling and R. A. Werner. 1999. On-line $\delta^{18}O$ measurement of organic and inorganic substances. *Rapid Communications in Mass Spectrometry* 13: 1685–1693.

Locatelli, C. 2008. "Metals". In: Y. Pico (ed.). *Wilson & Wilson's Comprehensive Analytical Chemistry—Food Contaminants and Residue Analysis*. Amsterdam: Elsevier.

Lott, M. J., J. D. Howa, L. A. Chesson and J. R. Ehleringer. 2015. Improved accuracy and precision in $\delta^{15}N_{AIR}$ measurements of explosives, urea, and inorganic nitrates by elemental analyzer/isotope ratio mass spectrometry using thermal decomposition. *Rapid Communications in Mass Spectrometry* 29: 1381–1388.

Nair, S., H. Geilmann, T. B. Coplen, H. Qi, M. Gehre, A. Schimmelmann and W. A. Brand. 2015. Isotopic disproportionation during hydrogen isotopic analysis of nitrogen-bearing organic compounds. *Rapid Communications in Mass Spectrometry* 29: 878–884.

Orellana, F. A., C. G. Gálvez, F. A. Orellana, C. G. Gálvez, M. T. Roldán, C. García-Ruiz, M. T. Roldán and C. García-Ruiz. 2013. Applications of laser-ablation-inductively-coupled plasma-mass spectrometry in chemical analysis of forensic evidence. *TrAC Trends in Analytical Chemistry* 42: 1–34.

Paoletti, C. and K. Esbensen. 2015. Distributional assumptions in food and feed commodities—development of fit-for-purpose sampling protocols. *Journal of AOAC International* 98: 295–300.

Petersen, L., P. Minkkinen and K. H. Esbensen. 2005. Representative sampling for reliable data analysis: Theory of Sampling. *Chemometrics and Intelligent Laboratory Systems* 77: 261–277.

Qi, H., T. B. Coplen and L. I. Wassenaar. 2011. Improved online $\delta^{18}O$ measurements of nitrogen- and sulfur-bearing organic materials and a proposed analytical protocol. *Rapid Communications in Mass Spectrometry* 25: 2049–2058.

Rees, G., S. D. Kelly, P. Cairns, H. Ueckermann, S. Hoelzl, A. Rossmann and M. J. Scotter. 2016. Verifying the geographical origin of poultry: The application of stable isotope and trace element (SITE) analysis. *Food Control* 67: 144–154.

Santamaria-Fernandez, R., J. G. Martínez-Sierra, J. M. Marchante-Gayón, J. I. García-Alonso and R. Hearn. 2009. Measurement of longitudinal sulfur isotope variations by laser ablation MC-ICP-MS in single human hair strands. *Analytical and Bioanalytical Chemistry* 394: 225–233.

Santamaria-Fernandez, R. and J. -C. Wolff. 2010. Application of laser ablation multicollector inductively coupled plasma mass spectrometry for the measurement of calcium and lead isotopes in packaging for discriminatory purposes. *Rapid Communications in Mass Spectrometry* 24: 1993–1999.

Santrock, J. and J. M. Hayes. 1987. Adaptation of the Unterzaucher procedure for determination of oxygen-18 in organic substances. *Analytical Chemistry* 59: 119–127.

Sjåstad, K. -E., T. Andersen and S. L. Simonsen. 2013. Application of laser ablation inductively coupled plasma multicollector mass spectrometry in determination of lead isotope ratios in common glass for forensic purposes. *Spectrochimica Acta Part B: Atomic Spectroscopy* 89: 84–92.

Tinggi, U., C. Reilly and C. Patterson. 1997. Determination of manganese and chromium in foods by atomic absorption spectrometry. *Food Chemistry* 60: 123–128.

Tinggi, U., C. Reilly, S. Hahn and M. Capra. 1992. Comparison of wet digestion procedures for the determination of cadmium and lead in marine biological tissues by Zeeman graphite furnace atomic absorption spectrometry. *Science of the Total Environment* 125: 15–23.

Yang, L. 2009. Accurate and precise determination of isotopic ratios by MC-ICP-MS: A review. *Mass Spectrometry Reviews* 28: 990–1011.

CHAPTER 3

Interpreting Stable Isotope Ratios in Plants and Plant-based Foods

James R. Ehleringer

3.1 Introduction

"We North Americans look like corn chips with legs."

Professor Todd Dawson, UC Berkeley in The Omnivore's Dilemma

Photosynthesis is one of the fundamental biochemical and physiological processes that distinguish plants from animals. Plants can be categorized as having one of three different *photosynthetic pathways*: C_3, C_4, or CAM. While C_3 photosynthesis is the most common pathway globally, many of our North American foods are dominated directly or indirectly by corn (maize), a C_4 pathway plant.

3.2 Our foods come from plants with different photosynthetic pathways

The C_3 pathway can be found in the most primitive plant-based foods, such as algae, and the most evolutionarily-advanced plants, the angiosperms, which comprise most of our food crops today (Ehleringer and Sandquist 2014). The term C_3 refers to the observation that the first fixed organic compound during photosynthesis is a molecule containing three carbon atoms. This mode of photosynthesis proceeds via the enzyme ribulose bisphosphate carboxylase-oxygenase (Rubisco) that combines ribulose bisphosphate, a 5-carbon molecule, with CO_2 to form two 3-carbon molecules (phosphoglyceric acid, PGA). Most of the tree crops, root crops, and vegetables that we consume today share this photosynthetic

Department of Biology, University of Utah, Salt Lake City, UT 84112, USA.
Email: jim.ehleringer@utah.edu

pathway. As shown in the leaf cross section in Figure 3.1, plants with C_3 photosynthesis have chloroplasts distributed in mesophyll cells that carry out both the light and dark reactions of photosynthesis.

In contrast, fewer plants utilize the C_4 photosynthetic pathway than the C_3 pathway (Sage 2016). Yet C_4 plants account for far greater productivity on a land-surface basis than you would expect on a percent species basis because of their high rates of photosynthesis. The term C_4 refers to the observation that the first fixed organic compound during photosynthesis is a molecule containing four carbon atoms. As foods that are commonly eaten, C_4 plants are confined to the grasses and a few leafy vegetables, most notably sugar cane and corn (Sage 2016). Not all grasses are C_4 plants, but only those commonly referred to as "warm season" grasses. Biochemically, C_4 photosynthesis is an evolutionarily-derived pathway expanding on the basic C_3 photosynthetic pathway. The C_3 photosynthetic pathway component of the C_4 pathway, including Rubisco, is largely restricted to interior vascular bundle sheath cells (Sage 2014). These vascular bundles are surrounded by a layer of mesophyll cells that contain the enzyme phosphoenol pyruvate carboxylase (PEPcase), which combines phosphoenol pyruvate and CO_2 to form a 4-carbon molecule as the initial photosynthetic reaction. PEPcase effectively acts as a CO_2 pump, increasing CO_2 levels around the Rubisco and making it more efficient. Thus, plants with C_4 photosynthesis have leaves that exhibit a "Kranz" or wreath-like anatomy (Figure 3.1), in which light and dark reactions of photosynthesis occur in interior bundle sheath portions of the leaf. Critical to the functioning of C_4 photosynthesis is the *spatial separation* of carboxylation activities into different cells.

Lastly, we have CAM photosynthesis, which refers to plants with Crassulacean Acid Metabolism as this pathway was first identified in plants from the Family Crassulaceae. Although CAM plants are far less important for global primary productivity than C_3 and C_4 plants, they do include a number of important foodstuffs, such as agave, pineapple, and vanilla. The

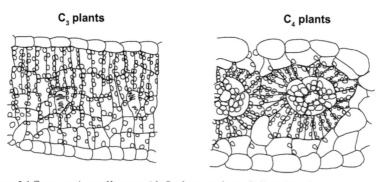

Figure 3.1 Cross-sections of leaves with C_3 photosynthesis (left) and C_4 photosynthesis (right). Note the difference in the distributions of chloroplast in the two different photosynthetic pathway types. Drawing is a copyright free graphic from http://plantecology.info.

biochemical function of CAM plants is the same as C_4 plants, except that both PEPcase and Rubisco activities take place in the same cell (Sage et al. 2014). This can occur because plants with CAM photosynthesis exhibit a *temporal separation* of carboxylation activities. PEPcase is active at night, when C_4 acid concentrations build up in cell vacuoles. During the daylight hours, PEPcase is inactive and the C_4 acids within the vacuole are decarboxylated. The CO_2 released is then combined with ribulose bisphosphate via Rubisco to form PGA. Anatomically, most CAM plants are succulent (containing extensive water storage cells) and the photosynthetic activities are restricted to mesophyll cells near the epidermal surface (Sage 2014).

3.3 Photosynthetic pathway differences result in differences in plant carbon isotope ratios

Differences in the initial photosynthetic carboxylation reactions among C_3, C_4, and CAM plants result in large differences in carbon isotope ratios of plant tissues (Farquhar et al. 1989). Thus, we can consider C_3 plants as one group and C_4 and CAM plants as a second group. As shown in Figure 3.2, it is clear that the carbon isotope ratios of C_3, C_4, and CAM plant tissues do not exhibit single values; instead there can be wide variations in plant carbon isotope ratios. It is also evident that there is essentially no overlap in the carbon isotope ratios of C_3 *vs.* C_4/CAM plants.

For the purposes of isotopic identification of different food sources, we see that it is possible to distinguish different food sources based on their carbon isotope ratio values. The more interesting question regarding carbon isotope ratios in plants will be, "What do variations in carbon isotope ratio values tell us about the environment in which the food was

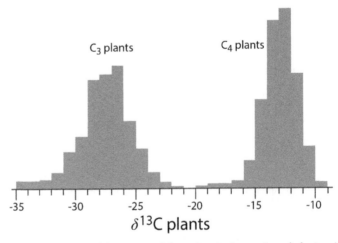

Figure 3.2 Relative frequency histograms of the carbon isotope ratios of plants with C_3 and C_4 photosynthesis. Drawing is adapted from Cerling et al. (2016).

cultivated?" Here, carbon dioxide sources, water stress, nutrient stress, and light environment play a role in determining variations in carbon isotope ratios of a single crop grown under different cultivation regimes. Farquhar et al. (1989) presented the fundamental equations describing the biochemical, environmental, and physiological bases for variations in plant carbon isotope ratios. From these theoretical foundations, we learn that the source of atmospheric carbon dioxide and the degree to which stomata remain open are key factors. While Farquhar et al. (1989) presented carbon isotope abundances in terms of carbon isotope discrimination (Δ), here we convert those relationships to carbon isotope ratio in δ-notation, the term used throughout this volume.

Variations in the $\delta^{13}C$ values of C_3 plants were first proposed by Farquhar et al. (1982) as Equation 3.1:

$$\delta^{13}C_p = \delta^{13}C_{atm} - a - (b-a) \times c_i/c_a \qquad \text{Equation 3.1}$$

where $\delta^{13}C_p$ is the $\delta^{13}C$ value of the plant, $\delta^{13}C_{atm}$ is the $\delta^{13}C$ value of the CO_2 in the atmosphere, a is the diffusional fractionation against ^{13}C in air, b is the net enzymatic fractionation against $^{13}CO_2$ by Rubisco, and c_i and c_a are the CO_2 concentrations inside the leaf and in the atmosphere, respectively. Since a, b, and c_a are constants, it is therefore variations in $\delta^{13}C_{atm}$ and c_i/c_a that drive the observed variations in $\delta^{13}C_p$.

The factors influencing the $\delta^{13}C_p$ of a C_4 plant are somewhat more complicated, since the CO_2 is first fixed by PEPcase in the mesophyll cells, decarboxylated in the bundle sheath cells, and again carboxylated via Rubisco in the bundle sheath cells (Farquhar 1983). However, the expression to describe the $\delta^{13}C_p$ value in a C_4 plant is simply a modification of the expression used for C_3 plants, Equation 3.2:

$$\delta^{13}C_p = \delta^{13}C_{atm} - a - (b_4 + b_3\phi - a) \times c_i/c_a \qquad \text{Equation 3.2}$$

where b_3 is the net enzymatic fractionation against $^{13}CO_2$ by Rubisco (similar to b above), b_4 is the net enzymatic fractionation against $^{13}CO_2$ by PEPcase, and ϕ is the fraction of decarboxylated CO_2 that diffuses out of the bundle sheath cells and back to the mesophyll cells.

The expression to describe CAM photosynthesis becomes a simplification of the equation to describe C_4 photosynthesis, since the stomata are typically closed during daylight hours and the Rubisco fractionation is not give the opportunity to express itself (Farquhar et al. 1989):

$$\delta^{13}C_p = \delta^{13}C_{atm} - a - (b_4 - a) \times c_i/c_a \qquad \text{Equation 3.3}$$

While Equation 3.3 predicts no variations in the $\delta^{13}C_p$ values of CAM plants, variation is in fact observed (Figure 3.2). This variation in $\delta^{13}C_p$ values arises because under well-watered conditions, many CAM plants will fix some proportion of their CO_2 using C_3 photosynthesis near the end of the diurnal cycle.

3.4 Genetic variations can lead to differences in carbon isotope ratios

We should expect to see genetically-based variation in ^{13}C composition of our foods as a result of both directed and natural selection. Genetic variation in $\delta^{13}C_p$ values is known to occur among annual crops, woody crops, and wild species (Dhanapal et al. 2015; Santiso et al. 2015; Verlinden et al. 2015). We can describe the relationship between carbon isotope ratio and plant physiology in terms relevant to understanding the basis of genetically-based $\delta^{13}C_p$ differences among taxa or within a single species. Consider an economic analogy to describe the basis of $\delta^{13}C_p$ variations—our food plants can differ in their photosynthesis rates and in their transpiration rates.

Within leaves, the c_i/c_a values reflect a balance between the CO_2 supply from the atmosphere through stomata on the leaf epidermis and the CO_2 demand from photosynthesis (Figure 3.3). The stomatal conductance, regulating the CO_2 supply rate, is influenced by relative humidity, light level, and water stress. In turn, the rate of photosynthesis is regulated by light levels (influencing the light reactions of photosynthesis) and the availability of CO_2 for Rubisco enzymatic fixation. In such a relationship of CO_2 supply rate and CO_2 demand (consumption rate) within a leaf, the operational point of supply *vs.* demand will be the c_i/c_a value. Over the long term, the $\delta^{13}C_p$ value represents the flux-based c_i/c_a average of these instantaneous physiological processes. We now know that genetic variation in $\delta^{13}C_p$ within a single C[3] species may span 3–4‰ (Hubick et al. 1986; Farquhar et al. 1989; Comstock and Ehleringer 1992; Ehleringer et al. 1993; Flanagan and Johnsen 1995; Lauteri et al. 1997). This naturally-occurring variation must be taken into consideration when determining, on the basis of ^{13}C composition, whether or not a foodstuff is authentic.

For example, the common bean (*Phaseolus vulgaris*—the green bean, dry bean, or string bean) exhibits significant ^{13}C genetic variations when grown under common field conditions (Figure 3.4). Early and late flowering beans differ in their flowering times by only 8–10 days, yet these maturation differences are significant enough to result in distinct carbon isotope ratio groupings. Genetic variations in ^{13}C values are known to have a high heritability (White et al. 1990; Zacharisen et al. 1999), with much of this ^{13}C variation positively correlated with leaf conductance to CO_2 and water vapor (Ehleringer 1990). To a consumer, the differences in ^{13}C values of the common bean could be the result of growing different cultivars or could be the result of crops grown under drought *vs.* irrigated conditions.

3.5 Drought histories are recorded by enrichment of carbon isotopes of C[3] plants

A major factor influencing carbon isotope ratios in C[3] plants is a partial stomatal closure in response to water stress, as a result of depleted soil

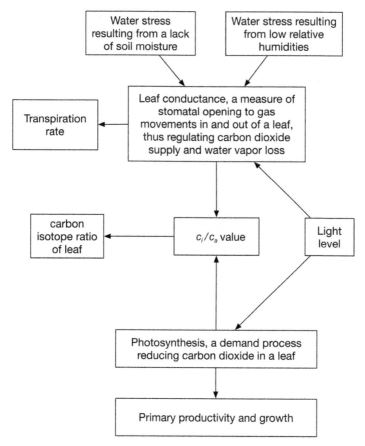

Figure 3.3 Interactions between stomatal conductance regulating the supply of CO_2 into the leaf, photosynthesis consuming CO_2 in the leaf, and intercellular CO_2 concentration (the operational point). Carbon isotope ratio is a long-term measure of the intercellular CO_2 concentration.

moisture contents (i.e., drought) or low atmospheric water vapor content (i.e., low relative humidity) (Farquhar et al. 1989; Ehleringer et al. 1993; Bowling et al. 2008). The effect of a partial stomatal closure is manifested as a decrease in c_i values (Equation 3.1). As c_i/c_a values decrease, $\delta^{13}C_p$ values will increase (Figure 3.5). Agricultural studies have shown that variations in carbon isotope ratios within a single crop species can differ 2–4‰ among plants cultivated under irrigated *vs.* rainfed conditions (White et al. 1990; Ehleringer et al. 1993; Adirejo et al. 2014; Bchir et al. 2016; Polania et al. 2016). By the time a food crop has arrived to the distributor or consumer, it is often difficult to distinguish between variations in $\delta^{13}C_p$ values associated with genetic factors *vs.* variations associated with cultivation under water deficits.

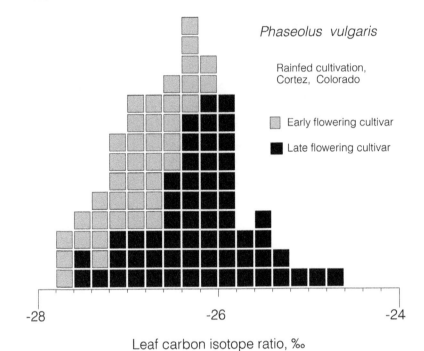

Figure 3.4 Carbon isotope ratio variation in different genetic lines of the common bean grown outdoors in Cortez, Colorado. Drawing is a copyright free graphic from http://plantecology. Based on data originally from Ehleringer et al. (1990).

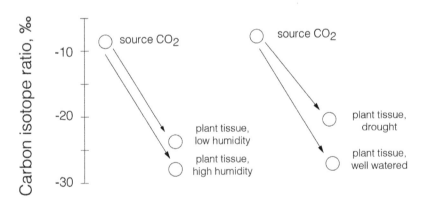

Figure 3.5 Changes in the carbon isotope ratios of a C_3 crop when grown under drought and non-drought conditions.

3.6 Shade and indoor growth are recorded depletions in carbon isotopes of plants

Plants cultivated in shade exhibit more negative $\delta^{13}C_p$ values than when grown under full or partial sunlight conditions (Ehleringer et al. 1986; Farquhar et al. 1989). For some crops this difference in cultivation practice has strong economic implications. For example, *shade-grown coffee* often commands a greater price than sun-grown coffee. While to the consumer this may be difficult to distinguish, shade-grown coffee can be easily identified on the basis of the $\delta^{13}C_p$ values (DaMatta 2004).

3.7 Different tissues can exhibit different ¹³C contents

Food scientists, using carbon isotopes to analyze foods, must be aware that there can be differences in the ^{13}C contents of biochemical components of an organ as well as ^{13}C differences among plant organs (Tcherkez et al. 2011). The theory presented in Section 3.4 pertains to the bulk $\delta^{13}C_p$ leaf values. Once the initial carbohydrates are formed during photosynthesis, subsequent biochemical reactions can result in variations of 2–10‰ in the ^{13}C contents of different compounds within a leaf (Badeck et al. 2005; Werner et al. 2011). These patterns are well recognized and often expressed as differences between the compound of interest and the bulk leaf value. Generally, lipids are ^{13}C depleted relative to the bulk leaf, while cellulose tends to be ^{13}C enriched relative to the bulk leaf. This becomes most apparent when carbohydrates from the photosynthetic cycle are broken down to 2-carbon molecules, such as acetate or acetyl-Co-A, which then form the substrates for synthesis reactions. As a result, lipids, starches, and proteins within tissues of a food crop may have distinctly different carbon isotope ratios (Tieszen and Fagre 1993; Hobbie and Werner 2004). Given that the chemical compositions of fruits, seeds, and roots are not necessarily the same as the biochemical compositions of leaves, it is not surprising that there can be 2–3‰ differences in the ^{13}C contents of different tissues (Badeck et al. 2005). There are, however, multiple hypotheses to explain why non-photosynthetic tissues, such as seeds and fruits, are generally ^{13}C enriched relative to leaves (Cernusak et al. 2009).

3.8 The soil and atmospheric water environment is recorded as variations in hydrogen and oxygen isotopes in plants

The hydrogen (δ^2H) and oxygen ($\delta^{18}O$) isotope ratios of plant organic matter and of water within plant tissues reflect the source water (soil or irrigation) δ-values and are additionally influenced by kinetic isotope fractionation factors and the δ-values of atmospheric water vapor (Flanagan et al. 1991; Roden and Ehleringer 1999; Roden et al. 2000; Cernusak et al. 2002). The $\delta^{18}O$ value of CO_2 has little, if any, impact on the $\delta^{18}O$ values of leaf water and plant organic matter. In general, the δ^2H and $\delta^{18}O$ values of xylem water in the roots and stem reflect the values of soil water, since there is no kinetic fractionation during water uptake through roots (Ehleringer and Dawson 1992; Ehleringer et al. 2000; Dawson et al. 2002).

As xylem water leaves the stem and enters an evaporating tissue (e.g., leaf, fruit, non-suberized stem), kinetic fractionation occur, resulting in plant tissue water that is isotopically enriched relative to source and xylem water as predicted by a modified Craig-Gordon model (Figure 3.6). The enrichment in leaf and fruit water δ^2H and $\delta^{18}O$ values results in tissue water values that fall to the right of the meteoric water line (Flanagan and Ehleringer 1991; Roden and Ehleringer 1999). There are several useful applications of this information in authenticating food sources. For example, consider authentication of fruit juices. Authentic juices will contain waters that do not plot on the meteoric water line, but are instead evaporatively enriched (Brause et al. 1984; Hammond 1996; Ogrinc et al. 2009; Chesson et al. 2010a). Reconstituted juices are likely to contain water that falls close to or on the meteoric water line. Similarly, fruits and vegetable should have waters with δ^2H and $\delta^{18}O$ values exhibiting evaporative enrichment (Dunbar and Wilson 1983; Fugel et al. 2005; Bong et al. 2008; Camin et al. 2011).

The water contained in beverages is also useful for authenticating the origins of wines (Breas et al. 1994; Rossmann et al. 1999; Calderone et al. 2004; Flamini and Panighel 2006; West et al. 2007) and beer (Chesson et al. 2010b). It is even possible to analyze the CO_2 in sparkling wines and beer to determine its authenticity, especially if there are large differences between the $\delta^{13}C$ values (Gonzalez-Martin et al. 1997; Calderone et al. 2007).

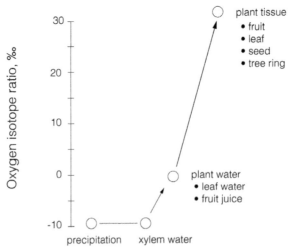

Figure 3.6 Changes in the oxygen isotope ratios of liquid water as precipitation, in the stem xylem, in the leaf, and in the fruit. Shown also is the change in the oxygen isotope ratio when the oxygen in water becomes fixed in organic compounds in a plant.

By knowing the δ^2H and $\delta^{18}O$ values of the source water and the relative humidity of the growth regime, it should be possible to make general predictions about the expected δ^2H and $\delta^{18}O$ values of fruit juices and leafy vegetables. Using this quantitative approach, it is feasible to verify broadly the authenticity of a fruit juice or leafy vegetable, if post-harvest evaporation has been minimal (Greule et al. 2015).

There is significant enrichment when the ^2H and ^{18}O compositions of labile tissue waters become permanently recorded in the organic matter of all components of plant tissues (Figure 3.6), including cellulose, sugars, protein, and lipids (DeNiro and Epstein 1989; Kornexl et al. 1999; Roden et al. 2000; Schmidt et al. 2001).

3.9 Soil nitrogen sources are recorded in plant nitrogen isotopes

Understanding plant nitrogen stable isotope values can be complicated, because plant nitrogen can be taken up by roots, through mycorrhizal fungal associations, or symbiotically through bacteria in root nodules. Plants that derive nitrogen from soils tend to take up either ammonium or nitrate ions, and the same soil nitrogen sources are used when mycorrhizal associations acquire soil nitrogen for their host plants. While most variations in plant nitrogen isotope ratios ($\delta^{15}N$) are associated with differences in the sources of soil nitrogen, symbiotic nitrogen fixation results in plant $\delta^{15}N$ values that are independent of the soil $\delta^{15}N$ values (Evans 2001; Robinson 2001; Craine et al. 2009). Within a plant, $\delta^{15}N$ values tend to be relatively homogeneous, but root-to-leaf differences can be as large as 3–7‰ in plants with ectomycorrhizal fungal associations and grown under high fertilizer regimes (Craine et al. 2015; Hobbie 2015). Thus, understanding the nature of roots and belowground processes is important when interpreting the $\delta^{15}N$ values of leafy vegetable, seed, and grain foods.

The $\delta^{15}N$ values of plants grown under non-fertilized conditions tend to range from −5 to +8‰, depending on soil conditions. Craine et al. (2015) summarized a range of studies, which showed that the $\delta^{15}N$ values of leaves tend to be similar to the $\delta^{15}N$ values of inorganic N in the soil. Thus, plant $\delta^{15}N$ values are a good indication of the soil N sources. However, for those plants with biological nitrogen fixation through symbiotic bacterial interactions in root nodules (such as legumes), plant $\delta^{15}N$ values tend to range from 0 to +3‰ and are independent of soil $\delta^{15}N$ values (Robinson 2001; Craine et al. 2015).

If soil ^{15}N values (mostly) influence plant ^{15}N values, what then drives the variations in soil ^{15}N values? Most studies have concluded that soil $\delta^{15}N$ values represent a balance between nitrogen inputs and nitrogen losses, especially gaseous losses that tend to enrich soil ^{15}N values because isotopically ^{15}N lighter gases are preferentially evolved during microbial degradation (Koba et al. 1998; Evans 2001; Hogberg and Read 2006; Craine et al. 2009; Craine et al. 2015; Hobbie 2015). In food systems, understanding the $\delta^{15}N$ values of fertilizers becomes key to understanding variations in plant $\delta^{15}N$ values.

3.10 Fertilizer nitrogen isotope ratios

Today, most nitrogen added to soil as fertilizer (ammonium, nitrate, urea) is derived commercially from atmospheric nitrogen by the Haber-Bosch process and, for this reason, the nitrogen isotope ratios of commercial synthetic fertilizers tend to be approximately 0‰ (Bateman et al. 2005; Bateman and Kelly 2007; Bateman et al. 2007). While there are many economic advantages to using synthetic fertilizers, biologically-based fertilizers are still used, especially in the production of organic foods (Chapter 12). These fertilizers tend to have more positive $\delta^{15}N$ values than synthetic fertilizers (Table 3.1), with the higher $\delta^{15}N$ values reflecting the trophic-level ^{15}N enrichments of animal tissues (Post 2002). This large difference in the $\delta^{15}N$ values of synthetic *vs.* organic fertilizers provides a basis for using plant $\delta^{15}N$ values to verify whether or not a food has been cultivated using an organic fertilizer (Bateman et al. 2005; Bateman and Kelly 2007; Bateman et al. 2007).

Table 3.1 Average values of nitrogen isotope ratios of different fertilizer types commonly used in crop cultivation. Data are presented as means ± one standard deviation and adapted from Bateman and Kelly (2007).

Fertilizer type	$\delta^{15}N_{2AIR}$ (‰)
Synthetic fertilizer	−0.2 ± 2.1
Seaweed-based fertilizer	+2.5 ± 1.5
Mammalian, non-manure fertilizer	+5.9 ± 1.0
Fish-based fertilizer	+7.1 ± 3.6
Manure-compost fertilizer	+8.1 ± 3.9

3.11 Sulfur isotopes in plants

Sulfur isotopes in plants exhibit limited ^{34}S depletion within an ecosystem (Peterson and Fry 1987; Krouse and Grinenko 1991; Tanz and Schmidt 2010; Song et al. 2015), although within trophic studies animals fed high protein diets exhibit an average enrichment of 2‰ in $\delta^{34}S$ values per trophic level (McCutchan et al. 2003). Across a broad spectrum of food resources, ranging from marine fish through cultivated corn, we can detect a large 15‰ range in $\delta^{34}S$ values, which can be attributed to microbial processes on both land and in the ocean, especially as sulfur is transformed among its oxidative states (Krouse and Grinenko 1991). Given an apparent limited sulfur isotope fractionation within plants, the $\delta^{34}S$ values of plants can be used as an initial indicator of soil $\delta^{34}S$ values.

Soil sulfur isotopes are known to vary geographically, particularly along marine-to-inland gradients and in the vicinity of SO_2 pollutants (Krouse and Grinenko 1991). As a consequence, $\delta^{34}S$ values have been used to assign geographical food origins (Tanz and Schmidt 2010). In combination with other stable isotopes, $\delta^{34}S$ values have been used to attribute geographical origins to several crops, such as coffee (Rodrigues et al. 2011), but sulfur isotope analysis has been more often applied to identify the origins of animal proteins and dairy products (Rossmann et al. 2000; Boner and Forstel 2004; Camin et al. 2007).

3.12 Strontium isotopes in plants

There is a rich history applying strontium isotope ratio ($^{87}Sr/^{86}Sr$) to ecological and food studies as a tracer of edaphic properties (Capo et al. 1998; Stewart et al. 1998). Strontium is taken up through plant roots as a substitute for calcium and the $^{87}Sr/^{86}Sr$ ratio thus becomes a diagnostic measure of the soil since there is no isotopic fractionation during uptake. Given that the $^{87}Sr/^{86}Sr$ ratio in geologic formations is a function of soil age, we find large geographical variations in $^{87}Sr/^{86}Sr$ ratios across the continents that can be used to assign possible provenance (Beard and Johnson 2000; Kelly et al. 2005; Chesson et al. 2012).

Geographically-based differences in plant $^{87}Sr/^{86}Sr$ ratios have been particularly useful in assigning region-of-origin to foods and food products. For example, consider that there are sufficient $^{87}Sr/^{86}Sr$ variations in the soils of the Hawaiian Islands to determine on which island coffee beans were cultivated. While the Hawaiian Islands are overall quite young in geologic terms, the ages of the individual islands are sufficiently different that soil— and thus coffee—$^{87}Sr/^{86}Sr$ ratios vary across cultivation sites (Rodrigues et al. 2011). When Sr isotope analyses of coffee beans were combined with $\delta^{13}C$, $\delta^{15}N$, and $\delta^{34}S$ values, coffee beans grown in each of the Hawaiian Islands could be fully distinguished from each other. Liu et al. (2014) extended the application of Sr isotopes to source coffee beans, showing that coffee from 14 countries in Africa, America, and Asia could be distinguished on the basis of $^{87}Sr/^{86}Sr$, $\delta^{11}B$, and $\delta^{18}O$ values. As the $^{87}Sr/^{86}Sr$ ratios are reflective of soil type and soil ages at the site of cultivation, a variety of other food crops have been geographically distinguished on the basis of Sr isotopes, including asparagus (Ariyama et al. 2008), onions (Hiraoka et al. 2016), wheat (Podio et al. 2013; Liu et al. 2016), peppers (Song et al. 2014), and rice (Song et al. 2014).

3.13 Conclusions

There are well-established theories to explain variations observed in the natural abundances of hydrogen, carbon, and oxygen isotopes in plants. Here we briefly introduced the theories that provide food scientists with sufficient information to predict variations in the isotopes of these elements in plant-based foods. While our understanding of natural isotope ratio variations in nitrogen, sulfur, and strontium is not yet as fully developed, it is possible to generate preliminary interpretations of variations in these isotope ratios of plant and plant-based foods and link these variations to growth condition and geography. The applications of stable isotope analysis to plant biology and food science provide a foundation for interpreting environmental, cultivation, and genetic-based variations in our plant-based foods.

References

Adirejo, A. L., O. Navaud, S. Munos, N. B. Langlade, T. Lamaze and P. Grieu. 2014. Genetic control of water use efficiency and leaf carbon isotope discrimination in sunflower (*Helianthus annuus* L.) subjected to two drought scenarios. *Plos ONE* 9: e101218.

Ariyama, K., M. Shinozaki and A. Kawasaki. 2008. Identification of Marchfeld asparagus using Sr isotope ratio measurements by MC-ICP-MS. *Analytical and Bioanalytical Chemistry* 390: 487–494.

Badeck, F. W., G. Tcherkez, S. Nogues, C. Piel and J. Ghashghaie. 2005. Post-photo synthetic fractionation of stable carbon isotopes between plant organs—a widespread phenomenon. *Rapid Communications in Mass Spectrometry* 19: 1381–1391.

Bateman, A. S. and S. D. Kelly. 2007. Fertilizer nitrogen isotope signatures. *Isotopes in Environmental and Health Studies* 43: 237–247.

Bateman, A. S., S. D. Kelly and T. D. Jickells. 2005. Nitrogen isotope relationships between crops and fertilizer: Implications for using nitrogen isotope analysis as an indicator of agricultural regime. *Journal of Agricultural and Food Chemistry* 53: 5760–5765.

Bateman, A. S., S. D. Kelly and M. Woolfe. 2007. Nitrogen isotope composition of organically and conventionally grown crops. *Journal of Agricultural and Food Chemistry* 55: 2664–2670.

Bchir, A., J. M. Escalona, A. Gallé, E. Hernández-Montes, I. Tortosa, M. Braham and H. Medrano. 2016. Carbon isotope discrimination ($\delta^{13}C$) as an indicator of vine water status and water use efficiency (WUE): Looking for the most representative sample and sampling time. *Agricultural Water Management* 167: 11–20.

Beard, B. L. and C. M. Johnson. 2000. Strontium isotope composition of skeletal material can determine the birth place and geographic mobility of humans and animals. *Journal of Forensic Sciences* 45: 1049–1061.

Boner, M. and H. Forstel. 2004. Stable isotope variation as a tool to trace the authenticity of beef. *Analytical and Bioanalytical Chemistry* 378: 301–310.

Bong, Y. S., K. S. Lee, W. J. Shin and J. S. Ryu. 2008. Comparison of the oxygen and hydrogen isotopes in the juices of fast-growing vegetables and slow-growing fruits. *Rapid Communications in Mass Spectrometry* 22: 2809–2812.

Bowling, D. R., D. E. Pataki and J. T. Randerson. 2008. Carbon isotopes in terrestrial ecosystem pools and CO_2 fluxes. *New Phytologist* 178: 24–40.

Brause, A. R., J. M. Raterman, D. R. Petrus and L. W. Doner. 1984. Fruits and fruit products: Verificaiton of authenticity of orange juice. *Journal of Association of Official Analytical Chemistry* 67: 535–539.

Breas, O., F. Reniero and G. Serrini. 1994. Isotope ratio mass spectrometry: Analysis of wines from different European countries. *Rapid Communications in Mass Spectrometry* 8: 967–970.

Calderone, G., C. Guillou, F. Reniero and N. Naulet. 2007. Helping to authenticate sparkling drinks with C-13/C-12 of CO_2 by gas chromatography-isotope ratio mass spectrometry. *Food Research International* 40: 324–331.

Calderone, G., N. Naulet, C. Guillou and F. Reniero. 2004. Characterization of European wine glycerol: Stable isotope carbon isotope. *Journal of Agricultural and Food Chemistry* 52: 5902–5906.

Camin, F., L. Bontempo, K. Heinrich, M. Horacek, S. D. Kelly, C. Schlicht, F. Thomas, F. J. Monahan, J. Hoogewerff and A. Rossmann. 2007. Multi-element (H, C, N, S) stable isotope characteristics of lamb meat from different European regions. *Analytical and Bioanalytical Chemistry* 389: 309–320.

Camin, F., P. M. Bontempo, S. Fabroni, W. Faedi, S. Magnani, B. G. M. Bonoli, M. R. Tabilio, S. Musmeci, A. Rossmann, S. D. Kelly and P. Rapisarda. 2011. Potential isotopic and chemical markers for characterising organic fruits. *Food Chemistry* 125: 1072–1082.

Capo, R. C., B. W. Stewart and O. A. Chadwick. 1998. Strontium isotopes as tracers of ecosystem processes: Theory and methods. *Geoderma* 82: 197–225.

Cerling, T. E., J. E. Barnette, G. J. Bowen, L. A. Chesson, J. R. Ehleringer, C. H. Remien, P. Shea, B. J. Tipple and J. B. West. 2016. Forensic stable isotope biogeochemistry. *Annual Review of Earth and Planetary Sciences* 44: 175–206.

Cernusak, L. A., J. S. Pate and G. D. Farquhar. 2002. Diurnal variation in the stable isotope composition of water and dry matter in fruiting *Lupinus angusifolius* under field conditions. *Plant, Cell and Environment* 25: 893–907.

Cernusak, L. A., G. Tcherkez, C. Keitel, W. K. Cornwell, L. S. Santiago, A. K. Knapp, M. M. Barbour, D. G. Williams, P. B. Reich, D. Ellsworth, T. E. Dawson, H. G. Griffiths, G. D. Farquhar and I. J. Wright. 2009. Why are non-photosynthetic tissues generally ^{13}C enriched compared with leaves in C$_3$ plants? Review and synthesis of current hypotheses. *Functional Plant Biology* 36: 199–213.

Chesson, L. A., G. J. Bowen and J. R. Ehleringer. 2010a. Analysis of the hydrogen and oxygen stable isotope ratios of beverage waters without prior water extraction using isotope ratio infrared spectroscopy. *Rapid Communications in Mass Spectrometry* 24: 3205–3213.

Chesson, L. A., B. J. Tipple, G. N. Mackey, D. P. Fernandez and J. R. Ehleringer. 2012. Strontium isotope ratios of tap water from the coterminous USA. *Ecosphere* 3: Art67.

Chesson, L. A., L. O. Valenzuela, S. P. O'Grady, T. E. Cerling and J. R. Ehleringer. 2010b. Links between purchase location and stable isotope ratios of bottled water, soda, and beer in the United States. *Journal of Agricultural and Food Chemistry* 58: 7311–7316.

Comstock, J. P. and J. R. Ehleringer. 1992. Correlating genetic variation in carbon isotopic composition with complex climatic gradients. *Proceedings of the National Academy of Sciences of the United States of America* 89: 7747–7751.

Craine, J. M., E. N. J. Brookshire, M. D. Cramer, N. J. Hasselquist, K. Koba, E. Marin-Spiotta and L. Wang. 2015. Ecological interpretations of nitrogen isotope ratios of terrestrial plants and soils. *Plant and Soil* 396: 1–26.

Craine, J. M., A. J. Elmore, M. P. Aidar, M. Bustamante, T. E. Dawson, E. A. Hobbie, A. Kahmen, M. C. Mack, K. K. McLauchlan, A. Michelsen, G. B. Nardoto, L. H. Pardo, J. Penuelas, P. B. Reich, E. A. Schuur, W. D. Stock, P. H. Templer, R. A. Virginia, J. M. Welker and I. J. Wright. 2009a. Global patterns of foliar nitrogen isotopes and their relationships with climate, mycorrhizal fungi, foliar nutrient concentrations, and nitrogen availability. *New Phytologist* 183: 980–992.

DaMatta, F. M. 2004. Ecophysiological constraints on the production of shaded and unshaded coffee: A review. *Field Crops Research* 86: 99–114.

Dawson, T. E., S. Mambelli, A. H. Plamboeck, P. H. Templer and K. P. Tu. 2002. Stable isotopes in plant ecology. *Annual Review of Ecology and Systematics* 33: 507–559.

DeNiro, M. J. and S. Epstein. 1989. Determination of the concentration and stable isotopic composition of oxygen in organic matter containing carbon, hydrogen, oxygen, nitrogen, and sulfur. *Analytical Chemistry* 61: 1887–1889.

Dhanapal, A. P., J. D. Ray, S. K. Singh, V. Hoyos-Villegas, J. R. Smith, L. C. Purcell, C. Andy King, P. B. Cregan, Q. Song and F. B. Fritschi. 2015. Genome-wide association study (GWAS) of carbon isotope ratio (δ^{13}C) in diverse soybean [*Glycine max* (L.) Merr.] genotypes. *Theoretical and Applied Genetics* 128: 73–91.

Dunbar, J. and A. T. Wilson. 1983. Oxygen and hydrogen isotopes in fruit and vegetable juices. *Plant Physiology* 72: 725–727.

Ehleringer, J. R. 1990. Correlations between carbon isotope discrimination and leaf conductance to water vapor in common beans. *Plant Physiology* 93: 1422–1425.

Ehleringer, J. R. and T. E. Dawson. 1992. Water uptake by plants—perspectives from stable isotope composition. *Plant Cell and Environment* 15: 1073–1082.

Ehleringer, J. R., C. B. Field, Z. F. Lin and C. Y. Kuo. 1986. Leaf carbon isotope and mineral composition in subtropical plants along an irradiance cline. *Oecologia* 70: 520–526.

Ehleringer, J. R., A. E. Hall and G. D. Farquhar (eds.). 1993. *Stable Isotopes and Plant Carbon/Water Relations*. Academic Press, San Diego.

Ehleringer, J. R., J. R. Roden and T. E. Dawson. 2000. Assessing ecosystem-level water relations through stable isotope ratio analyses. pp. 181–198. *In*: O. E. Sala, R. Jackson, H. A. Mooney and R. Howarth (eds.). *Methods in Ecosystem Science*. Springer Verlag, New York.

Ehleringer, J. R. and D. R. Sandquist. 2014. Photosynthesis: Physiological and ecological considerations. pp. 245–268. *In*: L. Taiz, E. Zeiger, I. M. Moller and A. Murphy (eds.). *Plant Physiology and Development*, Sixth Edition. Sinuer Associates, Sunderland, MA.

Ehleringer, J. R., J. W. White, D. A. Johnson and M. Brick. 1990. Carbon isotope discrimination, photosynthetic gas exchange, and transpiration efficiency in beans and range grasses. *Acta Oecologica* 11: 611–625.

Evans, R. D. 2001. Physiological mechanisms influencing plant nitrogen isotope composition. *Trends in Plant Science* 6: 121–126.

Farquhar, G. D. 1983. On the nature of carbon isotope discrimination in C_4 species. *Australian Journal of Plant Physiology* 10: 205–226.

Farquhar, G. D., J. R. Ehleringer and K. T. Hubick. 1989. Carbon isotope discrimination and photosynthesis. *Annual Review of Plant Physiology and Plant Molecular Biology* 40: 503–537.

Farquhar, G. D., M. H. O'Leary and J. A. Berry. 1982. On the relationship between carbon isotope discrimination and the intercellular carbon dioxide concentration in leaves. *Australian Journal of Plant Physiology* 9: 121–137.

Flamini, R. and A. Panighel. 2006. Mass spectrometry in grape and wine chemistry. Part II: The consumer protection. *Mass Spectrometry Reviews* 25: 741–774.

Flanagan, L. B., J. P. Comstock and J. R. Ehleringer. 1991. Comparison of modeled and observed environmental influences on the stable oxygen and hydrogen isotope composition of leaf water in *Phaseolus vulgaris* L. *Plant Physiology* 96: 588–596.

Flanagan, L. B. and J. R. Ehleringer. 1991. Effects of mild water stress and diurnal changes in temperature and humidity on the stable oxygen and hydrogen isotopic composition of leaf water in *Cornus stolonifera* L. *Plant Physiology* 97: 298–305.

Flanagan, L. B. and K. H. Johnsen. 1995. Genetic variation in carbon isotope discrimination and its relationship to growth under field conditions in full-sib families of *Picea mariana*. *Canadian Journal of Forestry Research* 25: 39–47.

Fugel, R., R. Carle and A. Schieber. 2005. Quality and authenticity control of fruit purees, fruit preparations and jams—a review. *Trends in Food Science & Technology* 16: 433–441.

Gonzalez-Martin, I., C. Gonzalez-Perez and E. Macias-Marques. 1997. Contribution to the study of the origin of CO_2 in Spanish sparkling wines by determination of the $^{13}C/^{12}C$ isotope ratio. *Journal of Agricultural and Food Chemistry* 45: 1149–1151.

Greule, M., A. Rossmann, H. L. Schmidt, A. Mosandl and F. Keppler. 2015. A stable isotope approach to assessing water loss in fruits and vegetables during storage. *Journal of Agricultural and Food Chemistry* 63: 1974–1981.

Hammond, D. A. 1996. Authenticity of fruit juices, jams and preserves. pp. 15–59. *In*: P. R. Ashurst and M. J. Dennis (eds.). *Food Authenticity*. Blackie Academic & Professional, London.

Hiraoka, H., S. Morita, A. Izawa, K. Aoyama, K. -C. Shin and T. B. Nakano. 2016. Tracing the geographical origin of onions by strontium isotope ratio and strontium content. *Analytical Sciences* 32: 781–788.

Hobbie, E. A. and R. A. Werner. 2004. Intramolecular, compound-specific, and bulk carbon isotope patterns in C_3 and C_4 plants: A review and synthesis. *New Phytologist* 161: 371–385.

Hobbie, S. E. 2015. Plant species effects on nutrient cycling: Revisiting litter feedbacks. *Trends in Ecology and Evolution* 30: 357–363.

Hogberg, P. and D. J. Read. 2006. Towards a more plant physiological perspective on soil ecology. *Trends in Ecology and Evolution* 21: 548–554.

Hubick, K. T., G. D. Farquhar and R. Shorter. 1986. Correlation between water-use efficiency and carbon isotope discrimination in diverse peanut (*Arachis*) germ plasm. *Australian Journal of Plant Physiology* 13: 803–816.

Kelly, S., K. Heaton and J. Hoogewerff. 2005. Tracing the geographical origin of food: The application of multi-element and multi-isotope analysis. *Trends in Food Science & Technology* 16: 555–567.

Koba, K., N. Tokuchi, T. Yoshioka, E. A. Hobbie and G. Iwatsubo. 1998. Natural abundance of nitrogen-15 in a forest soil. *Soil Science Society of America Journal* 62: 778–781.

Kornexl, B. E., M. Gehre, R. Hofling and R. A. Werner. 1999. Online $\delta^{18}O$ measurement of organic and inorganic substances. *Rapid Communications in Mass Spectrometry* 13: 1685–1693.

Krouse, H. R. and V. A. Grinenko (eds.). 1991. *Stable Isotopes Natural and Anthropogenic Sulphur in the Environment*. John Wiley & Sons, New York.

Lauteri, M., A. Scartazza, M. C. Guido and E. Brugnoli. 1997. Genetic variation in photosynthetic capacity, carbon isotope discrimination and mesophyll conductance in provenances of *Castanea sativa* adapted to different environments. *Functional Ecology* 11: 675–683.

Liu, H. C., Y. Wei, H. Lu, S. Wei, T. Jiang, Y. Zhang and B. Gui. 2016. Combination of the $^{87}Sr/^{86}Sr$ ratio and light stable isotopic values ($\delta^{13}C$, $\delta^{15}N$ and δD) for identifying the geographical origin of winter wheat in China. *Food Chemistry* 212: 367–373.

Liu, H. C., C. F. You, C. Y. Chen, Y. C. Liu and M. T. Chung. 2014. Geographic determination of coffee beans using multi-element analysis and isotope ratios of boron and strontium. *Food Chemistry* 142: 439–445.

McCutchan, J. H., W. M. Lewis, C. Kendall and C. C. McGrath. 2003. Variation in trophic shift for stable isotope ratios of carbon, nitrogen, and sulfur. *Oikos* 102: 378–390.

Ogrinc, N., K. Bat, I. J. Kosir, T. Golob and R. Kokkinofta. 2009. Characterization of commercial Slovenian and Cypriot fruit juices using stable isotopes. *Journal of Agricultural and Food Chemistry* 57: 6764–6769.

Peterson, B. J. and B. Fry. 1987. Stable isotopes in ecosystem studies. *Annual Review of Ecology and Systematics* 18: 293–320.

Podio, N. S., M. V. Baroni, R. G. Badini, M. Inga, H. A. Ostera, M. Cagnoni, E. A. Gautier et al. 2013. Elemental and isotopic fingerprint of Argentinean wheat. Matching soil, water and crop composition to differentiate provenance. *Journal of Agricultural and Food Chemistry* 61: 3763–3773.

Polania, J. A., C. Poschenrieder, S. Beebe and I. M. Rao. 2016. Effective use of water and increased dry matter partitioned to grain contribute to yield of common bean improved for drought resistance. *Frontiers in Plant Science* 7: Art660.

Post, D. M. 2002. Using stable isotopes to estimate trophic position: Models, methods, and assumptions. *Ecology* 83: 703–718.

Robinson, D. 2001. $\delta^{15}N$ as an integrator of the nitrogen cycle. *Trends in Ecology and Evolution* 16: 153–162.

Roden, J. S. and J. R. Ehleringer. 1999. Observations of hydrogen and oxygen isotopes in leaf water confirm the Craig-Gordon model under wide-ranging environmental conditions. *Plant Physiology* 120: 1165–1173.

Roden, J. S., G. G. Lin and J. R. Ehleringer. 2000. A mechanistic model for interpretation of hydrogen and oxygen isotope ratios in tree-ring cellulose. *Geochimica et Cosmochimica Acta* 64: 21–35.

Rodrigues, C., M. Brunner, S. Steiman, G. J. Bowen, J. M. F. Nogueira, L. Gautz, T. Prohaska and C. Máguas. 2011. Isotopes as tracers of the Hawaiian coffee-producing regions. *Journal of Agricultural and Food Chemistry* 59: 10239–10246.

Rossmann, A., G. Haberhauer, S. Holzl, P. Horn, F. Pichlmayer and S. Voerkelius. 2000. The potential of multielement stable isotope analysis for regional origin assignment of butter. *European Food Research and Technology* 211: 32–40.

Rossmann, A., F. Reniero, I. Moussa, H. L. Schmidt, G. Versini and M. H. Merle. 1999. Stable oxygen isotope content of water of EU data-bank wines from Italy, France and Germany. *Zeitschrift Fur Lebensmittel-Untersuchung Und-Forschung a-Food Research and Technology* 208: 400–407.

Sage, R. F. 2014. Photosynthetic efficiency and carbon concentration in terrestrial plants: The C_4 and CAM solutions. *Journal of Experimental Botany* 65: 3323–3325.

Sage, R. F. 2016. A portrait of the C_4 photosynthetic family on the 50th anniversary of its discovery: Species number, evolutionary lineages, and hall of fame. *Journal of Experimental Botany* 67: 4039–4056.

Sage, R. F., R. Khoshravesh and T. L. Sage. 2014. From proto-Kranz to C_4 Kranz: Building the bridge to C_4 photosynthesis. *Journal of Experimental Botany* 65: 3341–3356.

Santiso, X., L. López, K. J. Gilbert, R. Barreiro, M. C. Whitlock and R. Retuerto. 2015. Patterns of genetic variation within and among populations in *Arbutus unedo* and its relation with selection and evolvability. *Perspectives in Plant Ecology, Evolution and Systematics* 17: 185–192.

Schmidt, H. -L., R. A. Werner and A. Roßmann. 2001. [18]O pattern and biosynthesis of natural plant products. *Phytochemistry* 58: 9–32.

Song, B. -Y., M. K. Gautam, J. -S. Ryu, D. Lee and K. -S. Lee. 2015. Effects of bedrock on the chemical and Sr isotopic compositions of plants. *Environmental Earth Sciences* 74: 829–837.

Song, B. -Y., J. -S. Ryu, H. S. Shin and K. -S. Lee. 2014. Determination of the source of bioavailable Sr using [87]Sr/[86]Sr tracers: A case study of hot pepper and rice. *Journal of Agricultural and Food Chemistry* 62: 9232–9238.

Stewart, B. W., R. C. Capo and O. A. Chadwick. 1998. Quantitative strontium isotope models for weathering, pedogenesis and biogeochemical cycling. *Geoderma* 82: 173–195.

Tanz, N. and H.-L. Schmidt. 2010. δ^S34-value measurements in food origin assignments and sulfur isotope fractionations in plants and animals. *Journal of Agricultural and Food Chemistry* 58: 3139–3146.

Tcherkez, G., A. Mahe and M. Hodges. 2011. $^{12}C/^{13}C$ fractionations in plant primary metabolism. *Trends in Plant Science* 16: 499–506.

Tieszen, L. L. and T. Fagre. 1993. Carbon isotopic variability in modern archaeological maize. *Journal of Archaeological Science* 20: 25–40.

Verlinden, M. S., R. Fichot, L. S. Broeckx, B. Vanholme, W. Boerjan and R. Ceulemans. 2015. Carbon isotope compositions ($\delta^{13}C$) of leaf, wood and holocellulose differ among genotypes of poplar and between previous land uses in a short-rotation biomass plantation. *Plant, Cell & Environment* 38: 144–156.

Werner, R. A., N. Buchmann, R. T. W. Siegwolf, B. E. Kornexl and A. Gessler. 2011. Metabolic fluxes, carbon isotope fractionation and respiration—lessons to be learned from plant biochemistry. *New Phytologist* 191: 10–15.

West, J. B., J. R. Ehleringer and T. E. Cerling. 2007. Geography and vintage predicted by a novel GIS model of wine $\delta^{18}O$. *Journal of Agricultural and Food Chemistry* 55: 7075–7083.

White, J. W., J. A. Castillo and J. R. Ehleringer. 1990. Associations between productivity, root growth and carbon isotope discrimination in *Phaseolus vulgaris* under water deficit. *Australian Journal of Plant Physiology* 17: 189–198.

Zacharisen, M. H., M. A. Brick, A. G. Fisher, J. B. Ogg and J. R. Ehleringer. 1999. Relationships between productivity and carbon isotope discrimination among dry bean lines and F-2 progeny. *Euphytica* 105: 239–250.

CHAPTER 4

Introduction to Stable Isotopes in Food Webs

Timothy D. Jardine,[1], Keith A. Hobson[2] and David X. Soto[3]*

4.1 Introduction

"Things can be low on the food chain, but that doesn't mean they're lowly."

Gary Larson (b. 1950), American cartoonist and creator of The Far Side

Here we set the dinner table with an introduction to food webs. Prior chapters allowed the reader to grasp the fundamentals of isotope ratio measurements and how those are expressed in plants, so we can now begin to think about how these signals are picked up by animals from their diet. This chapter draws extensively on isotope applications in the ecological literature, as well as decades of captive rearing studies.

4.2 A brief history of isotopes in foodweb ecology

The forensic application of stable isotopes requires an understanding of the ways in which isotopes behave in food webs. Empirical and theoretical work on food webs began around the turn of the 20th century (Pimm et al. 1991), driven largely by a need for quantitative information that could predict yields in fisheries and agriculture. Food webs later became entrenched in the ecological literature, with the work of Elton (1927) and

[1] School of Environment and Sustainability, University of Saskatchewan, Saskatoon, SK, CANADA.
[2] Department of Biology, Western University, London, ON, CANADA.
Email: khobson6@uwo.ca
[3] Environment Canada, Saskatoon, SK, CANADA.
Email: David.Soto@ec.gc.ca
* Corresponding author: tim.jardine@usask.ca

Lindeman (1942), who visualized biomass pyramids and food cycles (Figure 4.1). Food webs can be described simply as "diagrams depicting which species in a community interact" (Pimm 1992); food webs depict both actors (structure) and actions (function), but both the occurrence and magnitude of interactions can change in space and time. Later, when consideration of food webs moved beyond production systems, into the domains of animal ecology and conservation biology, the temporal and spatial dynamic nature of food webs posed analytical challenges, because single-point assessments, using the tools of the day, namely short-term (visual) foraging observations or stomach content analyses, proved insufficient to capture the mixture of the energy pathways leading to an individual animal, especially if the individual or its diet was transient. What was needed was a recorder of information that could be retrieved from an organism that detailed its history and relationship with other organisms, in its ecological orbit, preferably on different time scales. Enter stable isotopes!

Stable isotopes of the major biological elements, especially those of carbon ($^{13}C/^{12}C$) and nitrogen ($^{15}N/^{14}N$), have aided greatly in reconstructing the past niche, habitat, and trophic level (food chain position) of individual organisms. We can look to laboratory experiments in the late 1970s for

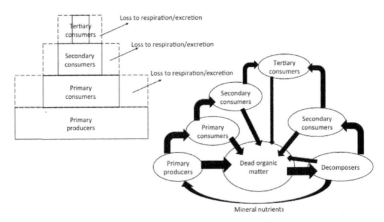

Figure 4.1 Simple conceptual diagrams of (left) biomass pyramids and (right) food cycles. Biomass is reduced at each step in the food chain because animals burn carbon (respiration) and lose nitrogen (excretion) as they carry out living functions. Primary producers include photosynthetic organisms (e.g., grasses, trees, algae); primary consumers are grazers (e.g., rabbits, grasshoppers, snails). Secondary consumers such as fish, amphibians and small mammals feed on primary consumers. Tertiary consumers are top predators (e.g., lions, eagles, polar bears, crocodiles) that are not preyed upon themselves. When plants and animals die, they contribute to a large reservoir of dead organic matter (e.g., leaf litter, carcasses). Detritivores feed on this dead organic matter and are themselves preyed upon by other predators. The return of mineral nutrients to primary producers via the decomposer pathway completes the cycle. All steps in these pyramids and cycles can potentially be traced with stable isotopes.

the germination of isotope food web applications. DeNiro and Epstein (1978, 1981a) reared a variety of animals on diets that differed isotopically, and found that while $^{13}C/^{12}C$ was largely conserved, $^{15}N/^{14}N$ increased relative to the diet. Field investigations confirmed this latter finding, with an average enrichment in $\delta^{15}N$ of 3.4‰ (Minagawa and Wada 1984), a number that continues to be supported today, albeit with caveats. Here, finally, were the necessary connections between the isotopic variations produced by plants (see Chapter 3) and the animals that fed upon them for both $^{13}C/^{12}C$ (Minson et al. 1975; Haines 1976) and $^{15}N/^{14}N$ compositions (Miyake and Wada 1967; Steele and Daniel 1978). In addition, these early days saw the potential for hydrogen isotopes ($^{2}H/^{1}H$) in food web ecology based on the pioneering work of Estep and Dabrowski (1980), in which they observed that $^{2}H/^{1}H$ values in diet influenced $^{2}H/^{1}H$ values of mouse tissues. After this paper, however, little research occurred until the 2000s, primarily due to uncertainty over the relative contribution of diet and water to tissue H (DeNiro and Epstein 1981b) and analytical issues such as the confounding effect of extrinsic H—hydrogen present due to interactions with external water sources—during sample processing (Wassenaar and Hobson 2000a).

While controlled laboratory studies were critical for determining the steady-state relationship between the isotope ratios of an animal and its diet, what remained untested was the rate at which animals approached the isotope value of a new diet when diets were switched. From a forensic perspective, this is a crucial element since it defines the period over which dietary information can be derived from the isotope analysis of tissues. Fry and Arnold (1982) were the first systematically to assess isotopic change following a diet switch, and found that rapidly growing brown shrimp (*Penaeus aztecus*) quickly approached isotopic equilibrium, with half-lives (Box 4.1) that ranged from four to 19 days and were a function of growth rate on different diets. While Fry and Arnold were measuring turnover in shrimp, Tieszen et al. (1983) were conducting a similar experiment with gerbils (*Meriones unguienlatus*). They calculated half-lives that ranged from six to 48 days but varied as a function of tissue type. Liver, with its high metabolic rate, exhibited the fastest turnover and metabolically inert hair took the longest to reach equilibrium (Tieszen et al. 1983). Hobson and Clark (1992) later showed rapid but variable turnover rates in quail (*Coturnix japonica*) tissues, ranking in pace from a 3-day half-life in liver to a 173-day half-life in bone collagen. Though growth rates were not reported in either of the latter studies, it can be assumed that growth was minimal because adult gerbils and quail were used. Later investigations by Hesslein et al. (1993) found that isotopic change in juvenile (growing) whitefish (*Coregonus nasus*) was almost entirely driven by the addition of new tissue, with only minimal contributions from tissue replacement by turnover. These very different outcomes with an aquatic invertebrate and fish, where growth drives isotopic change, and a terrestrial mammal and

Box 4.1 Data for blood of Pallas' long-tongued bats (*Glossophaga soricina*) from Mirón et al. 2006.

How fast do isotope ratios change?

After switching to a new diet or habitat, tissues will take on the new value as a function of their metabolic activity and growth. This can create problems in forensics if an animal in question has switched diets in the recent past.

How do I estimate half-life?

Controlled laboratory experiments are used that involve repeated isotope measurements (δt) over time (t). Data are then modelled according to equation 1. By iteration, a best-fit value of c can be determined and a half-life calculated using equation 2.

The measurement unit is typically a **half-life**, the number of days required to move halfway from the old value (δi) to the new value (δe).

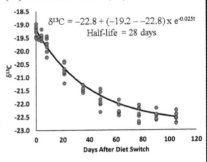

$$\delta^{13}C = -22.8 + (-19.2 - -22.8) \times e^{-0.025t}$$
Half-life = 28 days

1) $\delta t = \delta e + (\delta i - \delta e) \times e^{-ct}$

2) Half-life = $\ln(0.5)/c$

bird, where metabolism drives isotopic change, revealed fundamental differences across the animal kingdom that would be explored later.

Through the 1970s and 1980s, single element isotope analysis was the norm. This changed with the advent of continuous-flow isotope ratio mass spectrometry (CF-IRMS), which allowed the determination of $\delta^{13}C$ and $\delta^{15}N$ values from combustion of a single sample (Fry et al. 1992) and unlocked the potential for the construction of entire food webs in bivariate isotopic space, something that is taken for granted today (Figure 4.2). While early food web work occasionally included both $\delta^{13}C$ and $\delta^{15}N$ measurements, these two isotopes were often treated independently (e.g., Estep and Vigg 1985) or only considered a small number of trophic levels (Peterson et al. 1985). Ambrose and DeNiro (1985) showed the power of a dual-isotope approach in a savanna grassland setting, demonstrating that $\delta^{13}C$ values could differentiate between browsing and grazing mammals, and within each of those distinct food webs, predators could be separated from herbivores using $\delta^{15}N$ values. Dual-isotope bi-plots for comprehensive food webs containing both plants and animals originated in freshwater environments with Fry (1991) and in the oceans with Fry (1988) and Hobson and Welch (1992). Fry's (1991) diagrams showed sources from the land (terrestrial), water column (planktonic), and bottom (benthic) that underpinned food webs containing insects, crustaceans, amphibians, and

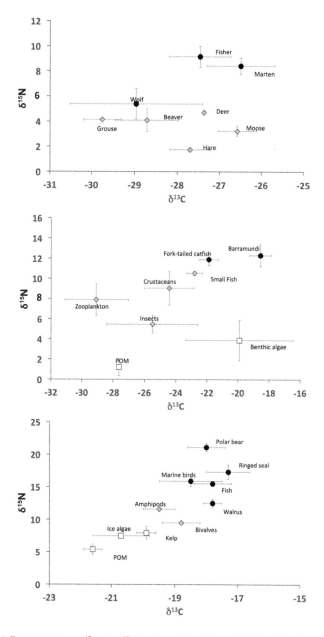

Figure 4.2 Representative $\delta^{15}N$ *vs.* $\delta^{13}C$ scatterplots of terrestrial (top), freshwater (middle), and marine (bottom) food webs for predators (solid circles), omnivores and grazers (shaded diamonds), and primary producers (open squares). The terrestrial scatterplot is for a forest-wetland ecosystem in central Canada. The freshwater scatterplot is for a series of waterholes in northern Australia (Jardine et al. 2013). The marine scatterplot is for Lancaster Sound in the High Arctic (Hobson and Welch 1992).

fish in a stream and lake. In coastal Newfoundland, Canada, Fry (1988) tracked particulate organic matter (POM, $\delta^{13}C \approx -21‰$, $\delta^{15}N \approx +5‰$) up to predatory fishes ($\delta^{13}C \approx -17‰$, $\delta^{15}N \approx +14‰$), representing approximately 2.5 trophic levels. In a high Arctic marine food web Hobson and Welch (1992) observed, based on $\delta^{13}C$ values, that ice algae and POM (largely plankton) were isotopically different and drove production in benthic areas and in the open water, respectively. From this base, $\delta^{15}N$ values progressively increased through invertebrates, fish, birds, and mammals, culminating in polar bears that were approximately 14‰ enriched in ^{15}N compared to the primary producers—representing *ca.* 5 trophic levels. The dual isotope approach rapidly became commonplace with greater analytical ease and further understanding of source and discrimination patterns between carbon and nitrogen isotopes. At present, other isotopes ($^2H/^1H$ or δ^2H, $^{18}O/^{16}O$ or $\delta^{18}O$, $^{34}S/^{32}S$ or $\delta^{34}S$) are increasingly providing food web information thanks to advances in IRMS techniques (e.g., Fry 2006; Wassenaar et al. 2015). Further insights into the chemical history of individual animals are now possible using compound specific isotope analyses (CSIA), and these approaches are slowly becoming commonplace. As the isotopic toolbox for food web studies continues to expand, multi-isotope and multi-compound diagrams will, no doubt, be seen more frequently.

This chapter will highlight developments in the past 25 years that have advanced our understanding of isotope behavior in food webs, with a view to providing guidance for those interested in forensic back-calculations of diet or habitat of questioned animal tissue: Where was the animal living when the tissue was formed? What was the animal eating when the tissue was formed? Some of the advancements in our understanding have improved the confidence with which stable isotope techniques can link animals to particular diets or habitats, while others have highlighted important caveats or limitations. It is only through careful consideration of these caveats that one can reliably interpret forensic evidence and for this evidence to contribute appropriately to the weight of evidence in any case.

4.3 Key considerations in studying food webs with isotopes

Tracing the composition of materials or their spatial origins using stable isotopes fundamentally depends on the existence of an isotopic structure in nature. All living organisms acquire their tissue isotopic compositions through food webs and it is important to know the basic processes that can alter the isotopic composition of consumers through their diets and so characterize them for forensic investigations.

There are three basic principles applied to the use of stable isotope measurements to infer prior diet and provenance (Hobson et al. 2010). The first is that the "isoscape" or "isotopic terrain" through which the material of interest moves or exchanges must be either known or inferred.

While sedentary animals such as livestock have a relatively constant isoscape and their provenance can be evaluated based on biogeochemical characteristics that confer source information (Chapter 6), wildlife are more complicated. Animals may move between areas with different physical characteristics (e.g., rainfall amounts and temperature ranges) that contain plant communities with different photosynthetic pathways (e.g., C_3, C_4, CAM; Chapter 3) or those growing on variable geological substrates containing distinct isotope ratios of trace elements [e.g., Pb (Vogel et al. 1990; Stewart et al. 2003)]. In many cases, isoscapes or the potential range of the animal are poorly defined. The second principle is that, even if the isoscape is well characterized, the isotopic composition of animals may not exactly match those of the baseline (food and water) due to isotopic discrimination through physiological processes, such as metabolic routing of specific macronutrients to tissues (Martinez del Rio et al. 2009; Wolf et al. 2015). The third principle is that the isotope value of a tissue sampled at a given time will be a function of dietary integration over the period in which it was grown. Metabolically active tissues will represent diet over a short period if they turn over quickly and over a long period if they turn over slowly (Hobson and Clark 1992). Other tissues like feather, hair, and nail are metabolically inert once they are formed and so it is possible to recover a longitudinal isotopic record of previous diet or habitat.

The successful use of a food web approach in assigning animals to geographic locations, therefore, requires an understanding of:

- baseline isotope patterns at the scale of interest,
- tissue-specific isotopic discrimination, and
- patterns of turnover.

The following sections will address each of these components in turn.

4.4 The spatial template: Isoscapes

For animal tissues, terrestrial and freshwater isotopic patterns have been used in assigning organisms to a particular geographic location. The scale of these investigations varies, from offshore *vs.* onshore feeding on the order of hundreds of meters by fish within lakes (Vander Zanden and Vadeboncoeur 2002), to regional movements of several kilometers by elephants across a forest-savanna transition (Cerling et al. 2009), to migrations by birds across continents and oceans that may cover thousands of kilometers (Hobson 2008). In all cases, biogeochemical processes impart distinct isotope ratios to food sources for animals that differ among areas. When these patterns vary systematically, we can refer to them as *isoscapes*, presenting a powerful tool to map the use of the landscape by wildlife or the likely origin of unlawfully acquired wildlife products.

4.4.1 Isoscapes derived from gradients in plant physiology (C_3/C_4/CAM)

The well-established differences in ^{13}C compositions among plants using C_3, C_4, and CAM photosynthesis have been described in Chapter 3. Early studies exploited these differences to investigate the provenance of animal tissues, including humans. The timing of domestication of corn/maize (a C_4 plant) was revealed by analyzing shifts in human bone collagen from lower $\delta^{13}C$ values (*ca.* −21‰) to higher $\delta^{13}C$ values (*ca.* −12‰) at archaeological sites in central United States (Van der Merwe and Vogel 1978). Modern applications allow identification of corn consumption by birds and their assignment to areas of dense corn-growing in the Midwest United States (Alisauskas et al. 1998; Wassenaar and Hobson 2000b). Plant photosynthetic pathways are, however, driven largely by climatic factors, and so can vary spatially across continents and/or with altitude. Thus, there has been interest in deriving isoscapes based on the expected distribution of C_3 and C_4 plants (e.g., Hattersley 1983).

Isoscapes modeled on the expected proportion of C_3 and C_4 plants across landscapes are based on plant physiology models that in turn make use of the climate conditions that favor each photosynthetic pathway (Still and Powell 2010). By assuming fixed endpoints for C_3 (−27‰) and C_4 (−12‰) plants, it is possible to estimate a foodweb $\delta^{13}C$ value that is spatially integrated as a mapped surface (GIS layer). These *foodweb surfaces* can then be modeled into a tissue carbon isotope ratio surface by applying known or estimated diet-to-tissue isotopic discrimination factors (see below). Using a combination of $\delta^{13}C$, $\delta^{15}N$, and δ^2H foodweb isoscapes, Hobson et al. (2012a) identified isotopically homogenous clusters in Africa to which animals could be assigned. Subsequently, the same general approach has been used to produce spatially explicit continuous *probability surfaces of origin* for Africa (Veen et al. 2014; Hobson et al. 2014a) and South America (Garcia-Perez and Hobson 2013; Hobson and Kardynal 2016). We anticipate that this approach will be applied to more continents and otherwise defined regions in both terrestrial and marine biomes. A great deal more work is, however, required to refine these tissue-specific isoscapes (Gutiérrez-Expósito et al. 2015).

Beyond categories based on differences in plant photosynthesis, enrichment in ^{13}C can also occur in woody plants (C_3 photosynthesis) under conditions of low water availability, providing further isotopic structure across landscapes with moisture gradients. For example, in southeast Queensland, Australia, leaf $\delta^{13}C$ values declined from *ca.* −26‰ in dry scrub (mean annual rainfall < 300 mm) to *ca.* −31‰ in subtropical rainforest (mean annual rainfall > 1600 mm) (Stewart et al. 1995). Awareness of these patterns allowed Marra et al. (1998) to assign American Redstarts (*Setophaga ruticilla*) to wet and dry habitats on the Jamaican wintering grounds. Any $\delta^{13}C$ isoscape that models expected average plant community values due to a variety of biogeographic and physiological processes will potentially assist with assignment of animal origins.

4.4.2 Isoscapes derived from gradients in plant N uptake (water stress, agriculture)

Plant $\delta^{15}N$ compositions can also be estimated from climate variables such as rainfall amounts that are known to influence plant metabolism (Craine et al. 2009). Nitrogen isotope ratios are, however, inherently more difficult to model as an isoscape due to numerous other factors such as agriculture and anthropogenic landscape changes and modification of soils (Nadelhoffer and Fry 1994; Pardo and Nadelhoffer 2010). Generally, high $\delta^{15}N$ values in aquatic organisms are associated with human-dominated landscapes under heavy nitrogen loading (Cabana and Rasmussen 1996), and high $\delta^{15}N$ values in terrestrial organisms with arid regions (Grocke et al. 1997). The nutrient source of N in water and microbial processes such as denitrification can, however, complicate characterization of the N isotopic baseline (Mayer and Wassenaar 2012).

Additionally, because trophic position strongly influences animal tissue $\delta^{15}N$ compositions, assigning organisms to $\delta^{15}N$ isoscapes requires a good understanding of isotopic discrimination. This is especially true for predatory animals because increases in $\delta^{15}N$ values occur with each step in the food chain (see below).

4.4.3 Isoscapes derived from sulfur biogeochemistry

Sulfate loading to the landscape and subsequent reduction leads to strong isotope effects, and a corresponding broad range of $\delta^{34}S$ values. Food web applications of $\delta^{34}S$ measurements are limited in part because of low %S in organic materials that creates analytical challenges. Generally, $\delta^{34}S$ values of sulfates are highly variable in freshwaters (range −40 to +50‰; Nehlich 2015) and freshwater organisms (Hesslein et al. 1991). Because of the frequent and strong distinction between marine and freshwater/terrestrial environments (Godbout et al. 2010), $\delta^{34}S$ values can be further used to constrain assignment in a multi-isotope approach (Haché et al. 2014; Hobson and Kardynal 2016). For example, a low $\delta^{34}S$ value would normally negate any association with a marine environment. There is also the potential for discrimination between different habitat types in the freshwater/terrestrial environment. Hebert and Wassenaar (2005) found that mallard feathers grown in wetlands embedded in agricultural landscapes of the Canadian prairies had much lower $\delta^{34}S$ values than those grown in boreal forests. Additional evidence suggests that S isotope pathways differ between sediments (low $\delta^{34}S$) and the water column (high $\delta^{34}S$) in lakes (Croisetiere et al. 2009), and early work with $\delta^{34}S$ measurements showed that this isotope could effectively differentiate between vegetation types in estuaries (Peterson et al. 1985).

4.4.4 Isoscapes derived from δ²H and δ¹⁸O in precipitation and surface waters

It has long been recognized that δ^2H and $\delta^{18}O$ values in precipitation in North America and other continents show broad patterns, with a general gradient of relatively enriched values in the hotter, lower latitude regions—like southern Mexico—to more depleted values in the cooler, higher latitude regions, such as northern Canada (Sheppard et al. 1969; Taylor 1974). Isotopic compositions of δ^2H and $\delta^{18}O$ in precipitation are strongly correlated and are generally described by Equation 4.1, known as the Global Meteoric Water Line (GMWL; Craig 1961).

$$\delta^2H = 8 \times \delta^{18}O + 10 \qquad\qquad \text{Equation 4.1}$$

This strong correlation owes its origin to fractionation during precipitation occurring at equilibrium, resulting in equal effects on H and O. The kinetic fractionation that occurs during evaporation is, however, stronger for the more diffusive $^2H^1H^{16}O$ water molecule, resulting in a slope for surface waters that is lower than the GMWL in dry regions. This effect is known as *deuterium excess* and is governed largely by humidity.

The continental patterns in precipitation isotopic variation, populated largely by data from the long-term International Atomic Energy Agency (IAEA) Global Network of Isotopes in Precipitation database, are now described for most regions of the world (Terzer et al. 2013; www.iaea. org/water). While this program was never intended to create regional isoscapes for these elements, it has nonetheless provided the backbone for the use of such isoscapes in animal and other material tracking (Hobson and Wassenaar 2008).

Yapp and Epstein (1982) provided the first major linkage between the abiotic precipitation isoscapes and plant material. Cormie et al. (1994) then showed an excellent correlation between δ^2H values in deer bone collagen and the amount-weighted average growing-season precipitation ^2H, which was further affirmed at continental scales for bird feathers (Chamberlain et al.1997; Hobson and Wassenaar 1997). Rescaling functions for ^2H, derived from known-origin individuals, now exist for many animal species (Hobson 2008), including various avian species (Hobson 2008; Hobson et al. 2012b, 2014a; Procházka et al. 2013; Rogers et al. 2012; Solovyeva et al. 2016; Sullins et al. 2016) and their insect prey (Wassenaar and Hobson 1998; Hobson et al. 1999, 2012c,d; Brattström et al. 2010; Flockhart et al. 2013) (Figure 4.3). Slopes of these regressions range from 0.3 to 1.4, with songbirds and waterfowl typically showing the strongest correlations and slopes closest to 1.

Few measurements of ^{18}O have been conducted on tissues of migratory organisms due, primarily, to the only recent availability of the facilitating technology—Thermal Conversion Elemental Analyzers (TC/EA) (Qi et al. 2011). Fourel et al. (1998), however, showed a strong relationship between

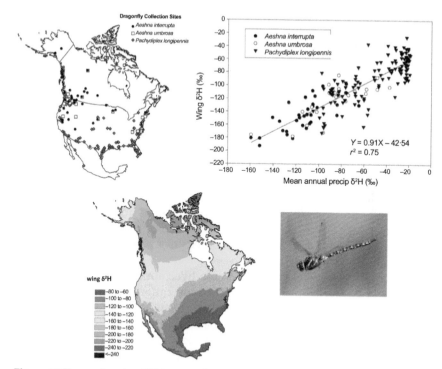

Figure 4.3 Dragonfly wing δ^2H isoscape for North America. Reprinted from Hobson et al. 2012c with permission. Copyright 2012, John Wiley & Sons, Inc.

the δ^2H and $\delta^{18}O$ values of chitin in the wings of Monarch Butterflies (*Danaus plexippus*), demonstrating that the meteoric relationship between these two elements can be passed through the foodweb. Similar results have since been reported for dragonflies (Hobson et al. 2012c,d), which expand the possibility to infer environmental and spatial information by measuring both isotopes in tissues of migratory animals. The current challenge is that oxygen can be fixed in animal tissues through diet, air, and drinking water and it is not clear how departures from the meteoric relationship should be interpreted (Pietsch et al. 2011; Hobson and Koehler 2015). Nonetheless, heat stress in animals and evapotranspiration in local environments are expected to change the relationship between tissue δ^2H and $\delta^{18}O$ values in animal tissues and so may be useful in identifying those individuals coming from specific habitats like deserts.

4.4.5 Isoscapes derived from strontium isotopes ($^{87}Sr/^{86}Sr$) and other radiogenic elements

Using radiogenic isotope systems to track animal origins has the potential advantage that there appears to be little, or no, isotopic fractionation from geologic sources through food webs and into tissues. As such, isotopes of radiogenic elements such as strontium ($^{87}Sr/^{86}Sr$) are being increasingly used in animal migration research. While these applications have been dominated, to date, by studies in aquatic systems (Hobson et al. 2010), the lack of Sr isotope fractionation between dietary inputs and tissues in terrestrial animals such as reindeer, elephant, mammoth, and mastodon bone and tusks (Aberg 1995; Vogel et al. 1990; Koch et al. 1995; Hoppe et al. 1999) highlights the potential for this isotope to identify geographical regions that an animal has inhabited, based on different regional geologies (Britton et al. 2009). Regional variations in $^{87}Sr/^{86}Sr$ ratios are largely a function of the age of the crust (Beard and Johnson 2000; Bataille et al. 2012; Bataille and Bowen 2012), as rocks that are older have higher $^{87}Sr/^{86}Sr$ ratios than younger rocks. While this generalization holds at large scales, significant variation in $^{87}Sr/^{86}Sr$ ratios also occurs on finer spatial scales, largely due to variation in rock type (Kennedy et al. 2000; Hoppe et al. 1999; Barnett-Johnson et al. 2008).

The use of other trace elements in foodweb applications has met with mixed success. Concentrations and isotope ratios of Hg ($^{202}Hg/^{198}Hg$, $^{201}Hg/^{198}Hg$, and $^{199}Hg/^{198}Hg$) can discriminate between sources in the marine environment and appear to separate coastal from oceanic foraging (Senn et al. 2010; Day et al. 2012); however, systematic patterns on broader scales have yet to be documented. Calcium isotopes ($^{44}Ca/^{40}Ca$ and $^{42}Ca/^{40}Ca$) are measurable in a variety of hard tissues and may be influenced by trophic level (Clementz et al. 2003) but require sophisticated instrumentation, and exhibit rather limited isotope ratio ranges (approximately $-3‰$ to $+1‰$) relative to instrumental precision (Reynard et al. 2010). Similarly, chlorine isotopes ($^{37}Cl/^{35}Cl$) have a relatively narrow range of values (approximately $-6‰$ to $0‰$). While chlorine has shown some promise in assessing exposure of organisms to industrial effluents at small scales (Dube et al. 2005), it is unknown whether broad-scale patterns exist. Stable isotopes of lead ($^{206}Pb/^{204}Pb$, $^{207}Pb/^{204}Pb$ and $^{208}Pb/^{204}Pb$) first gained attention in their use to track origins of elephant ivory (Vogel et al. 1990). Both geology and anthropogenic influences such as gasoline combustion (Scheuhammer and Templeton 1998) affect the distribution of lead isotope values across the landscape, and the recent employment of $^{206}Pb/^{207}Pb$ values in an isoscape framework (Keller et al. 2016) suggests potential use in forensic wildlife applications.

4.4.6 Marine isoscapes

The development of marine isoscapes could greatly assist in tracking rearing areas for marine mammals and fish stocks, but the vastness of the world's oceans and their complex biogeochemistry have so far limited isoscape applications. Early work by Best and Schell (1996) and Schell et al.

(1998) used gradients in $\delta^{13}C$ and $\delta^{15}N$ values measured in zooplankton to examine the movements of migratory Bowhead Whale (*Balaena mysticetus*) and Southern Right Whale (*Eubalaena australis*). Because plankton $\delta^{13}C$ becomes strongly negative as a function of latitude (Rau et al. 1982), the feeding origins of seabirds can also be estimated in the southern hemisphere (Quillfeldt et al. 2005; Cherel and Hobson 2007). Elsewhere, researchers have broadly inferred inshore and offshore feeding of marine vertebrates, based on relatively ^{13}C depleted pelagic *vs.* benthic marine food webs (Hobson et al. 1994; Reich et al. 2007; Karnovsky et al. 2012). Geographical marine gradients for other elements (i.e., $^{18}O/^{16}O$), recorded in hard biological materials such as otoliths, can potentially aid researchers by providing insights into where wild fish species have ranged throughout their lives (Trueman et al. 2012). Only recently have large-scale spatial applications of these patterns allowed display of marine isoscapes [i.e., contour plots for $\delta^{13}C$ and $\delta^{15}N$ (Graham et al. 2010)].

4.5 Isotopic discrimination and physiological considerations

Once an isoscape, or an expected isotopic signal, has been established, the next stage is to consider how that baseline signal is transferred to animals as an indication of potential origin—this is not as simple as bulk diet = bulk consumer. For many elements, only slight changes in stable isotope ratios occur between trophic levels after they are fixed by plant tissues. Different tissues (within the same animal) can, however, have widely varying isotope ratios because of their rate of turnover and their macromolecular composition (protein, lipid, carbohydrate). This range can be as large as 6‰ within an animal for bulk measurements (e.g., Smyntek et al. 2007), and up to 25‰ among individual amino acids or fatty acids (e.g., Bell et al. 2007; Lorrain et al. 2009). Furthermore, the physiological status of the animal at the time of death (e.g., fasting, breeding, weaning) can also influence isotope ratios because of differential allocation of nutrients and associated isotopic discrimination among tissues. Stable nitrogen isotopes in the tissues of consumers primarily provide a means of tracing protein pathways derived from diet since this element is largely absent in lipids and carbohydrates. Alternatively, $\delta^{13}C$ values represent all three macromolecules, and if lipids dominate the mass of tissue, then this can result in a $\delta^{13}C$ value that is significantly lower than the diet that was consumed to synthesize the lipid.

4.5.1 Elemental turnover

A fundamental question surrounding most isotope studies of animals is, "What period of feeding do the observed isotope data represent?" or, "Does a tissue isotope ratio represent foraging by the animal over a period of days, weeks, months prior to death or sampling, or its entire lifetime?"

Both growth (anabolism) and exchange (catabolism) contribute to tissue isotopic change following a change in diet. Rapidly growing tissues will change quickly because of dilution of the 'old' isotope ratio with tissue built from protein and lipid from the new diet. Similarly, tissues that are constantly broken down and rebuilt (even if they exhibit no net growth) can also show rapid change. Tieszen et al. (1983) first hypothesized that because oxygen consumption declines as body size increases, carbon turnover slows in tissues of animals with larger body size. This suggested that larger animals, which are often used for food (e.g., ungulates, cetaceans), are likely to have tissue isotope ratios that reflect a long-term diet. Before considering whether this hypothesis holds we must, however, first consider differences between endotherms (organisms that generate heat internally) and ectotherms (organisms that obtain heat from ambient sources).

For a given body size, endotherms have an order of magnitude higher field metabolic rate (Nagy et al. 1999) and food consumption rate (Farlow 1976) compared with ectotherms. As a result, large, slow-growing ectotherms have very slow rates of isotopic change (Figure 4.4), even in metabolically active tissues such as liver and blood plasma. Average isotope half-lives for elements in liver of birds and mammals are *ca.* 2 days with interquartile ranges of only *ca.* 2 days, whereas the corresponding values for fish are *ca.* 15 days and an interquartile range as high as *ca.* 40 days

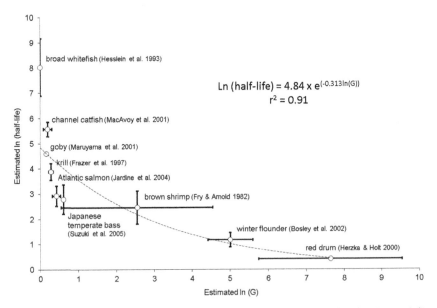

Figure 4.4 Half-life (calculated according to Box 4.1) *vs.* growth rate (calculated as G, month⁻¹) for fishes and crustaceans.

(Boecklen et al. 2011). Recent examples include American alligators (*Alligator mississippiensis*) with plasma half-lives of 60 days (Rosenblatt and Heithaus 2013) and leopard sharks (*Triakis semifasciata*) with plasma half-lives of *ca.* 35 days (Kim et al. 2012). These values are much higher than the averages for blood plasma of 2 days and 15 days for birds and mammals, respectively (Boecklen et al. 2011). Murray and Wolf (2012) report a half-life of 23 days for desert tortoise (*Gopherus agassizii*) plasma, which is much slower than similarly sized birds and mammals. Even relatively rapidly growing juvenile ectotherms can have relatively slow turnover rates.

Due to the fundamental differences between endotherms and ectotherms, recent meta-analyses that predict tissue isotopic half-life as a function of body mass treat these physiological categories differently (Thomas and Crowther 2015; Vander Zanden et al. 2015). While ectotherms (both invertebrates and vertebrates) show strong positive relationships between half-life and body mass, consistent with original predictions by Tieszen et al. (1983), the relationships for birds and mammals are weak (Vander Zanden et al. 2015). This may be due, in part, to the limited size range of birds and mammals used in diet-switch studies. Turnover rates are generally rapid for endotherms regardless of body size.

These meta-analyses produce general predictions to estimate half-life for a given body size of animals. For example, carbon and nitrogen in muscle from a 50 g mammal would have a predicted half-life of 29.4 days (Vander Zanden et al. 2015), very near the original estimate of 27.6 days calculated by Tieszen et al. (1983). Thomas and Crowther (2015) produced more general estimates across the animal kingdom that range from a half-life of 3–9 days in a 1 mg animal (e.g., an insect) to 115–321 days in a 100 kg animal (e.g., a domestic pig), with temperature also contributing to variation. However, there remains a great deal of uncertainty around these general equations. Recognizing these differences in the temporal integration of isotope values, within and among tissues, allows the refinement of seasonal movements or switches in diet (Cerling et al. 2009; Hobson and Bond 2012) and will also allow better estimation of the history of an animal under forensic investigation.

While these experiments have yielded useful information for C and N turnover, far less is known about turnover of H, O, and S, and especially trace elements. Soto et al. (2013a) found that fast-growing fish had high-energy demands, increasing the incorporation of ^{2}H-depleted products from dietary lipids into tissue-H. This produced a relatively high metabolic water contribution relative to that of the environmental water during tissue synthesis. Work to date suggests that S turnover occurs at approximately the same pace as that of C and N (Hesslein et al. 1993; MacAvoy et al. 2001; Harrison et al. 2011). While experimental data for turnover rates of trace elements are limited, Kennedy et al. (2000) found that different bony tissues within individual juvenile Atlantic salmon (*Salmo salar*) in the same river had the same Sr isotopic compositions (e.g., otoliths, vertebrae), suggesting similar turnover rates for different tissue types.

4.5.2 Isotopic discrimination

Upon reaching isotopic equilibrium with a new food source, animal tissues also demonstrate an isotopic diet-to-tissue *offset*. This is variously referred to as *discrimination* or *trophic enrichment*, and for an isotopic species X (e.g., ^{15}N) this offset is denoted as ΔX according to Equation 4.2

$$\Delta_t = \Delta_d + \Delta X \qquad\qquad \text{Equation 4.2}$$

where Δ_t is the isotope ratio of the tissue and Δ_d is the isotope ratio of the diet. Each isotope is subject to biochemical processes within organisms that lead to variable ΔX.

4.5.2.1 $\delta^{13}C$

Initial work with $\delta^{13}C$ in food webs showed progressive enrichment in ^{13}C from the base to the top of the food web (McConnaughey and McRoy 1979). This led to the conclusion that, because respired CO_2 was isotopically light, $\delta^{13}C$ measurements could potentially be used as an indicator of trophic level (Rau et al. 1983). In contrast, laboratory studies have shown that, on average, discrimination of ^{13}C from diet to consumer is limited. Published reports of these controlled studies reveal mean values that differ only slightly from zero (Table 4.1), with wide variations among species, diets, and tissues, ranging from −8.8 to +6.1‰ (Caut et al. 2009). As a taxonomic

Table 4.1 Isotopic discrimination factors (mean ± SE) as summarized in literature reviews of controlled dietary studies.

Reference	Isotope					
	$\delta^{13}C$	n	$\delta^{15}N$	n	$\delta^{34}S$	n
Minagawa and Wada 1984	.	.	3.40 ± 0.27‰ SE	16		.
Vander Zanden and Rasmussen 2001	0.47 ± 0.07‰ SE	42	2.92 ± 0.30‰ SE	35	.	.
Post 2002	0.39 ± 0.13‰ SE	107	3.4 ± 0.13‰ SE	56	.	.
McCutchan et al. 2003	0.44 ± 0.12‰ SE	111	2.0 ± 0.20‰ SE	83	0.4 ± 0.52‰ SE	13
Vanderklift and Ponsard 2003	.	.	2.54 ± 0.11‰ SE	134	.	.
Caut et al. 2009	0.75 ± 0.11‰ SE	290	2.75 ± 0.10‰ SE	268	.	.
Nehlich 2015	.	.	.	.	0.5 ± 0.5‰ SE	29

group, fish tend to have the highest $\Delta^{13}C$ (Caut et al. 2009). Among tissues, there is some evidence to suggest that keratins such as hair and feather have values at the upper end of the range (Dalerum and Angerbjorn 2005). For example, mice raised on constant diets have hair $\delta^{13}C$ that is consistently higher than muscle, blood, and liver (Wolf et al. 2015), whereas several bird species show feather $\delta^{13}C$ values that are higher than other tissues (reviewed in Kurle et al. 2013).

4.5.2.2 $\delta^{15}N$

In contrast to $\delta^{13}C$ values, $\delta^{15}N$ values clearly increase along food chains. The magnitude of change with each trophic level ranges from +2 to +5‰ (Table 4.1), although this "average" (bulk protein) value masks variation amongst individual amino acids. For some "source" amino acids (e.g., phenylalanine), nitrogen will largely be incorporated into the protein pool of the consumer with little to no enrichment (Popp et al. 2007; Chikaraishi et al. 2014). Other "trophic" amino acids (e.g., glutamic acid), however, show considerable enrichment. The net discrimination observed for bulk $\delta^{15}N$ values in consumers will reflect the degree to which the diet meets the amino acid requirement of the consumer (Robbins et al. 2005). Like $\delta^{13}C$ values, there is considerable variability in $\delta^{15}N$ values among tissues and taxa, which has been the subject of much investigation (Vanderklift and Ponsard 2003). Organisms that excrete ammonia (fishes and aquatic invertebrates) appear to have lower $\Delta^{15}N$ than those that excrete urea (mammals) or uric acid (birds and terrestrial invertebrates) (Post 2002; Vanderklift and Ponsard 2003). Some of this variation may, however, be due simply to the confounding influence of tissue type. For example, muscle $\Delta^{15}N$ differs little among birds, mammals, and fishes and is similar to the whole bodies of invertebrates, but the three former taxa have different $\Delta^{15}N$ for liver (Caut et al. 2009).

4.5.2.3 $\delta^{34}S$

The uptake of sulfur and the associated $\delta^{34}S$ values from diet to consumer involves the amino acids cysteine and methionine, both of which can be limiting in food sources. Available evidence suggests that, on average, $\Delta^{34}S$ is near zero (Fry 1988), which may reflect the efficient conversion of these dietary amino acids into consumer protein. McCutchan et al. (2003) report a mean $\Delta^{34}S$ value of 0.4‰ from controlled laboratory studies, and Nehlich (2015) compiled more recent data to obtain a similar estimate of 0.5‰ (Table 4.1). These meta-analyses show wide variation among species, diets, and tissue types. Estimates of $\Delta^{34}S$ range from a low of −3.2‰ for a tiger moth caterpillar (subfamily Arctiinae) feeding on cottonwood leaves, to a high of +7.3‰ for a Buckeye butterfly pupae (*Junonia coenia*) feeding on Plantago (McCutchan et al. 2003). Some of this variation could arise from analytical error associated with the dietary samples because low S plant material is particularly difficult to measure by IRMS. Two closely related fish species (*Salvelinus fontinalis* and *Oncorhynchus mykiss*) that were fed high protein diets also had considerably different $\Delta^{34}S$ under the same laboratory conditions (+1.6‰ and +4.0‰, respectively; McCutchan et al. 2003) pointing to other factors responsible for variation (described below).

4.5.2.4 δ^2H

In contrast to the isotopes discussed above, the hydrogen isotopic composition of animal tissues are influenced, not only by solid food, but also water (e.g., Hobson et al. 1999; Podlesak et al. 2008; Jardine et al. 2009), which makes the interpretation and estimation of isotopic discrimination more complex. The study of Birchall et al. (2005) first suggested that H isotopes could reflect trophic level, because a trophic enrichment was observed in consumers relative to their diet. These authors, and others (Soto et al. 2011; Topalov et al. 2013), did not, however, find strong correlations between δ^2H and $\delta^{15}N$ values. Consequently, this isotopic behaviour seems to indicate that trophic enrichment between consumer and diet is not related by similar mechanisms of metabolic discrimination that occur with ^{15}N. Experimental studies have determined that the accumulative influence of ambient water-H into consumer tissues at each trophic level can be the cause of the reflected trophic enrichment in ^2H, at least in aquatic systems. The proportion of tissue-H derived from ambient water can also vary remarkably among organisms (Solomon et al. 2009; Soto et al. 2013b; Wilkinson et al. 2015). This influence in aquatic food webs has been called a *trophic compounding effect*.

At the base of the food web (primary producers), photosynthetic processes discriminate in favor of the more abundant (lighter) isotope, hence plants generally have δ^2H values much more negative than source water (Estep and Hoering 1980). Transpiration in land plants, however, retains the heavier isotope (deuterium) preferentially and is responsible for the large differences in δ^2H values between aquatic and terrestrial primary producers. These differences have been used as the basis for tracing terrestrial and autochthonous inputs in aquatic ecosystems (Doucett et al. 2007).

4.5.2.5 $\delta^{18}O$

The oxygen isotope composition of animal tissues is related to that of source water, and poorly to that of dietary sources (Ehleringer et al. 2008; Nielson and Bowen 2010; Soto et al. 2013b; Schilder et al. 2015). This strong water effect can, however, sometimes be distorted in animals such as cats that, depending on the diet, do not require drinking water [non-obligate drinkers (Pietsch et al. 2011)]. Very few studies have determined the contribution of dietary sources to the final tissue-O, with estimates of 10% in aquatic crustaceans (Nielson and Bowen 2010). Other sources for tissue-O include atmospheric and dissolved O_2 for terrestrial and aquatic organisms, respectively. Water from internal tissues, whose isotopic composition is mainly influenced by ambient water, ultimately undergoes a net isotopic fractionation of *ca.* +15‰ during tissue synthesis. These complications, together with the strong correlation between δ^2H and $\delta^{18}O$ values, mean that oxygen isotopes alone may provide little foodweb information.

4.5.3 Factors affecting discrimination

The current consensus is that researchers should strive to use the most conservative value associated with their specific organism of interest. Various meta-analyses have, however, suggested some broad patterns across various taxa that could assist in choosing appropriate values for isotopic discrimination in the absence of controlled laboratory studies.

4.5.3.1 Diet quality

Much of the variation in isotopic discrimination, observed across studies, may be related to the quality of the diet offered to the animal; for example, the direct routing of high quality animal protein, without recycling, can lead to lower discrimination (Poupin et al. 2011). Quality can be defined in a variety of ways, ranging from protein content, N content and C:N ratio, all of which reflect a balance between amino acid intake, assimilation and excretion that can affect ^{15}N discrimination (Robbins et al. 2005). Diets vary widely in these endpoints, ranging from <1 to ~15% N and from <4 to >100 C:N. Although these measures largely reflect quantity they fail to accurately characterize protein quality and nutritionists, therefore, use the term *biological value* to describe the percent of absorbed protein that is retained by the animal. This is a better measure of protein quality, because it relates the amino acid composition of the diet to the needs of the consumer. A diet may be rich in protein but low in a particular, limiting, amino acid such as methionine (Florin et al. 2011)—diets can thus be expressed as the percentage of the most limiting amino acid. Studies that vary the quality of diet of a single species under controlled conditions have isolated the effects of this variable from other confounding factors, such as tissue type and taxon, and found that $\Delta^{15}N$ increases with decreasing diet quality (Florin et al. 2011).

4.5.3.2 Diet heterogeneity

Calculations of isotopic discrimination require an estimate of the isotope value of the bulk diet and yet formulated diets can contain ingredients that vary in digestibility, and the assimilated material can differ isotopically from the bulk diet (Codron et al. 2011). Further, formulated diets may be uniformly digested, but selectively partitioned into different tissues (Codron et al. 2012): a phenomenon referred to as *metabolic* or *isotopic routing* (Martinez del Rio et al. 2009). Finally, the carbon skeletons of lipids can be used to build non-essential amino acids that are then assimilated into proteins (Wolf et al. 2015). This means that animals fed lipid-rich diets can have proteinaceous tissues such as hair that are more ^{13}C-depleted than would be predicted from the protein in the diet (Wolf et al. 2015).

4.5.3.3 Physiological status

While the previous sections presented some of the taxonomic and tissue-specific factors responsible for variation in Δ^{15}N, variation also exists within a species for a given tissue. For example, Hobson et al. (1993) showed that Ross's Goose (*Chen rossii* or *Anser rossii*) fasting and undergoing significant protein catabolism during incubation experience an increase in body δ^{15}N values. Similarly, fasting penguins exhibit enrichment in ^{15}N across multiple tissues (Cherel et al. 2005). Mammals that have recently been weaned are enriched in ^{15}N relative to adults, indicative of their source of nutrition (mother's milk) (Hobson and Quirk 2014; Cherel et al. 2015). Therefore, information about the age of the animal in question can be helpful in determining an appropriate isotopic discrimination to apply (Kurle et al. 2014). Knowledge of these physiological processes is, therefore, necessary when using tissue δ^{15}N values to back-calculate the diet or habitat of the consumer, and also highlights the need for probabilistic approaches that provide an estimate, with associated uncertainty.

4.6 Hypothetical case studies

To conclude this chapter we give two examples of how food web processes lead to isotope ratios in the tissues of an animal, focusing on wild, rather than domesticated species (for the latter see Chapter 6). Because a primary issue in both food and wildlife forensics is determining origin (Espinoza et al. 2012), we consider how knowledge of food web processes can help eliminate possible origins using isotopes (Figure 4.5), and, when combined with robust probabilistic approaches, can be used to quantify the likelihood of assignment to a particular origin.

4.6.1 Marine case study

First, consider whale meat purchased from a market in Asia. Although genetic techniques can identify the species of whale (Baker et al. 2007), an isotope approach is necessary to identify the region of the ocean where the whale was foraging prior to slaughter. Suppose the isotope ratios of that whale meat were δ^{13}C = −29‰, δ^{15}N = +8‰, δ^{34}S = +16‰, δ^{2}H = −40‰, δ^{18}O = +5‰. What do these isotope ratios represent? ^{13}C is particularly

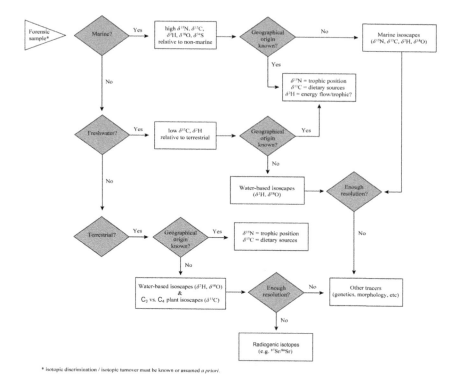

Figure 4.5 Basic overview of bulk isotope approaches to narrowing origins of a forensic sample using food web principles.

promising because of known latitudinal gradients in ocean plankton (Rau et al. 1982) that likely formed the base of the whale's food chain (or its diet, depending on the species). The $\delta^{13}C$ value can immediately narrow the search area for the origin of this whale, but consideration must be given to the role of *de novo* lipid synthesis in altering the tissue value. Extracting lipids and re-analyzing the lipid-free sample might yield a $\delta^{13}C$ value of −28‰. Combining the $\delta^{13}C$ and $\delta^{15}N$ values could pinpoint a bounded area in the southern ocean that spans approximately 5 degrees latitude (Graham et al. 2010). ^{34}S may offer little resolving power because seawater sulfate has a narrow range of values around +21‰ (Rees et al. 1978). This leads to equally narrow ranges, centred on +17‰, in offshore food webs (Nehlich et al. 2013). Stable hydrogen and oxygen isotope ratios are also unlikely to be particularly useful in this exercise because values in the oceans are relatively homogenous over extensive regions (Bowen 2010). The values for these isotopes are, however, more negative near the poles (Bowen 2010), so their inclusion, with appropriate measurement protocols (Wassenaar and Hobson 2000a), may provide support for an interpretation based on the ^{13}C data and more specifically in relation to origins from Antarctica. The inclusion of isotopic "fingerprints" using CSIA of known source populations could also provide higher resolution geo-location, but such data are currently very limited.

4.6.2 Terrestrial case study

Next, consider a case of *bush-meat* obtained at a market in southern Africa, with a small number of hairs remaining on the meat. Analysis of the hairs finds: $\delta^{13}C = -12‰$, $\delta^{15}N = +18‰$, $\delta^{34}S = +5‰$, $\delta^2H = +15‰$. The African continent is characterized by strong gradients in foodweb $\delta^{13}C$, $\delta^{15}N$, and δ^2H values (Hobson et al. 2012a), which can be used to estimate the most likely origin of the hair.

Although hair $\delta^{18}O$ varies across the landscape (Bowen 2010), there is a current lack of rescaling functions ($\delta^{18}O_{tissue}$ *vs.* $\delta^{18}O_{water}$) for this element and, since ^{18}O co-varies with ^2H, the focus will be on hydrogen. Human hair ^2H is strongly correlated with δ^2H values of drinking water but is partly confounded by an isotopically homogenous "supermarket" diet, with typical $\delta^2H = -115‰$ (Ehleringer et al. 2008). Since dietary-H accounts for >75% of body-H, the homogenous dietary value leads to a shallow slope (0.27) in the δ^2H_{hair} *vs.* δ^2H_{water} relationship (Ehleringer et al. 2008). Wild animals do not (in general) eat at the supermarket, and their diet δ^2H values are likely to co-vary with the δ^2H value of local water, leading to tighter and steeply sloped relationships between δ^2H_{tissue} and δ^2H_{water}. For example, a bird eating beetles in an area with high δ^2H values will obtain H enriched in ^2H from both water and diet because beetle chitin ^2H is also tightly linked to local precipitation (van Hardenbroek et al. 2012) (see also Figure 4.3 with a slope of 0.91 and R^2 of 0.75 for dragonflies). For mammals, while rabbit hair δ^2H values strongly correlate with local river water, feline hair δ^2H values do not, potentially because of physiological effects or large home ranges (Pietsch et al. 2011). Therefore, caution is needed when back-calculating the expected value of local precipitation that was used to grow the hair on the bush-meat sample. A reasonable choice of rescaling function would be rabbit hair. Applying appropriate probabilistic approaches (see Chapter 5) identifies two possible regions for the origin of this meat, one in the Sahel and the other in western South Africa (Hobson et al. 2012a). The addition of other techniques can then be used to further discriminate between and within these regions.

4.7 Combining tools

Given the uncertainties associated with the isotope food web approach to food forensics, there is a need to expand the toolbox with complementary approaches. Although the use of multiple isotopes can refine the search

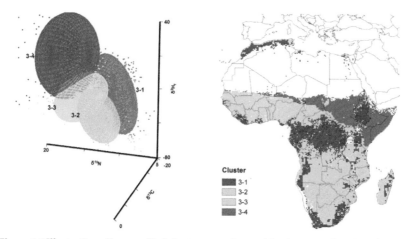

Figure 4.6 Illustration of how multiple isotopes can be combined to create location clusters in an isoscape (adapted from Hobson et al. 2012a).

area that has been identified using a single isotope (Hobson et al. 2015; Figure 4.6), there may remain considerable uncertainty, either because the identified area is large or because there are multiple possible source regions that match the isotope profile of the tissue. The use of prior information, such as species distributions, combined with Maximum Entropy Principle models (MaxEnt; Merow et al. 2013; Pekarsky et al. 2015) or Bayesian frameworks (VanWilgenburg and Hobson 2011) can also eliminate areas with low probability of origin. Population genetics can be combined with isotopes for animal assignment, provided sufficient genetic structure exists in the species of interest (Chabot et al. 2012), and other tracers such as fatty acid profiles (DePeters et al. 2013) can further discriminate among potential source populations or between wild and farm-raised animals. Application of these combined techniques, validated with robust field, and laboratory experiments, will further elucidate origins for a sample in question.

References

Aberg, G. 1995. The use of natural strontium isotopes as tracers in environmental studies. *Water, Air and Soil Pollution* 79: 309–322.

Alisauskas, R. T., E. E. Klaas, K. A. Hobson and C. D. Ankney. 1998. Stable-carbon isotopes support use of adventitious color to discern winter origins of lesser snow geese. *Journal of Field Ornithology* 69: 262–268.

Ambrose, S. H. and M. J. DeNiro. 1986. The isotopic ecology of East African mammals. *Oecologia* 69: 395–406.

Baker, C. S., J. G. Cooke, S. Lavery, M. L. Dalebout, R. L. Brownell Jr., Y. -U. Ma, N. Funahashi and C. Carraher. 2007. Estimating the number of whales entering trade using DNA profiling and capture-recapture analysis of market products. Publications, Agencies and Staff of the U.S. Department of Commerce. Paper 87.

Barnett-Johnson, R., F. C. Ramos, T. Pearson, C. B. Grimes and R. B. MacFarlane. 2008. Tracking natal origins of salmon using isotopes, otoliths, and landscape geology. *Limnology and Oceanography* 53: 1633–1642.

Bataille, C. P. and G. J. Bowen. 2012. Mapping $^{87}Sr/^{86}Sr$ variations in bedrock and water for large scale provenance studies. *Chemical Geology* 304: 39–52.

Bataille, C. P., J. Laffoon and G. J. Bowen. 2012. Mapping multiple source effects on the strontium isotopic signatures of ecosystems from the circum-Caribbean region. *Ecosphere* 3: art 118.

Beard, B. L. and C. M. Johnson. 2000. Strontium isotope composition of skeletal material can determine the birth place and geographic mobility of humans and animals. *Journal of Forensic Science* 45: 1049–1061.

Bell, J. G., T. Preston, R. J. Henderson, F. Strachan, J. E. Bron, K. Cooper and D. J. Morrison. 2007. Discrimination of wild and cultured European sea bass (*Dicentrarchus labrax*) using chemical and isotopic analysis. *Journal of Agricultural and Food Chemistry* 55: 5934–5941.

Best, P. B. and D. M. Schell. 1996. Stable isotopes in Southern Right Whale (*Eubalaena australis*) baleen as indicators of seasonal movements, feeding and growth. *Marine Biology* 124: 483–494.

Birchall, J., T. C. O'Connell, T. H. E. Heaton and R. M. Hedges. 2005. Hydrogen isotope ratios in animal body protein reflect trophic level. *Journal of Animal Ecology* 74: 877–881.

Boecklen, W. J., C. T. Yarnes, B. A. Cook and A. C. James. 2011. On the use of stable isotopes in trophic ecology. *Annual Review of Ecology, Evolution, and Systematics* 42: 411–440.

Bosley, K. L., D. A. Witting, R. C. Chambers and S. C. Wainright. 2002. Estimating turnover rates of carbon and nitrogen in recently metamorphosed winter flounder *Pseudopleuronectes americanus* with stable isotopes. *Marine Ecology Progress Series* 236: 233–240.

Bowen, G. J. 2010. Isoscapes: Spatial pattern in isotopic biogeochemistry. *Annual Reviews in Earth and Planetary Science* 38: 161–187.

Brattström, O., S. Bensch1, L. I. Wassenaar, K. A. Hobson and S. Åkesson. 2010. Understanding the migration ecology of European red admirals (*Vanessa atalanta*) using stable hydrogen isotopes. *Ecography* 33: 720–729.

Britton, K., V. Grimes, J. Dau and M. P. Richards. 2009. Reconstructing faunal migrations using intra-tooth sampling and strontium and oxygen isotope analyses: A case study of modern caribou (*Rangifer tarandus granti*). *Journal of Archaeological Science* 36: 1163–1172.

Cabana, G. and J. B. Rasmussen. 1996. Comparison of aquatic food chains using nitrogen isotopes. *Proceedings of the National Academy of Sciences USA* 93: 10844–10847.

Caut, S., E. Angulo and F. Courchamp. 2009. Variation in discrimination factors (^{15}N and ^{13}C): The effect of diet isotopic values and applications for diet reconstruction. *Journal of Applied Ecology* 46: 443–453.

Cerling, T. E., G. Wittenmyer, H. B. Rasmussen, F. Vollrath, C. E. Cerling, T. J. Robinson and I. Douglas-Hamilton. 2009. Stable isotopes in elephant hair document migration patterns and diet changes. *Proceedings of the National Academy of Sciences USA* 103: 371–373.

Chabot, A. M., K. A. Hobson, S. L. Van Wilgenburg and G. J. McQuat. 2012. Assigning breeding ground origins to migrant birds: Advances using a novel Bayesian approach combining genetic and stable isotope markers. *PLoS ONE* 7: e43627.

Chamberlain, C. P., J. D. Blum, R. T. Holmes, X. Feng, T. W. Sherry and G. R. Graves. 1997. The use of isotope tracers for identifying populations of migratory birds. *Oecologia* 109: 132–141.

Cherel, Y., K. A. Hobson, F. Bailleul and R. Groscolas. 2005. Nutrition, physiology, and stable isotopes: New information from fasting and moulting penguins. *Ecology* 86: 2881–2888.

Cherel, Y. and K. A. Hobson. 2007. Geographical variation in stable carbon isotope signatures of marine predators: A tool to investigate their foraging areas in the Southern Ocean. *Marine Ecology Progress Series* 329: 281–287.

Cherel, Y., K. A. Hobson and C. Guinet. 2015. Milk isotopic values demonstrate that nursing fur seal pups are a full trophic level higher than their mothers. *Rapid Communications in Mass Spectrometry* 29: 1485–1490.

Chikaraishi, Y., S. A. Steffan, N. O. Ogawa, N. F. Ishikawa, Y. Sasaki, M. Tsuchiya and N. Ohkouchi. 2014. High-resolution food webs based on nitrogen isotopic composition of amino acids. *Ecology and Evolution* 4: 2423–2449.

Clementz, M. T., P. Holden and P. L. Koch. 2003. Are calcium isotopes a reliable monitor of trophic level in marine settings? *International Journal of Osteoarchaeology* 13: 29–36.

Codron, D., J. Codron, M. Sponhemier, S. M. Bernasconi and M. Clauss. 2011. When animals are not quite what they eat: Diet digestibility influences ^{13}C-incorporation rates and apparent discrimination in a mixed-feeding herbivore. *Canadian Journal of Zoology* 89: 453–465.

Codron, D., M. Sponheimer, J. Codron, I. Newton, J. L. Lanham and M. Clauss. 2012. The confounding effects of source isotopic heterogeneity on consumer-diet and tissue-tissue stable isotope relationships. *Oecologia* 169: 939–953.

Cormie, A. B., H. P. Schwarcz and J. Gray. 1994. Relationship between the hydrogen and oxygen isotopes of deer bone and their use in the estimation of relative humidity. *Geochimica et Cosmochimica Acta* 60: 4161–4166.

Craig, H. 1961. Isotopic variations in meteoric waters. *Science* 133: 1702–1703.

Craine, J. M., A. J. Elmore, M. P. M. Aidar, M. Bustamante, T. E. Dawson, E. A. Hobbie, A. Kahmen et al. 2009. Global patterns of foliar nitrogen isotopes and their relationships with climate, mycorrhizal fungi, foliar nutrient concentrations, and nitrogen availability. *New Phytologist* 183: 980–992.

Croisetiere, L., L. Hare, A. Tessier and G. Cabana. 2009. Sulphur stable isotopes can distinguish trophic dependence on sediments and plankton in boreal lakes. *Freshwater Biology* 54: 1006–1015.

Dalerum, F. and A. Angerbjorn. 2005. Resolving temporal variation in vertebrate diets using naturally occurring stable isotopes. *Oecologia* 144: 647–658.

Day, R. D., D. G. Roseneau, S. S. Vander Pol, K. A. Hobson, O. F. X. Donard, R. S. Pugh, A. J. Moors and P. R. Becker. 2012. Regional, temporal, and species patterns of mercury in Alaskan seabird eggs: Mercury sources and cycling or food web effects? *Environmental Pollution* 166: 226–232.

DeNiro, M. J. and S. Epstein. 1978. Influence of diet on the distribution of carbon isotopes in animals. *Geochimica et Cosmochimica Acta* 42: 495–506.

DeNiro, M. J. and S. Epstein. 1981a. Influence of diet on the distribution of nitrogen isotopes in animals. *Geochimica et Cosmochimica Acta* 45: 341–351.

DeNiro, M. J. and S. Epstein. 1981b. Hydrogen isotope ratios of mouse tissues are influenced by a variety of factors other than diet. *Science* 214: 1374–1376.

DePeters, E. J., B. Puschner, S. J. Taylor and J. A. Rodzen. 2013. Can fatty acid and mineral compositions of sturgeon eggs distinguish between farm-raised vs. wild white (*Acipenser transmontanus*) sturgeon origins in California? *Forensic Science International* 229: 128–132.

Doucett, R. R., J. C. Marks, D. W. Blinn, M. Caron and B. A. Hungate. 2007. Measuring terrestrial subsidies to aquatic food webs using stable isotopes of hydrogen. *Ecology* 88: 1587–1592.

Dube, M. G., G. A. Benoy, S. Blenkinsopp, J. -M. Ferone, R. B. Brua and L. I. Wassenaar. 2005. Application of multi-stable isotope (^{13}C, ^{15}N, [34]S, [37]Cl) assays to assess spatial separation of fish (Longnose sucker *Catostomus catostomus*) in an area receiving complex effluents. *Water Quality Research Journal of Canada* 40: 275–287.

Ehleringer, J. R., G. J. Bowen, L. A. Chesson, A. G. West, D. W. Podlesak and T. E. Cerling. 2008. Hydrogen and oxygen isotope ratios in human hair are related to geography. *Proceedings of the National Academy of Sciences USA* 105: 2788–2793.

Elton, C. S. 1927. *Animal Ecology*. London, Sidgwick and Jackson.

Estep, M. F. and H. Dabrowski. 1980. Tracing food webs with stable hydrogen isotopes. *Science* 209: 1537–1538.

Estep, M. F. and T. C. Hoering. 1980. Biogeochemistry of the stable hydrogen isotopes. *Geochimica et Cosmochimica Acta* 44: 1197–1206.

Estep, M. F. and S. Vigg. 1985. Stable carbon and nitrogen isotope tracers of trophic dynamics in natural populations and fisheries of the Lahontan Lake system, Nevada. *Canadian Journal of Fisheries and Aquatic Sciences* 42: 1712–1719.

Farlow, J. O. 1976. A consideration of the trophic dynamics of a Late Cretaceous large-dinosaur community. *Ecology* 57: 841–857.

Flockhart, T., L. I. Wassenaar, T. Martin, K. A. Hobson, M. Wunder and D. R. Norris. 2013. Tracking multi-generational colonization of the breeding grounds by monarch butterflies in eastern North America. *Proceeding of the Royal Society of London B* 280: 20131087.

Florin, S. T., L. A. Felicetti and C. T. Robbins. 2011. The biological basis for understanding and predicting dietary-induced variation in nitrogen and sulphur isotope discrimination. *Functional Ecology* 25: 519–526.

Fourel, F., T. Merren, J. Morrison, L. I. Wassenaar and K. A. Hobson. 1998. Application of EA Pyrolysis-IRMS δD and $\delta^{18}O$ analysis to animal migration patterns. Micromass UK Limited Application Note 300.

Frazer, T. K., R. M. Ross, L. B. Quetin and J. P. Montoya. 1997. Turnover of carbon and nitrogen during growth of larval krill, *Euphasia superba* Dana: A stable isotope approach. *Journal of Experimental Marine Biology and Ecology* 212: 259–275.

Fry, B. 1988. Food web structure on Georges Bank from stable C, N, and S isotopic compositions. *Limnology and Oceanography* 33: 1182–1190.

Fry, B. 1991. Stable isotope diagrams of freshwater food webs. *Ecology* 72: 2293–2297.

Fry, B. 2006. *Stable Isotopes in Ecology*. Springer-Verlag, New York.

Fry, B. and C. Arnold. 1982. Rapid $^{13}C/^{12}C$ turnover during growth of brown shrimp (*Penaeus aztecus*). *Oecologia* 54: 200–204.

Fry, B., W. Brand, F. J. Mersch, K. Tholke and R. Garritt. 1992. Automated analysis system for coupled $\delta^{13}C$ and $\delta^{15}N$ measurements. *Analytical Chemistry* 64: 288–291.

Garcia-Perez, B. and K. A. Hobson. 2013. A multi-isotope (δ^2H, $\delta^{13}C$, $\delta^{15}N$) approach to establishing migratory connectivity of Barn Swallow (*Hirundo rustica*). *Ecosphere* 5: art 21.

Godbout, L., M. Trudel, J. R. Irvine, C. C. Wood, M. J. Grove, A. K. Schmitt and K. D. McKeegan. 2010. Sulfur isotopes in otoliths allow discrimination of anadromous and non-anadromous ecotypes of sockeye salmon (*Oncorhynchus nerka*). *Environmental Biology of Fishes* 89: 521–532.

Graham, B. S., P. L. Koch, S. D. Newsome, K. W. McMahon and D. Aurioles. 2010. Using isoscapes to trace the movements and foraging behavior of top predators in oceanic ecosystems. *In*: J. B. West, G. Bowen, T. Dawson and K. Tu (eds.). *Isoscapes: Understanding Movement, Pattern, and Process on Earth Through Isotope Mapping*. Springer-Verlag, New York.

Grocke, D. R., H. Bocherens and A. Mariotti. 1997. Annual rainfall and nitrogen-isotope correlation in macropod collagen: application as a palaeoprecipitation indicator. *Earth and Planetary Science Letters* 153: 279–285.

Gutiérrez-Expósito, C., F. Ramírez, I. Afán, M. G. Forero and K. A. Hobson. 2015. A deuterium feather isoscape for sub-Saharan Africa. *PLoS ONE* 10: e0135938.

Haché, S., K. A. Hobson, E. M. Bayne, S. L. Van Wilgenburg and M. A. Villard. 2014. Tracking natal dispersal in a coastal population of a migratory songbird using feather stable isotope (δ^2H, $\delta^{34}S$) tracers. *PLoS ONE* 9: e94437.

Haines, E. B. 1976. Relation between the stable carbon isotope composition of fiddler crabs, plants and soils in a salt marsh. *Limnology and Oceanography* 21: 880–883.

Harrison, S. M., O. Schmidt, A. P. Moloney, S. D. Kelly, A. Rossmann, A. Schellenberg, F. Camin, M. Perini, J. Hoogewerff and F. J. Monahan. 2011. Tissue turnover in ovine muscles and lipids as recorded by multiple (H, C, O, S) stable isotope ratios. *Food Chemistry* 124: 291–297.

Hattersley, P. W. 1983. The distribution of C_3 and C_4 grasses in Australia in relation to climate. *Oecologia* 57: 113–128.

Hebert, C. E. and L. I. Wassenaar. 2005. Feather stable isotopes in Western North American waterfowl: Spatial patterns, underlying factors, and management applications. *Wildlife Society Bulletin* 33: 92–102.

Herzka, S. Z. and G. J. Holt. 2000. Changes in isotopic composition of red drum (*Sciaenops ocellatus*) larvae in response to dietary shifts: Potential applications to settlement studies. *Canadian Journal of Fisheries and Aquatic Sciences* 57: 137–147.

Hesslein, R. H., M. J. Capel, D. E. Fox and K. Hallard. 1991. Stable isotopes of sulfur, carbon, and nitrogen as indicators of trophic level and fish migration in the lower MacKenzie River basin, Canada. *Canadian Journal of Fisheries and Aquatic Sciences* 48: 2258–2265.

Hesslein, R. H., K. A. Hallard and P. Ramlal. 1993. Replacement of sulfur, carbon, and nitrogen in tissue of growing broad whitefish (*Coregonus nasus*) in response to a change in diet traced by δ^{34}S, δ^{13}C, and δ^{15}N. *Canadian Journal of Fisheries and Aquatic Sciences* 50: 2071–2076.

Hobson, K. A. 2008. Applying isotopic methods to tracking animal movements. *In*: K. A. Hobson and L. I. Wassenaar (eds.). *Tracking Animal Migration using Stable Isotopes*. Academic Press, London.

Hobson, K. A. and R. G. Clark. 1992. Assessing avian diets using stable isotopes. I: Turnover of ^{13}C in tissues. *Condor* 94: 181–188.

Hobson, K. A. and H. E. Welch. 1992. Determination of trophic relationships within a high Arctic marine food web using δ^{13}C and δ^{15}N analysis. *Marine Ecology Progress Series* 84: 9–18.

Hobson, K. A. and L. I. Wassenaar. 1997. Linking wintering and breeding grounds of neotropical migrant songbirds using stable hydrogen isotopic analysis of feathers. *Oecologia* 109: 142–148.

Hobson, K. A. and A. Bond. 2012. Extending an indicator: Year-round information on seabird trophic ecology using multiple-tissue stable-isotope analyses. *Marine Ecology Progress Series* 461: 233–243.

Hobson, K. A. and T. Quirk. 2014. Effect of age and ration on diet-tissue isotopic (Δ^{13}C, Δ^{15}N) discrimination in Striped Skunks (*Mephitis mephitis*). *Isotopes in Environmental and Health Studies* 50: 300–306.

Hobson, K. A. and G. Koehler. 2015. On the use of stable-oxygen isotope (δ^{18}O) measurements for tracking avian movements in North America. *Ecology and Evolution* 5: 799–806.

Hobson, K. A. and K. Kardynal. 2016. An isotope (δ^{34}S) filter and geolocator results constrain a dual feather isoscape (δ^{2}H, δ^{13}C) to identify the wintering grounds of North American Barn Swallows. *Auk* 133: 86–98.

Hobson, K. A., R. T. Alisauskas and R. G. Clark. 1993. Stable-nitrogen isotope enrichment in avian tissues due to fasting and nutritional stress: Implications for isotopic analyses of diet. *The Condor* 95: 388–394.

Hobson, K. A., J. F. Piatt and J. Pitocchelli. 1994. Using stable isotopes to determine seabird trophic relationships. *Journal of Animal Ecology* 63: 786–798.

Hobson, K. A., L. Atwell and L. I. Wassenaar. 1999. Influence of drinking water and diet on the stable-hydrogen isotope ratios of animal tissues. *Proceedings of the National Academy of Sciences USA* 96: 8003–8006.

Hobson, K. A., L. I. Wassenaar and O. R. Taylor. 1999. Stable isotopes (δD and δ^{13}C) are geographic indicators of natal origins of monarch butterflies in eastern North America. *Oecologia* 120: 397–404.

Hobson, K. A., R. Barnett-Johnson and T. Cerling. 2010. Using isoscapes to track animal migration. *In*: J. B. West, G. Bowen, T. Dawson and K. Tu (eds.). *Isoscapes: Understanding Movement, Pattern, and Process on Earth through Isotope Mapping*. Springer-Verlag, New York.

Hobson, K. A., S. L. Van Wilgenburg, L. I. Wassenaar, R. L. Powell, C. J. Still and J. M. Craine. 2012a. A multi-isotope (δ^{13}C, δ^{15}N, δ^{2}H) feather isoscape to assign Afrotropical migrant birds to origins. *Ecosphere* 3: art44.

Hobson, K. A., S. L. Van Wilgenburg, L. I. Wassenaar and K. Larson. 2012b. Linking hydrogen (δ^{2}H) isotopes in feathers and precipitation: Sources of variance and consequences for assignment to global isoscapes. *PLoS ONE* 7: e35137.

Hobson, K. A., D. X. Soto, D. R. Paulson, L. I. Wassenaar and J. Matthews. 2012c. A dragonfly (δ^{2}H) isoscape for North America: A new tool for determining natal origins of migratory aquatic emergent insects. *Methods in Ecology and Evolution* 3: 766–772.

Hobson, K. A., R. C. Anderson, D. X. Soto and L. I. Wassenaar. 2012d. Isotopic evidence that dragonflies (*Pantala flavescens*) migrating through the Maldives come from the northern Indian subcontinent. *PLoS ONE* 7: e52594.

Hobson, K. A., S. L. Van Wilgenburg, Y. Ferrand, F. Gossman and C. Bastat. 2013. A stable isotope (δ^2H) approach to deriving origins of harvested Woodcock (*Scolopax rusticola*) taken in France. *European Journal of Wildlife Research* 59: 881–892.

Hobson, K. A., S. L. Van Wilgenburg, T. Wesolowski, M. Maziarz, R. G. Biljsma, A. Grendelmeier and J. W. Mallord. 2014. A multi-isotope (δ^2H, δ^{13}C, δ^{15}N) approach to establishing migratory connectivity in Palearctic-Afrotropical migrants: An example using Wood Warblers *Phylloscopus sibilatrix*. *Acta Ornithologica* 49: 57–69.

Hobson, K. A., L. K. Blight and P. Arcese. 2015. Human-induced long-term shifts in gull diet from marine to terrestrial sources in North America's coastal Pacific: More evidence from more isotopes (δ^2H, δ^{34}S). *Environmental Science and Technology* 49: 10834–10840.

Hoppe, K. A., P. L. Koch, R. W. Carlson and S. D. Webb. 1999. Tracking mammoths and mastadons: Reconstruction of migratory behavior using strontium isotope ratios. *Geology* 27: 439–442.

Jardine, T. D., D. L. MacLatchy, W. L. Fairchild, R. A. Cunjak and S. B. Brown. 2004. Rapid carbon turnover during growth of Atlantic salmon (*Salmo salar*) smolts in sea water, and evidence for reduced food consumption by growth-stunts. *Hydrobiologia* 527: 63–75.

Jardine, T. D., K. A. Kidd and R. A. Cunjak. 2009. An evaluation of deuterium as a food source tracer in temperate streams of eastern Canada. *Journal of the North American Benthological Society* 28: 885–893.

Jardine, T. D., R. J. Hunt, S. J. Faggotter, D. Valdez, M. A. Burford and S. E. Bunn. 2013. Carbon from periphyton supports fish biomass in waterholes of a wet-dry tropical river. *River Research and Applications* 29: 560–573.

Karnovsky, N. J., K. A. Hobson and S. J. Iverson. 2012. From Lavage to lipids: Innovations and limitations in estimating diets of seabirds. *Marine Ecology Progress Series* 451: 263–284.

Keller, A. T., L. A. Regan, C. C. Lundstrom and N. W. Bower. 2016. Evaluation of the efficacy of spatiotemporal Pb isoscapes for provenancing of human remains. *Forensic Science International* 261: 83–92.

Kennedy, B. P., J. D. Blum, C. L. Folt and K. H. Nislow. 2000. Using natural strontium isotopic signatures as fish markers: Methodology and application. *Canadian Journal of Fisheries and Aquatic Sciences* 57: 2280–2292.

Kim, S. L., C. Martinez del Rio, D. Casper and P. L. Koch. 2012. Isotopic incorporation rates for shark tissues from a long-term captive feeding study. *The Journal of Experimental Biology* 215: 2495–2500.

Koch, P. L., J. Heisinger, C. Moss, R. W. Carlson, M. L. Fogel and A. K. Behrensmeyer. 1995. Isotopic tracking of change in diet and habitat use in African elephants. *Science* 267: 1340–1343.

Kurle, C. M., M. E. Finkelstein, K. R. Smith, D. George, D. Ciani, P. L. Koch and D. R. Smith. 2013. Discrimination factors for stable isotopes of carbon and nitrogen in blood and feathers from chicks and juveniles of the California condor. *The Condor* 115: 492–500.

Kurle, C. M., P. L. Koch, B. Tershy and D. A. Croll. 2014. The effects of sex, tissue type, and dietary components on stable isotope discrimination factors (Δ^{13}C and Δ^{15}N) in mammalian omnivores. *Isotopes in Environmental and Health Studies* 50: 307–321.

Lindeman, R. L. 1942. The trophic dynamic aspect of ecology. *Ecology* 23: 399–417.

Lorrain, A., B. Graham, F. Menard, B. Popp, S. Bouillon, P. van Breugel and Y. Cherel. 2009. Nitrogen and carbon isotope values of individual amino acids: A tool to study foraging ecology of penguins in the Southern Ocean. *Marine Ecology Progress Series* 391: 293–306.

MacAvoy, S. E., S. A. Macko and G. C. Garman. 2001. Isotopic turnover in aquatic predators: Quantifying the exploitation of migratory prey. *Canadian Journal of Fisheries and Aquatic Sciences* 58: 923–932.

Marra, P. P., K. A. Hobson and R. T. Holmes. 1998. Linking winter and summer events in a migratory bird using stable carbon isotopes. *Science* 282: 1884–1886.

Martinez del Rio, C., N. Wolf, S. A. Carleton and L. Z. Gannes. 2009. Isotopic ecology ten years after a call for more laboratory experiments. *Biological Reviews* 84: 91–111.

Maruyama, A., Y. Yamada, B. Rusuwa and M. Yuma. 2001. Change in the stable nitrogen isotope ratio in the muscle tissue of a migratory goby, *Rhinogobius* sp., in a natural setting. *Canadian Journal of Fisheries and Aquatic Sciences* 58: 2125–2128.

Mayer, B. and L. I. Wassenaar. 2012. Isotopic characterization of nitrate sources and transformations in Lake Winnipeg and its contributing rivers, Manitoba, Canada. *Journal of Great Lakes Research* 38: 135–146.

McConnaughey, T. and C. P. McRoy. 1979. Food-web structure and the fractionation of carbon isotopes in the Bering Sea. *Marine Biology* 53: 257–262.

McCutchan, J. H. Jr., W. M. Lewis, Jr., C. Kendall and C. C. McGrath. 2003. Variation in trophic shift for stable isotope ratios of carbon, nitrogen, and sulfur. *Oikos* 102: 378–390.

Merow, C., M. J. Smith and J. A. Silander, Jr. 2013. A practical guide to MaxEnt for modelling species' distributions: What it does, and why inputs and settings matter. *Ecography* 36: 1058–1069.

Minagawa, M. and E. Wada. 1984. Stepwise enrichment of ^{15}N along food chains: Further evidence and relation between ^{15}N and animal age. *Geochimica et Cosmochimica Acta* 48: 1135–1140.

Minson, D. J., M. M. Ludlow and J. H. Troughton. 1975. Differences in natural carbon isotope ratios of milk and hair from cattle grazing tropical and temperate pastures. *Nature* 256: 602.

Mirón, L. L., L. Gerardo Herrera, N. Ramirez and K. A. Hobson. 2006. Effect of diet quality on carbon and nitrogen turnover and isotopic discrimination in blood of a New World nectarivorous bat. *The Journal of Experimental Biology* 209: 541–548.

Miyake, Y. and E. Wada. 1967. The abundance ratio of ^{15}N/^{14}N in marine environments. *Records of Oceanographic Works in Japan* 9: 37–53.

Murray, I. W. and B. O. Wolf. 2012. Tissue carbon incorporation rates and diet-to-tissue discrimination in ectotherms: Tortoises are really slow. *Physiological and Biochemical Zoology* 85: 96–105.

Nadelhoffer, K. J. and B. Fry. 1994. Nitrogen isotopes studies in forest ecosystems. *In*: K. Lajtha and R. H. Michener (eds.). *Stable Isotopes in Ecology and Environmental Science*. Blackwell Scientific, Oxford.

Nagy, K. A., I. A. Girard and T. K. Brown. 1999. Energetics of free-ranging mammals, reptiles, and birds. *Annual Reviews in Nutrition* 19: 247–277.

Nehlich, O. 2015. The application of sulphur isotope analyses in archaeological research: A review. *Earth-Science Reviews* 142: 1–17.

Nehlich, O., J. H. Barrett and M. P. Richards. 2013. Spatial variability in sulphur isotope values of archaeological and modern cod (*Gadus morhua*). *Rapid Communications in Mass Spectrometry* 27: 2255–2262.

Nielson, K. E. and G. J. Bowen. 2010. Hydrogen and oxygen in brine shrimp chitin reflect environmental water and dietary isotopic composition. *Geochimica et Cosmochimica Acta* 74: 1812–1822.

Pardo, L. H. and K. J. Nadelhoffer. 2010. Using nitrogen isotope ratios to assess terrestrial ecosystems at regional and global scales. *In*: J. B. West, G. Bowen, T. Dawson and K. Tu (eds.). *Isoscapes: Understanding Movement, Pattern, and Process on Earth through Isotope Mapping*. Springer-Verlag, New York.

Pekarsky, S., A. Angert, B. Haese, M. Werner, K. A. Hobson and R. Nathan. 2015. Enriching the isotopic toolbox for migratory connectivity analysis: A new approach for migratory species breeding in remote or unexplored areas. *Diversity and Distributions* 21: 416–427.

Peterson, B. J., R. W. Howarth and R. H. Garritt. 1985. Multiple stable isotopes used to trace the flow of organic matter in estuarine food webs. *Science* 227: 1361–1363.

Pietsch, S. J., K. A. Hobson, L. I. Wassenaar and T. Tütken. 2011. Tracking cats: Problems with placing feline carnivores on δ^{18}O, δD isoscapes. *PLoS ONE* 6: e24601.

Pimm, S. L. 1992. *Food Webs*. Chapman and Hall, New York.

Pimm, S. L., J. H. Lawton and J. E. Cohen. 1991. Food web patterns and their consequences. *Nature* 350: 669–674.

Podlesak, D. W., A. -M. Torregrossa, J. R. Ehleringer, M. D. Dearing, B. H. Passey and T. E. Cerling. 2008. Turnover of oxygen and hydrogen isotopes in the body water, CO_2, hair, and enamel of a small mammal. *Geochimica et Cosmochimica Acta* 72: 19–35.

Popp, B. N., B. S. Graham, R. J. Olson, C. C. S. Hannides, M. J. Lott, G. A. Lopez-Ibarra, F. Galvan-Magana and B. Fry. 2007. Insight into the trophic ecology of Yellowfin Tuna, *Thunnus albacares*, from compound-specific nitrogen isotope analysis of proteinaceous amino acids. *In*: T. E. Dawson and R. T. W. Siegwolf (eds.). *Stable Isotopes as Indicators of Ecological Change*. Academic Press, London.

Post, D. M. 2002. Using stable isotopes to estimate trophic position: Models, methods, assumptions. *Ecology* 83: 703–718.

Poupin, N., C. Bos, F. Mariotti, J. -F. Huneau, D. Tome and H. Fouillet. 2011. The nature of dietary protein impacts the tissue-to-diet ^{15}N discrimination factors in laboratory rats. *PLoS ONE* 6: e28046.

Procházka, P., S. van Wilgenburg, J. Neto, R. Yosef and K. A. Hobson. 2013. Using stable hydrogen isotopes (δ^2H) and ring recoveries to trace natal origins in a Eurasian passerine with a migratory divide. *Journal of Avian Biology* 44: 1–10.

Qi, H., T. B. Coplen and L. I. Wassenaar. 2011. Improved online $\delta^{18}O$ measurements of nitrogen- and sulfur-bearing organic materials and a proposed analytical protocol. *Rapid Communications in Mass Spectrometry* 25: 2049–2058.

Quillfeldt, P., R. A. R. McGill and R. W. Furness. 2005. Diet and foraging areas of Southern Ocean seabirds and their prey inferred from stable isotopes: Review and case study of Wilson's storm-petrel. *Marine Ecology Progress Series* 295: 295–304.

Rau, G. H., R. E. Sweeney and I. R. Kaplan. 1982. Plankton $^{13}C/^{12}C$ ratio changes with latitude: Differences between northern and southern oceans. *Deep Sea Research* 29: 1035–1039.

Rau, G. H., A. J. Mearns, D. R. Young, R. J. Olson, H. A. Schafer and I. R. Kaplan. 1983. Animal $^{13}C/^{12}C$ correlates with trophic level in pelagic food webs. *Ecology* 64: 1314–1318.

Rees, C. E., W. J. Jenkins and J. Monster. 1978. The sulphur isotopic composition of ocean water sulphate. *Geochimica et Cosmochimica Acta* 42: 377–381.

Reich, K. J., K. A. Bjorndal and A. B. Bolten. 2007. The "lost years" of green turtles: Using stable isotopes to study cryptic life stages. *Biology Letters* 3: 712–714.

Reynard, L. M., G. M. Henderson and R. E. M. Hedges. 2010. Calcium isotope ratios in animal and human bone. *Geochimica et Cosmochimica Acta* 74: 3735–3750.

Robbins, C. T., L. A. Felicetti and M. Sponheimer. 2005. The effect of dietary protein quality on nitrogen isotope discrimination in mammals and birds. *Oecologia* 144: 534–540.

Rogers, K. M., L. I. Wassenaar, D. X. Soto and J. A. Bartle. 2012. A feather-precipitation hydrogen isoscape model for New Zealand: Implications for eco-forensics. *Ecosphere* 3: art62.

Rosenblatt, A. E. and M. R. Heithaus. 2013. Slow isotope turnover rates and low discrimination values in the American alligator: Implications for interpretation of ectotherm stable isotope data. *Physiological and Biochemical Zoology* 86: 137–148.

Schell, D. M., B. A. Barnett and K. Vinette. 1998. Carbon and nitrogen isotope ratios in zooplankton of the Bering, Chukchi and Beaufort Seas. *Marine Ecology Progress Series* 162: 11–23.

Scheuhammer, A. M. and D. M. Templeton. 1998. Use of stable isotope ratios to distinguish sources of lead exposure in wild birds. *Ecotoxicology* 7: 37–42.

Schilder, J., C. Tellenbach, M. Möst, P. Spaak, M. van Hardenbroek, M. J. Wooller and O. Heiri. 2015. The stable isotopic composition of *Daphnia ephippia* reflects changes in $\delta^{13}C$ and $\delta^{18}O$ values of food and water. *Biogeosciences* 12: 3819–3830.

Senn, D. B., E. J. Chesney, J. D. Blum, M. S. Bank, A. Maage and J. P. Shine. 2010. Stable isotope (N, C, Hg) study of methylmercury sources and trophic transfer in the northern Gulf of Mexico. *Environmental Science and Technology* 44: 1630–1637.

Sheppard, S. M., R. L. Nielsen and H. P. Taylor. 1969. Oxygen and hydrogen isotope ratios of clay minerals from porphyry copper deposits. *Economic Geology* 64: 755–777.

Smyntek, P. M., M. A. Teece, K. L. Schulz and S. J. Thackeray. 2007. A standard protocol for stable isotope analysis of zooplankton in aquatic food web research using mass balance correction models. *Limnology and Oceanography* 52: 2135–2146.

Solomon, C. T., J. J. Cole, R. R. Doucett, M. L. Pace, N. D. Preston, L. E. Smith and B. C. Weidel. 2009. The influence of environmental water on the hydrogen stable isotope ratio in aquatic consumers. *Oecologia* 161: 313–324.

Solovyeva, D., K. A. Hobson, N. Kharitonova, J. Newton, J. W. Fox, V. Afansyev and A. D. Fox. 2016. Combining stable hydrogen (δ^2H) isotopes and geolocation to assign Scaly-sided Mergansers to moult river catchments. *Journal of Ornithology* 157: 663–669.

Soto, D. X., L. I. Wassenaar, K. A. Hobson and J. Catalan. 2011. Effects of size and diet on stable hydrogen isotope values (δD) in fish: Implications for tracing origins of individuals and their food sources. *Canadian Journal of Fisheries and Aquatic Sciences* 68: 2011–2019.

Soto, D. X., K. A. Hobson and L. I. Wassenaar. 2013a. The influence of metabolic effects on stable hydrogen isotopes in tissues of aquatic organisms. *Isotopes in Environmental and Health Studies* 49: 305–311.

Soto, D. X., L. I. Wassenaar and K. A. Hobson. 2013b. Stable hydrogen and oxygen isotopes in aquatic food webs are tracers of diet and provenance. *Functional Ecology* 27: 535–543.

Steele, K. W. and R. M. Daniel. 1978. Fractionation of nitrogen isotopes by animals: A further complication to the use of variations in the natural abundance of ^{15}N for tracer studies. *Journal of Agricultural Science* 90: 7–9.

Stewart, G. R., M. H. Turnbull, S. Schmidt and P. D. Erskine. 1995. ^{13}C natural abundance in plant communities along a rainfall gradient: A biological indicator of water availability. *Australian Journal of Plant Physiology* 22: 51–55.

Stewart, R. E. A., P. A. Outridge and R. A. Stern. 2003. Walrus life history movements reconstructed from lead isotopes in annual layers of teeth. *Marine Mammal Science* 19: 806–818.

Still, C. J. and R. L. Powell. 2010. Continental-scale distributions of vegetation stable carbon isotope ratios. *In*: J. B. West, G. J. Bowen, T. E. Dawson and K. P. Tu (eds.). *Isoscapes: Understanding Movements, Pattern and Process on Earth through Isotope Mapping*. Springer, New York.

Sullins, D. S., W. C. Conway, D. A. Haukos, K. A. Hobson, L. I. Wassenaar, C. E. Comer and I. -K. Hung. 2016. American woodcock migratory connectivity indicated by hydrogen isotopes. *Journal of Wildlife Management* 80: 510–526.

Suzuki, K. W., A. Kasai, K. Nakayama and M. Tanaka. 2005. Differential isotopic enrichment and half-life among tissues in Japanese temperate bass (*Lateolabrax japonicus*) juveniles: Implications for analyzing migration. *Canadian Journal of Fisheries and Aquatic Sciences* 62: 671–678.

Taylor, H. P., Jr. 1974. An application of oxygen and hydrogen isotope studies to problems of hydrothermal alteration and ore deposition. *Economic Geology* 69: 843–883.

Terzer, S., L. I. Wassenaar, L. J. Araguás-Araguás and P. K. Aggarwal. 2013. Global isoscapes for δ^{18}O and δ^2H in precipitation: Improved prediction using regionalized climatic regression models. *Hydrology and Earth System Sciences Discussions* 10: 7351–7393.

Thomas, S. M. and T. W. Crowther. 2015. Predicting rates of isotopic turnover across the animal kingdom: A synthesis of existing data. *Journal of Animal Ecology* 84: 861–870.

Tieszen, L. L., T. W. Boutton, K.G. Tesdahl and N.A. Slade. 1983. Fractionation and turnover of stable carbon isotopes in animal tissues: Implications for δ^{13}C analysis of diet. *Oecologia* 57: 32–37.

Topalov, K., A. Schimmelmann, P. D. Polly, P. E. Sauer and M. Lowry. 2013. Environmental, trophic, and ecological factors influencing bone collagen δ^2H. *Geochimica et Cosmochimica Acta* 111: 88–104.

Trueman, C. N., K. M. MacKenzie and M. R. Palmer. 2012. Identifying migrations in marine fishes through stable-isotope analysis. *Journal of Fish Biology* 81: 826–847.

Vanderklift, M. A. and S. Ponsard. 2003. Sources of variation in consumer-diet δ^{15}N enrichment: A meta analysis. *Oecologia* 136: 169–182.

Vander Zanden, M. J. and J. B. Rasmussen. 2001. Variation in δ^{15}N and δ^{13}C trophic fractionation: Implications for food web studies. *Limnology and Oceanography* 46: 2061–2066.

Vander Zanden, M. J. and Y. Vadeboncoeur. 2002. Fishes as integrators of benthic and pelagic food webs in lakes. *Ecology* 83: 2152–2161.

Vander Zanden, M. J., M. K. Clayton, E. K. Moody, C. T. Solomon and B. C. Weidel. 2015. Stable isotope turnover and half-life in animal tissues: A literature synthesis. *PLoS ONE* 10: e0116182.

Van der Merwe, N. J. and J. C. Vogel. 1978. [13]C content of human collagen as a measure of prehistoric diet in woodland North America. *Nature* 276: 815–816.

Van Hardenbroek, M., D. R. Grocke, P. E. Sauer and S. A. Elias. 2012. North American transect of stable hydrogen and oxygen isotopes in water beetles from a museum collection. *Journal of Palaeolimnology* 48: 461–470.

Van Wilgenburg, S. L. and K. A. Hobson. 2011. Combining stable-isotope (δD) and band recovery data to improve probabilistic assignment of migratory birds to origin. *Ecological Applications* 21: 1340–1351.

Veen, T., M. B. Hjernquist, S. L. Van Wilgenburg, K. A. Hobson, E. Folmer, L. Font and M. Klaassen. 2014. Identifying the African wintering grounds of hybrid flycatchers using a multi-isotope (δ^2H, δ^{13}C, δ^{15}N) assignment approach. *PLoS ONE* 9: e98075.

Vogel, J. C., B. Eglington and J. M. Auret. 1990. Isotope fingerprints in elephant bone and ivory. *Nature* 346: 747–749.

Wassenaar, L. and K. A. Hobson. 1998. Natal origins of migratory Monarch Butterflies at wintering colonies in Mexico: New isotopic evidence. *Proceedings of the National Academy of Sciences* 95: 15436–15439.

Wassenaar, L. I. and K. A. Hobson. 2000a. Improved method for determining the stable-hydrogen isotopic composition (δD) of complex organic materials of environmental interest. *Environmental Science and Technology* 34: 2354–2360.

Wassenaar, L. I. and K. A. Hobson. 2000b. Stable-carbon and hydrogen isotope ratios reveal breeding origins of red-winged blackbirds. *Ecological Applications* 10: 911–916.

Wassenaar, L. I., K. A. Hobson and L. Sisti. 2015. An online temperature-controlled vacuum-equilibration preparation system for the measurement of δ^2H values of non-exchangeable-H and of δ^{18}O values in organic materials by isotope-ratio mass spectrometry. *Rapid Communications in Mass Spectrometry* 29: 397–407.

Wilkinson, G. M., J. J. Cole and M. L. Pace. 2015. Deuterium as a food source tracer: Sensitivity to environmental water, lipid content, and hydrogen exchange. *Limnology and Oceanography: Methods* 13: 213–223.

Wolf, N., S. D. Newsome, J. Peters and M. L. Fogel. 2015. Variability in the routing of dietary proteins and lipids to consumer tissues influences tissue-specific isotopic discrimination. *Rapid Communications in Mass Spectrometry* 29: 1448–1456.

Yapp, C. J. and S. Epstein. 1982. A re-examination of cellulose carbon-bound hydrogen δD measurements and some factors affecting plant-water D/H relationships. *Geochimica et Cosmochimica Acta* 46: 955–965.

CHAPTER 5

Data Analysis Interpretation
Forensic Applications and Examples

Hannah B. Vander Zanden[1,]* and *Lesley A. Chesson*[2]

5.1 Introduction

"Essentially, all models are wrong, but some are useful."

(Box and Draper 1987)

Following food sampling and isotope ratio analysis, forensic practitioners must next reach some sort of conclusion about the collected data. An investigator assigns meaning and then determines the significance and implications of the findings using data analysis and interpretation. Unfortunately, there is no *one-size-fits-all* approach for evaluating collected data, or evidence, in forensic investigations, including investigations that collect isotope ratio data. This issue was highlighted in a report on the capacity of forensic sciences in the United States, released by the National Academy of Sciences (National Research Council 2009). Forensic evidence is frequently used in the legal system to support definitive conclusions about matches between specimens, between specimens and individuals, and between specimens and potential sources. However, as noted in the report:

> "The simple reality is that the interpretation of forensic evidence is not always based on scientific studies to determine its validity. This is a serious problem. Although research has been done in some disciplines, there is a notable dearth of peer-reviewed, published studies establishing the scientific bases and validity of many forensic methods."

[1] University of Utah, Global Change & Sustainability Center, Department of Geology & Geophysics, 115 South 1460 East, Salt Lake City, UT 84112, USA.
[2] IsoForensics, Inc., 421 Wakara Way, Suite 100, Salt Lake City, UT 84108, USA.
  Email: lesley@isoforensics.com
* Corresponding author: h.vanderzanden@utah.edu

The report contained recommendations for the forensic science community, including: standardized terminology and reporting; more and better research; good practices and standards; and quality control, assurance, and improvement. Since the publication of the 2009 report, several studies have addressed the recommendation for standardized reporting of forensic evidence [e.g., (Howes et al. 2013; Howes et al. 2014; Simmross 2014; Biedermann et al. 2015)]. The European Network of Forensic Science Institutes (ENFSI) recently released its *Guideline for Evaluative Reporting in Forensic Science* (ENFSI 2015), which provided forensic practitioners with a recommended framework for evaluative reports.

Analytical forensic disciplines—such as isotope chemistry—are ideally suited to address several recommendations of the 2009 National Academy of Sciences report. The field has standardized terminology and good practices, while quality assurance, quality control, and improvements to these activities are primary concerns (see Chapters 1 and 2). As described in other chapters of this volume, there exists a plethora of high quality research on the use of isotope ratio analysis to examine foods, beverages, and their flavorings. In addition, isotope data are amenable for use in an evaluative reporting framework, like that described by the 2015 ENFSI report.

This chapter principally concerns the interpretation of data collected via isotope ratio mass spectrometry (IRMS) and allied techniques in the context of food forensic applications. While isotopic composition data may complement other measurement results (e.g., elemental concentrations, fatty acid profiles), we address analysis and interpretation approaches that may be most informative with data consisting of isotope ratios. First, we consider the importance of analytical accuracy and data quality in order to make conclusive inferences. Next, we describe the common statistical approaches and methods for exploring food isotopic data. Finally, we introduce two examples of investigating food authenticity and origin using statistical approaches common to ecological studies. Our goal is to provide the reader with the necessary foundation and impetus to use data analysis and interpretation methods more effectively in addressing food forensic questions.

5.2 Reporting of isotopic evidence in forensic casework

Effective isotope data interpretation require accurate and precise measurement results, which are products of multiple processes (Carter and Barwick 2011), as presented within the framework of a laboratory's quality assurance plan in Figure 5.1, including validation of isotope data normalization and corrections, evaluation of potential biases, and review of long-term quality control data collected in the laboratory. These processes ultimately support a variety of reporting activities. For example, technical

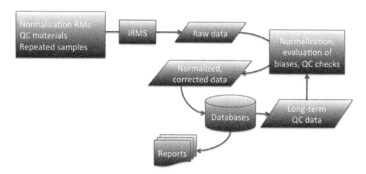

Figure 5.1 Schematic representation of a quality assurance program, the practices used to ensure that a laboratory is producing data that are accurate, precise, and that meet the requirements required for effective isotope data analysis and interpretation.

reports describe the process or results of scientific research; these reports are often used as the basis for publications in the peer-reviewed literature. Measurement results are typically presented in data reports, which allow the individual submitting the sample to then apply the results as needed.

Particularly relevant to the judicial system are *evaluative reports*, which not only present measurement results, but also critically examine the collected evidence in the context of prior scientific research. As noted in the ENFSI guide

"Evaluative reports for use in court should be produced when two conditions are met:

1. The forensic practitioner has been asked by a *mandating authority* or party to examine and/or compare material (typically recovered trace material with reference material from known potential sources).
2. The forensic practitioner seeks to evaluate findings with respect to particular competing *propositions* set by the specific case circumstances or as indicated by the mandating authority" (ENFSI 2015).

For most cases, opposing prosecution and defense positions provide the competing propositions.

One of the most common approaches used to evaluate competing propositions in forensic science employs a likelihood ratio (LR), simply the ratio of two probabilities or "the probability of the evidence under two competing hypotheses" (Aitken and Taroni 2004). Ideally, these competing hypotheses are mutually exclusive; for example

$P(E \mid H_p)$: the prosecution claims that the evidence was observed given that the questioned material came from the same source as the control sample

$P(E \mid H_D)$: the defense claims that the evidence was observed given that the questioned material came from a source other than the control sample

The likelihood ratio is then calculated as

$$LR = \frac{P(E|H_P)}{P(E|H_D)}$$

If the LR is greater than 1, the evidence supports the "prosecutor's hypothesis" that the questioned and control specimens shared a common origin. If the LR is less than 1, the evidence supports the "defense's hypothesis" that the observations were due to chance. A likelihood ratio near 1 indicates that the evidence favors neither the prosecution's nor the defense's position. Thus, the magnitude of the LR may be used in a qualitative context as the value of the evidence for the prosecution.

A likelihood ratio is generally easier to interpret and explain to the court than other, more exploratory data analysis and interpretation techniques, such as multivariate hypothesis testing. Groups have recommended verbal equivalents of LRs to further clarify reports such that a LR of 5,000 becomes "moderately strong evidence" (Association of Forensic Science Providers 2009). In addition, the single number provided by the LR can be considered a straightforward modifier to the odds of the two opposing propositions in a system of Bayesian inference. Calculation of a LR is the preferred method for evaluative reporting described in the ENFSI guide (ENFSI 2015). There have been multiple reports published on the use of likelihood ratios to evaluate evidence (Aitken and Lucy 2004; Aitken and Taroni 2004; Nordgaard and Rasmusson 2012), including some focused specifically on cases where isotope data were collected (Pierrini et al. 2007; Farmer et al. 2009; Carter et al. 2014).

In instances where no competing propositions exist and evaluative reporting is not needed, a variety of other statistical approaches can be used to interpret collected isotope ratio data. This is often the case in food studies where investigators are initially interested in examining patterns and trends in measurement results and then potentially using these to understand something about food authenticity or origin—e.g., isotopic differences in meat from swine raised on different diets or in different geographic locations. A review of approaches typically used for exploring, analyzing, and interpreting the isotopic compositions of food is presented in the next section.

Regardless of the type of reporting, a forensic practitioner should include in any technical, data, or evaluative report a statement of *measurement uncertainty*: "Nonnegative parameter characterising the dispersion of the quantity values being attributed to a measurand, based on the information used" (JCGM 2012). Generally, measurement uncertainty represents the value after "±" and provides the report's reader information on measurement precision. International standard ISO/IEC 17025 (ISO 2005) requires calculation of measurement uncertainty. However, a study on applications of stable isotope analysis in ecology found that approximately one-fifth of the 330 ecological papers surveyed failed to report any form of analytical error associated with IRMS data (Jardine and Cunjak 2005). Even within the papers that did report analytical error (81%), there was considerable variation in the terminology and approaches used to quantify and describe the error.

Uncertainty calculations typically consider at a minimum measurement *repeatability*, or variability between repeated analyses within an analytical sequence, and *reproducibility*, or variability between analytical sequences, between labs, and/or over time. Data collected as part of the laboratory's quality assurance program (Figure 5.1) provide the information needed for these evaluations. Additional sources of uncertainty can include *inherent variability* (inter-sample variability)—such as that associated with different individuals of a population (Barnes et al. 2008; Vander Zanden et al. 2012) or different batches of manufacture (Benson et al. 2009; Gentile et al. 2009)— and *data processing*, such as blank correction (Ohlsson 2013) or δ-value normalization (Skrzypek et al. 2010; Skrzypek 2013). Related specifically to data processing, the Forensic Isotope Ratio Mass Spectrometry (FIRMS) Network presented simple rules for combining uncertainties in its Good Practice Guide (Carter and Barwick 2011); more recently, a spreadsheet template for calculating measurement uncertainty in δ-values has been published (Dunn et al. 2015). The inherent variability associated with different groups or specimens is typically the basis of matches made between specimens, between specimens and individuals, and between specimens and potential sources (Gentile et al. 2011).

5.3 Common approaches for food isotope data analysis and interpretation

As noted above, technical reports written by forensic laboratories describe the results of scientific research, often forming the framework for a peer-reviewed publication. Unlike evaluative reporting—where likelihood ratios are most commonly used to reach conclusions about collected data—a variety of statistical approaches have been applied to analyze and interpret isotope data in the food literature. To date we are aware of only a limited number of studies published describing the application of the LR approach to food, to determine specifically the geographical origin of wine and olive oil, respectively (Martyna et al. 2014; Własiuk et al. 2015). Using the meat literature as an example library (see Chapter 6), the approaches used to analyze and interpret food isotope data can be generally divided into two types: *exploratory*, in which the characteristics of a data set are summarized, and *inferential*, in which an underlying data distribution is inferred from the collected measurement results.

It must be noted that data derived from the chemical analysis of biological material—such as food—are often non-parametric in nature and thus parametric statistics, such as *t*-tests and ANOVA (analysis of variance), should not be directly applied to these data sets. In some cases data might require transformation (e.g., logarithmic or square root) prior to statistical analysis and interpretation. Alternatively, forensic practitioners can use robust statistics that are fit for analysis of measurement results drawn from a variety of data distributions—for example, median values and box plots.

5.3.1 Exploratory data analysis techniques

Common exploratory data analysis techniques in food forensics include the *t*-test or one-way ANOVA, used to test for significant differences between groups, as well as the Pearson's correlation coefficient, which measures the linear correlation between two variables, and linear regression, which determines the relationship between two variables. These data analysis techniques are potentially useful in some court settings, especially when accompanied by data visualizations. Common visualization tools include box plots, histograms, and scatter plots; scatter plots are often used to display relationships between different isotopic compositions—e.g., the hydrogen isotopic compositions of fish tissues, such as otoliths (earbones) or muscle, and the water in which the fish lived (Whitledge et al. 2006). Scatter plots can also be used with multivariate ellipse-based metrics and their estimation via Bayesian inference, which provides a robust way to compare the isotopic composition of sample groups, even when groupings have different sample sizes (Jackson et al. 2011; Oulhote et al. 2011).

One visualization method frequently applied to multivariate data sets is *principal component analysis* (PCA), an unsupervised data analysis method that reduces the number of variables to a few, uncorrelated, linear combinations of data—or *principal components*—emphasizing characteristics in the data set. The first principal component accounts for as much variability observed in the data set as possible, with each following principal component accounting for as much of the remaining variability as possible. PCA is particularly useful for providing analysts with an overview of complex, multivariate data sets (Bro and Smilde 2014). As such, PCA is often used in studies that measure multiple isotope ratios or those that collect element content data along with isotopic compositions. For example, a 2015 study on chicken from four provinces of China (Zhao et al. 2016) found that approximately 75% of the total elemental and isotopic variability observed in the poultry samples could be explained by the first four principal components.

Statistics that are useful for grouping samples when pre-defined classes are not available include *k*-means clustering. In this technique, samples are partitioned in a specific number (*k*) of groups that are as homogenous as possible with respect to the measurands by minimizing the distance to data points and the cluster centers. A recent study on the traceability of shrimp used *k*-means clustering to group samples based on carbon and nitrogen isotopic compositions as well as elemental concentration (Ortea and Gallardo 2015). The *k*-means classification approach—in conjunction with discriminant function analysis (see Section 5.3.2)—could correctly classify ~ 90% of samples as farmed or wild-caught based on the stable isotopic compositions of carbon and nitrogen (Ortea and Gallardo 2015).

5.3.2 Discriminant function analysis

Like PCA, discriminant function analysis (DA) reduces multi-dimensional data to combinations that emphasize patterns within the data set, specifically by modeling differences among *classes* of data. However, PCA does not explicitly take class differences into account; that is, it does not incorporate foreknowledge about groups. In general, the goal of DA is to describe a mathematical function that can be applied to classify an unknown food sample to the correct class—dietary group, production method, region-of-origin, etc.—when discrete classes are known *a priori*. Common DA approaches includes canonical discriminant analysis (CDA), linear discriminant analysis (LDA), partial least-squares discriminant analysis (PLS-DA), and quadratic discriminant analysis (QDA).

A 2007 study on European lamb meat used LDA to assess the reliability of DA to classify samples to twelve geographic regions using hydrogen, carbon, nitrogen, and sulfur isotopic compositions (Camin et al. 2007). Samples of authentic, known-origin lamb meat were used as a training set and the model was evaluated with leave-one-out cross validation. By using LDA, 78% of the samples were assigned to the correct group, i.e., the correct geographical origin (Camin et al. 2007).

A word of caution is needed on using DA techniques: there are certain criteria that should be met before DA is applied; otherwise, the technique simply makes the observed data fit predefined classes. As noted in a recent study of coffee (Carter et al. 2015a), the level of source discrimination provided by DA was exaggerated when compared to other approaches. It is recommended that the number of observations be significantly greater than the number of independent variables (> 5:1) and the number of observations in the smallest group be greater than the number of independent variables (Carter et al. 2015a). It may still be possible to narrow the range of origins if large sample sets are not available, simply by using X-Y and geo-spatial plots. As an alternative to DA, a recent paper advocated class-modeling methods for food forensics, since the technique does not automatically consider the existence of other classes for grouping purposes (Rodionova et al. 2016). At least one meat study has used class modeling—specifically soft independent modeling of class analogy (SIMCA)—to authenticate beef cattle diet using a variety of isotopic compositions and other molecular biomarkers (i.e., fatty acids, β-carotene, lutein, and α-tocopherol) (Osorio et al. 2013).

There are additional statistical techniques that—despite infrequent application in the published literature—may be useful in food forensics. These include *kernel density estimation* (KDE), an exploratory and non-parametric data analysis method that can be used to visualize the distribution underlying a continuous variable (see, for example, de Rijke et al. 2016), and *support vector machines* (SVM) (or support vector networks), which consist of both supervised or unsupervised learning models that classify samples into groups. At least one study on salmon has used an *artificial neural network* (ANN) to identify organic farmed fish based on measurements of carbon and nitrogen isotope ratios with fatty acid content (Molkentin et al. 2007). In this application, the ANN was applied to model the potentially complex and non-linear patterns in the measured data. Similar to DA, however, these techniques perform best with very large data sets; otherwise they are prone to forcing observations to fit predefined groups.

5.4 Likelihood-based data analysis approaches in food forensics

Applications of stable isotope data to food forensics are generally focused on two types of inquiry: authenticity and origin. Authenticity studies seek to determine if food items meet claims regarding content and production methods or if two samples resemble one another, whereas origin studies investigate the geographic source location. The type of information that isotope analysis of different elements can provide is covered in other chapters in this book, as certain isotopes may be more appropriate to authenticity *vs.* origin questions. In both categories of inquiry, however, the goal is to provide probabilistic statements about the characteristics of the material to determine either a likely group membership or a likely geographic origin. We next introduce two likelihood-based approaches used in ecological studies that will also be effective in food forensics.

The first category for food sample assignment is a *discrete approach*, in which there are a limited number of groups to which an unknown sample may belong. The discrete example below uses DA (described above) to characterize the probability of belonging to a particular group and relies on comparative samples to define *a priori* groupings that are categorical in nature. This approach will often be relevant to authenticity questions, though the categories could also be geographic in nature if a potential geographic range is divided into discrete regions. While verification of authenticity of food products is one of the most important issues in food quality control, the efficacy of a discrete approach will rely on the design of the classes. That is, the groups must be isotopically distinct and also characterize the extent of the unknown samples, as an unknown sample from a unique group that is not part of the training data cannot be classified accurately. Some examples in which a discrete approach would be appropriate include distinguishing between organic *vs.* synthetic fertilizer use; livestock diets (e.g., grass-fed or corn diet); pure *vs.* adulterated products (e.g., sugar added); or local *vs.* non-local origin.

The second approach for food sample assignment is a *continuous approach*, which uses an isotopic distribution across space to determine the likelihood of a material originating from any number of geographic locations. These isotopic landscapes, or *isoscapes*, can be generated by interpolation of isotopic measurements, often based on their co-variance with additional parameters such as latitude and altitude (West et al. 2010). Whereas these surfaces will be discretized pixels (dependent on the isoscape resolution) across the geographic domain considered, this approach is termed *continuous* because possible origins can be considered within the entire spatial range.

The most readily available and commonly used isoscapes are those of the δ^2H and $\delta^{18}O$ compositions of precipitation, which exhibit predictable patterns across continental scales (Craig 1961; Dansgaard 1964; Bowen et al. 2005). Plants and animals incorporate environmental water and propagate these baseline patterns through food webs; H and O isotopes thus serve as a tracer of the location where the tissue was synthesized (Rubenstein and Hobson 2004; Hobson and Wassenaar 2008). Relating precipitation values to other types of biological material requires calibration samples to determine the isotopic relationship between the baseline model and the sample material in order to create a rescaling function. Rescaling functions are often calculated as a simple scalar offset or a linear regression, owing to the strong relationship between organic materials and the isotopic composition of environmental water. Across samples collected from large spatial ranges, correlations between the isotope ratios of water and a diversity of organic materials have been observed, including wildlife tissues, such as bird feathers (Bowen et al. 2005; Hobson et al. 2012), and foods, such as olive oil (Chiocchini et al. 2016), beef (Chesson et al. 2011), and alcoholic drinks (West et al. 2007; Carter et al. 2015b). It is also possible directly to interpolate stable isotope values from the material of interest (often through Kriging approaches), in which case a rescaling function is not necessary (Hobson et al. 1999; McMahon et al. 2013; Vander Zanden et al. 2015). A direct interpolation requires samples of known origin from across the spatial range of interest, and the resulting isoscape will not be applicable to other sample types. For example, Hobson et al. (1999) created a hydrogen isoscape from monarch butterfly wings (*Danaus plexippus*) based on samples of known origin from across eastern North America. However, collecting a spatially exhaustive sample set may be a major limitation of using this approach in food forensic applications.

Prior probabilities that incorporate additional data sources can be incorporated in both approaches for food assignment. For example, group membership can be weighted to proportions in the training data or in other observations (Royle and Rubenstein 2004; Vander Zanden et al. 2015), while geographic assignments can incorporate spatial priors such as known animal production ranges, species distributions, or population densities (Wunder et al. 2005; Van Wilgenburg and Hobson 2011; Flockhart et al. 2013; Guillemain et al. 2014). The outcomes of both the discrete and continuous approaches consist of posterior probabilities that provide an estimate of the strength of the support for belonging to a group or originating from a particular location. We suggest that an odds ratio framework can be used to create decision-making thresholds and determine the potential reliability of assignments, and we present examples of this framework in the context of both examples.

5.4.1 Discrete approach

Discriminant function analysis and related methods that invert analysis of variance-type models use a likelihood-based approach to first characterize the isotope distributions or probability density for a limited number of categories and then provide a posterior probability of belonging to each group. A training data set will always be necessary to define the groupings before unknown samples can be assessed for group membership. Analysis of variance can be informative to indicate whether particular independent variable(s) will be useful in DA, as significant differences among groups are necessary to provide revealing outcomes. Discriminant analysis assumes a normal distribution in the sample data, which may not always apply to all food isotope data. Linear discriminant analysis assumes homogeneity in the variance among groups, while quadratic discriminant analysis relaxes this assumption and can be used when variance is heterogeneous among groups. Most statistical software packages include built-in functions for performing DA that adhere to these assumptions, although fully Bayesian models could be developed to work with data that are not normally distributed.

Because the utility of a discriminant model will be based on the classification accuracy, it is important to evaluate the performance with samples of known group membership. To perform this validation step, it is preferable to divide the training data into two equal groups in order to train the model with the first group and then assess the assignment accuracy with the remaining half. A commonly used method for evaluating a discriminant function analysis is *leave-one-out cross-validation* (also called jackknifed validation), which excludes one observation at a time when formulating a discriminant function with the remaining data and then classifies the excluded observation. This metric essentially tests for outliers, which are less likely to be classified accurately when not included in the training data. Therefore, it is preferable to have a sufficiently large training data set that can be split for calibration and validation steps to evaluate accuracy instead of the leave-one-out approach.

To demonstrate an application of DA, we employ a data set that consists of $\delta^{13}C$ and $\delta^{15}N$ values to characterize dietary differences in barn-raised commercial chickens and free-range homegrown chickens. In this data set, oven-dried (but not de-fatted) breast muscle was analyzed for 32 barn-raised chickens purchased in grocery stores and 27 free-range chickens obtained from local households in Brazil. The original study demonstrated significant differences in stable isotopic composition between the two groups (Coletta et al. 2012). We also explore the utility of these data to authenticate the nutritional history of 35 unknown chicken breast samples from a separate study (Chesson et al., unpublished data).

The Brazilian chicken data were divided equally to use half as the calibration data (to train the model) and half as the validation data. Because the variance was not equal between the two groups, QDA was used. In addition, it was assumed that it was equally likely that a sample could have originated from either group, so null priors were used. The validation data resulted in a classification accuracy of 97%, with just one barn-raised sample being erroneously assigned to the free-range group, indicating that unknown samples collected in Brazil could likely be assigned with high accuracy, as seen in Figure 5.2.

We also explored the assignment outcomes using chicken breast data for samples of unknown production status obtained at U.S. grocery stores (Chesson et al., unpublished data), even though the training data did not include samples from the U.S. We urge caution in overstating the applicability of the Brazilian chicken results to other samples. In this exploratory exercise, however, all of the U.S. chickens were assigned to the barn-raised commercial group with a probability ≥ 0.96 (Figure 5.2), which was expected under the assumption that commercial chicken diets will be similar between the two countries and that the majority of chicken breast available in conventional grocery stores is commercially raised.

Odds ratios can be used to evaluate the strength of the support for assigning the samples to one group over another. This metric compares the odds of the posterior probability and the odds of a null model based on random chance. In this case, the null model would predict that each group is equally likely, and the odds of assigning it to one group *vs.* the other is 0.5/0.5 or 1. If we use a posterior probability threshold of 0.8, then the odds of the isotope-based assignment would be 0.8/0.2 or 4. An odds ratio

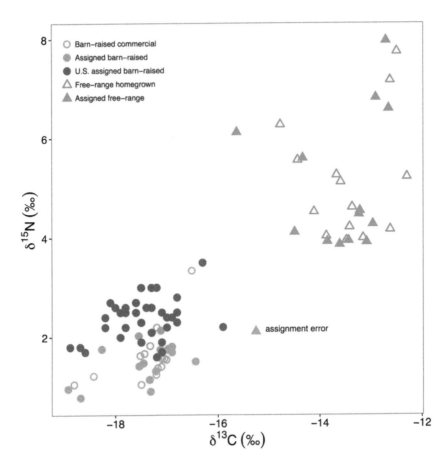

Figure 5.2 An example of a discrete approach to characterize production conditions from the carbon and nitrogen stable isotope compositions of chicken breast muscle, using training data from Coletta et al. (2012). The training data were randomly divided into two groups, such that the samples used for calibration are represented with open symbols, while the accuracy was evaluated with the samples represented by solid symbols. Solid gold circles and solid green triangles were classified accurately, while a single gold triangle was a barn-raised commercial sample that was erroneously assigned to the free-range group, resulting in a classification accuracy of 97%. Solid gray circles represent chicken of unknown production conditions purchased in the U.S. (Chesson et al., unpublished data), which were all assigned to the barn-raised commercial group using this discriminant analysis model.

of 4:1 means that the use of the isotope data to determine the production conditions of the chickens is four times more informative than random chance. Because the minimum observed posterior probability was 0.96, the use of the isotope-based assignment is actually at least 24 times more informative than random. Even if the posterior probability remains the same among different analyses, the odds ratios can change with the total number of possible groups. For example, three classification groups would reduce the odds of the null model to 0.33/0.66 or 0.5, and maintaining a posterior probability threshold of 0.8 with odds of 0.8/0.2 (or 4) would result in an odds ratio of 4/0.5 or 8:1. Ultimately, users of discrete sample assignment approaches must select an odds ratio that incorporates an appropriate level of risk for the specifics of their application.

5.4.2 Continuous approach

Investigations in which potential sources are not divided into discrete classes or regions, but rather considered across a contiguous geographic range or isoscape, provides a second, probabilistic approach to evaluating food isotope data that we illustrate here. Isotope-based geographic assignments evaluate the probability for each point in space, or pixel of the isoscape, to be the true origin of a sample. The continuous surface geospatial analytical approach has been utilized frequently in ecological studies to determine the geographic origin of migratory animals (Hobson and Wassenaar 2008; Hobson et al. 2009; Hobson et al. 2012; Flockhart et al. 2013; Cryan et al. 2014; Nelson et al. 2015). In contrast, isoscapes have been applied infrequently in food forensic studies to date, including a single study each on wine (West et al. 2007), honey (Chesson et al. 2014), and olive oil (Chiocchini et al. 2016), and none have used the analytical approaches presented here to determine the origin of unknown samples.

In the continuous approach, it is necessary to use a relevant isoscape that predicts the isotopic composition of samples across space, such as the commonly employed isoscapes of δ^2H and $\delta^{18}O$ compositions in precipitation. Amount-weighted mean annual precipitation maps are available at waterisotopes.org. Additionally, the web-accessible IsoMAP cyber-infrastructure (http://isomap.org) allows users to create precipitation isoscapes that are relevant to the time period and region of study, the ability to perform assignments, and tools for project documentation and archival.

A critical component before beginning the assignment process is to determine the isotopic relationship between the isoscape (e.g., environmental water) and the component of interest, and this must be evaluated on collections of known-origin samples. These relationships often take the form of a fixed offset or simple linear regression to rescale the isoscape to values relevant to the sample composition (Bowen et al. 2005; West et al. 2007; Chesson et al. 2011; Hobson et al. 2012; Bowen et al. 2014; Chiocchini et al. 2016). However, if the baseline isoscape was derived directly from samples of the study compound over a wide spatial distribution, then no further calibration would be necessary. Thus, as in the discrete approach, training data of known-origin are needed to calibrate the model, either to build the rescaling relationship or the isoscape itself.

The analytical framework used by IsoMAP as well as many recent studies of animal migration employs a semi-parametric Bayesian framework to evaluate the probability that each possible site in the geographic range is the true origin of a sample of unknown origin (Wunder et al. 2005; Wunder 2010; Bowen et al. 2014). The continuous surface assignment model can also incorporate variance estimates, such as those associated with the isoscape, laboratory measurement, and heterogeneity among and between samples, and building in a hierarchical error structure can provide flexibility over a single estimate of bulk variance (Wunder 2010).

Outputs from IsoMAP and similar algorithms are gridded maps that provide the probability of origin values and that sum to 1 across the entire surface (Bowen et al. 2014; Vander Zanden et al. 2015). Given the small posterior probability values, the maps are often relativized to the largest posterior probability to rescale the maps for ease of visualization. As in the discrete approach, the strength of the evidence favoring one location relative to any other can be evaluated through odds ratios (OR), or more specifically:

$$OR = \frac{\dfrac{P(i)}{1 - P(i)}}{\dfrac{P(max)}{1 - P(max)}}$$

where $P(i)$ is the probability in a given pixel and $P(max)$ is the cell of highest probability in the raster. In the example below, we consider 2:1 and 19:1 odds ratios to select likely regions of origin. Predicting assignment accuracy also requires known-origin samples, and ideally half of the training data would be used in the calibration and the other for validation.

We demonstrate an application of continuous surface assignment using hamburger (cooked beef patty) samples collected at outlets of a local and a national restaurant chain in Salt Lake City, Utah, USA (Chapter 6). The δ^2H composition of animal tissue is expected to relate to the local precipitation where it is raised, and previously, the intrinsic δ^2H composition of de-fatted beef samples from known-origin cattle herds within the continental U.S. was demonstrated to be significantly and strongly correlated to that of precipitation with the equation $\delta^2H_{beef} = 0.82 \times \delta^2H_{water} - 80$ (Chesson et al. 2011). We can, therefore, rescale the δ^2H isoscape of mean annual precipitation in the U.S. (Bowen and Revenaugh 2003) to a beef δ^2H isoscape in order to perform geographic assignment. We incorporated a pooled error term that consisted of four sources of uncertainty: (1) the standard deviation for 10 replicate analyses of a single hamburger patty sample as a representation of analytical error equal to 1.6‰ (Chesson et al. 2008), (2) the standard deviation of hamburger patties purchased at the same restaurant on five consecutive days to represent sample heterogeneity equal to 5.7‰ (Chesson et al., unpublished data), (3) the standard deviation of the residuals in the regression relating precipitation to beef δ^2H composition as a representation in rescaling error equal to 8.7‰ (Chesson et al. 2011), and (4) a spatially variant error associated with the precipitation isoscape estimates calculated as a component of the isoscape modeling process with a mean of 4.3‰ across the spatial range (Bowen and Revenaugh 2003).

The geographic assignment model was evaluated within the spatial domain of the contiguous U.S., and posterior probability maps (Figures 5.3a, b) were evaluated at odds ratio thresholds of 2:1 and 19:1 (Figures 5.3c, d). Known-origin samples were not available in the hamburger data set to test the accuracy of assignments for similar samples, but the odds ratio framework provides theoretical predictions of accuracy, such that a 19:1 threshold is expected to result in an inaccurate assignment at a rate of 5%. Regardless of the threshold selected, the maps demonstrate clear differences in the likely origin of the beef samples, such that the beef used in the local restaurant chain originated in the western region of the U.S., while the beef used in the national restaurant originated in the Midwestern U.S. It is important to note that the total area of the likely region increases

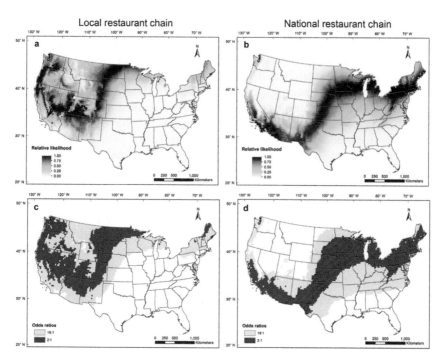

Figure 5.3 Posterior probability maps scaled to the maximum likelihood (a, b) and regions of likely origin using 2:1 and 19:1 odds ratios (c, d) for two different hamburger samples (cooked beef patty) purchased in Salt Lake City, Utah, USA at a local restaurant chain (a, c) or a national restaurant chain (b, d) (data from Chapter 6).

from the 2:1 threshold to the 19:1 threshold, but the predicted accuracy also increases. This trade-off between accuracy and precision is something investigators must evaluate in selecting an acceptable level of risk relevant to the study (Vander Zanden et al. 2014).

5.5 Conclusion

The numerous applications and increasing prevalence of stable isotope measurements in the field of food forensics have revealed that analysis of these biogeochemical markers can be a powerful tool to reveal information about sample history and origin. However, generating inferences from stable isotope data may require consideration on several levels, including sample collection, data quality, and statistical analysis. While standardized reporting of forensic evidence continues to pose challenges, evaluative reports that examine the evidence in the context relative to the judicial system are particularly important (ENFSI 2015).

Much of the stable isotope food forensic literature has been concerned with exploring and describing the isotopic differentiation between groupings of samples. The statistical analyses in these studies often document the markers or combinations of markers that might be most useful in separating potential groups of interests, yet few studies have applied such characterizations to classify unknown samples in authenticity and origin inquiries. We introduce discrete and continuous approaches that are used frequently in the ecological literature for the determination of group membership and geographic origin studies. These probabilistic approaches are natural extensions for the descriptive studies that have comprised the bulk of isotopic food forensic publications, and they provide a rigorous framework for evaluating the strength of the evidence using odds ratios.

We expect that use of probabilistic approaches will become one of the more powerful data interpretation methods used by forensic scientists to authenticate samples or yield geographic-based inferences of sample origin. Particularly for geographic origin studies, the availability of isoscapes to characterize the isotopic variation in plants and water (Bowen et al. 2005; West et al. 2006; West et al. 2008; Bowen 2010; Still and Powell 2010) as well as the proliferation of isotopic data and models to characterize other environmental gradients (e.g., Somes et al. 2010; Bataille and Bowen 2012; McMahon et al. 2013) will continue to contribute to the utility of this approach in the future.

References

Aitken, C. G. G. and D. Lucy. 2004. Evaluation of trace evidence in the form of multivariate data. *Journal of the Royal Statistical Society: Series C (Applied Statistics)* 53: 109–122.
Aitken, C. G. G. and F. Taroni. 2004. *Statistics and the Evaluation of Evidence for Forensic Scientists.* 2nd ed. Statistics in Practice. Chichester, England; Hoboken, N.J.: Wiley.

Association of Forensic Science Providers. 2009. Standards for the formulation of evaluative forensic science expert opinion. *Science & Justice* 49: 161–164.

Barnes, C., S. Jennings, N. V. C. Polunin and J. E. Lancaster. 2008. The importance of quantifying inherent variability when interpreting stable isotope field data. *Oecologia* 155: 227–235.

Bataille, C. P. and G. J. Bowen. 2012. Mapping $^{87}Sr/^{86}Sr$ variations in bedrock and water for large scale provenance studies. *Chemical Geology* 304–305: 39–52.

Benson, S. J., C. J. Lennard, P. Maynard, D. M. Hill, A. S. Andrew and C. Roux. 2009. Forensic analysis of explosives using isotope ratio mass spectrometry (IRMS)—Discrimination of ammonium nitrate sources. *Science & Justice* 49: 73–80.

Biedermann, A., J. Vuille, F. Taroni and C. Champod. 2015. The need for reporting standards in forensic science. *Law, Probability and Risk* 14: 169–173.

Bowen, G. J. 2010. Statistical and geostatistical mapping of precipitation water isotope ratios. pp. 139–160. *In*: J. B. West, G. J. Bowen, T. E. Dawson and K. P. Tu (eds.). *Isoscapes: Understanding Movement, Pattern, and Process on Earth through Isotope Mapping*. Dordrecht: Springer Netherlands.

Bowen, G. J., Z. Liu, H. B. Vander Zanden, L. Zhao and G. Takahashi. 2014. Geographic assignment with stable isotopes in IsoMAP. *Methods in Ecology and Evolution* 5: 201–206.

Bowen, G. J. and J. Revenaugh. 2003. Interpolating the isotopic composition of modern meteoric precipitation. *Water Resources Research* 39: 1299–1311.

Bowen, G. J., L. I. Wassenaar and K. A. Hobson. 2005. Global application of stable hydrogen and oxygen isotopes to wildlife forensics. *Oecologia* 143: 337–348.

Box, G. E. P. and N. R. Draper. 1987. *Empirical Model-Building and Response Surfaces*. Wiley Series in Probability and Mathematical Statistics. New York: Wiley.

Bro, R. and A. K. Smilde. 2014. Principal component analysis. *Analytical Methods* 6: 2812–2831.

Camin, F., L. Bontempo, K. Heinrich, M. Horacek, S. D. Kelly, C. Schlicht, F. Thomas, F. Monahan, J. Hoogewerff and A. Rossmann. 2007. Multi-element (H, C, N, S) stable isotope characteristics of lamb meat from different European regions. *Analytical and Bioanalytical Chemistry* 389: 309–320.

Carter, J. F. and V. Barwick (eds.). 2011. Good practice guide for isotope ratio mass spectrometry. FIRMS Network.

Carter, J. F., S. Doyle, B. -L. Phasumane and N. NicDaeid. 2014. The role of isotope ratio mass spectrometry as a tool for the comparison of physical evidence. *Science & Justice* 54: 327–334.

Carter, J. F., H. S. A. Yates and U. Tinggi. 2015a. Isotopic and elemental composition of roasted coffee as a guide to authenticity and origin. *Journal of Agricultural and Food Chemistry* 63: 5771–5779.

Carter, J. F., H. S. A. Yates and U. Tinggi. 2015b. A global survey of the stable isotope and chemical compositions of bottled and canned beers as a guide to authenticity. *Science & Justice* 55: 18–26.

Chesson, L. A., D. W. Podlesak, A. H. Thompson, T. E. Cerling and J. R. Ehleringer. 2008. Variation of hydrogen, carbon, nitrogen, and oxygen stable isotope ratios in an American diet: Fast food meals. *Journal of Agricultural and Food Chemistry* 56: 4084–4091.

Chesson, L. A., B. J. Tipple, J. D. Howa, G. J. Bowen, J. E. Barnette, T. E. Cerling and J. R. Ehleringer. 2014. Stable isotopes in forensics applications. pp. 285–317. *In*: T. E. Cerling (ed.). *Treatise on Geochemistry*. Elsevier.

Chesson, L. A., L. O. Valenzuela, G. J. Bowen, T. E. Cerling and J. R. Ehleringer. 2011. Consistent predictable patterns in the hydrogen and oxygen stable isotope ratios of animal proteins consumed by modern humans in the USA. *Rapid Communications in Mass Spectrometry* 25: 3713–3722.

Chiocchini, F., S. Portarena, M. Ciolfi, E. Brugnoli and M. Lauteri. 2016. Isoscapes of carbon and oxygen stable isotope compositions in tracing authenticity and geographical origin of Italian extra-virgin olive oils. *Food Chemistry* 202: 291–301.

Coletta, L. D., A. L. Pereira, A. A. D. Coelho, V. J. M. Savino, J. F. M. Menten, E. Correr, L. C. França and L. A. Martinelli. 2012. Barn vs. free-range chickens: Differences in their diets determined by stable isotopes. *Food Chemistry* 131: 155–160.

Craig, H. 1961. Isotopic variations in meteoric waters. *Science* 133: 1702–1703.

Cryan, P. M., C. A. Stricker and M. B. Wunder. 2014. Continental-scale, seasonal movements of a heterothermic migratory tree bat. *Ecological Applications* 24: 602–616.

Dansgaard, W. 1964. Stable isotopes in precipitation. *Tellus* 16: 436–468.

Dunn, P. J. H., D. Malinovsky and H. Goenaga-Infante. 2015. Calibration strategies for the determination of stable carbon absolute isotope ratios in a glycine candidate reference material by elemental analyser-isotope ratio mass spectrometry. *Analytical and Bioanalytical Chemistry* 407: 3169–3180.

ENFSI. 2015. *ENFSI guideline for evaluative reporting in forensic science.*

Farmer, N., W. Meier-Augenstein and D. Lucy. 2009. Stable isotope analysis of white paints and likelihood ratios. *Science & Justice* 49: 114–119.

Flockhart, D. T. T., L. I. Wassenaar, T. G. Martin, K. A. Hobson, M. B. Wunder and D. R. Norris. 2013. Tracking multi-generational colonization of the breeding grounds by monarch butterflies in eastern North America. *Proceedings of the Royal Society B: Biological Sciences* 280: 20131087.

Gentile, N., L. Besson, D. Pazos, O. Delémont and P. Esseiva. 2011. On the use of IRMS in forensic science: Proposals for a methodological approach. *Forensic Science International* 212: 260–271.

Gentile, N., R. T. W. Siegwolf and O. Delémont. 2009. Study of isotopic variations in black powder: Reflections on the use of stable isotopes in forensic science for source inference. *Rapid Communications in Mass Spectrometry* 23: 2559–2567.

Guillemain, M., S. L. V. Wilgenburg, P. Legagneux and K. A. Hobson. 2014. Assessing geographic origins of Teal (*Anas crecca*) through stable-hydrogen (δ^2H) isotope analyses of feathers and ring-recoveries. *Journal of Ornithology* 155: 165–172.

Hobson, K. A., H. Lormée, S. L. Van Wilgenburg, L. I. Wassenaar and J. M. Boutin. 2009. Stable isotopes (δD) delineate the origins and migratory connectivity of harvested animals: The case of European woodpigeons. *Journal of Applied Ecology* 46: 572–581.

Hobson, K. A., D. X. Soto, D. R. Paulson, L. I. Wassenaar and J. H. Matthews. 2012. A dragonfly (δ^2H) isoscape for North America: A new tool for determining natal origins of migratory aquatic emergent insects. *Methods in Ecology and Evolution* 3: 766–772.

Hobson, K. A., S. L. Van Wilgenburg, L. I. Wassenaar and K. Larson. 2012. Linking hydrogen (δ^2H) isotopes in feathers and precipitation: Sources of variance and consequences for assignment to isoscapes. *PLoS ONE* 7: e35137.

Hobson, K. A. and L. I. Wassenaar. 2008. Tracking animal migration with stable isotopes. *In*: K. A. Hobson and L. I. Wassenaar (eds.). Terrestrial Ecology Series. Elseiver/Academic Press.

Hobson, K. A., L. I. Wassenaar and O. R. Taylor. 1999. Stable isotopes (δD and δ^{13}C) are geographic indicators of natal origins of monarch butterflies in eastern North America. *Oecologia* 120: 397–404.

Howes, L. M., K. P. Kirkbridge, S. F. Kelty, R. Julian and N. Kemp. 2013. Forensic scientists' conclusions: How readable are they for non-scientist report-users? *Forensic Science International* 231: 102–112.

Howes, L. M., K. A. Martire and S. F. Kelty. 2014. Response to recommendation 2 of the 2009 NAS Report—Standards for formatting and reporting expert evaluative opinions: Where do we stand? *Forensic Science Policy & Management: An International Journal* 5: 1–14.

ISO. 2005. *ISO/IEC 17025:2005(E) General requirements for the competence of testing and calibration laboratories.* Switzerland: ISO Copyright Office.

Jackson, A. L., R. Inger, A. C. Parnell and S. Bearhop. 2011. Comparing isotopic niche widths among and within communities: SIBER—Stable Isotope Bayesian Ellipses in R. *Journal of Animal Ecology* 80: 595–602.

Jardine, T. D. and R. A. Cunjak. 2005. Analytical error in stable isotope ecology. *Oecologia* 144: 528–533.

JCGM. 2012. International vocabulary of metrology—Basic and general concepts and associated terms (VIM 3rd edition), JCGM 200:2012 (JCGM 200:2008 with minor corrections). Paris: Bureau International des Poids et Mesures.

Martyna, A., G. Zadora, I. Stanimirova and D. Ramos. 2014. Wine authenticity verification as a forensic problem: An application of likelihood ratio test to label verification. *Food Chemistry* 150: 287–295.

McMahon, K. W., L. L. Hamady and S. R. Thorrold. 2013. A review of ecogeochemistry approaches to estimating movements of marine animals. *Limnology and Oceanography* 58: 697–714.

Molkentin, J., H. Meisel, I. Lehmann and H. Rehbein. 2007. Identification of organically farmed Atlantic salmon by analysis of stable isotopes and fatty acids. *European Food Research and Technology* 224: 535–543.

National Research Council. 2009. Strengthening Forensic Science in the United States: A Path Forward. Washington, D.C.: National Academies Press. http://www.nap.edu/catalog/12589.

Nelson, D. M., M. Braham, T. A. Miller, A. E. Duerr, J. Cooper, M. Lanzone, J. Lemaître and T. Katzner. 2015. Stable hydrogen isotopes identify leapfrog migration, degree of connectivity, and summer distribution of Golden Eagles in eastern North America. *The Condor* 117: 414–429.

Nordgaard, A. and B. Rasmusson. 2012. The likelihood ratio as value of evidence—more than a question of numbers. *Law, Probability and Risk* 11: 303–315.

Ohlsson, K. E. A. 2013. Uncertainty of blank correction in isotope ratio measurement. *Analytical Chemistry* 85: 5326–5329.

Ortea, I. and J. M. Gallardo. 2015. Investigation of production method, geographical origin and species authentication in commercially relevant shrimps using stable isotope ratio and/or multi-element analyses combined with chemometrics: An exploratory analysis. *Food Chemistry* 170: 145–153.

Osorio, M. T., G. Downey, A. P. Moloney, F. T. Röhrle, G. Luciano, O. Schmidt and F. J. Monahan. 2013. Beef authentication using dietary markers: Chemometric selection and modelling of significant beef biomarkers using concatenated data from multiple analytical methods. *Food Chemistry* 141: 2795–2801.

Oulhote, Y., B. Le Bot, S. Deguen and P. Glorennec. 2011. Using and interpreting isotope data for source identification. *Trends in Analytical Chemistry* 30: 302–312.

Pierrini, G., S. Doyle, C. Champod, F. Taroni, D. Wakelin and C. Lock. 2007. Evaluation of preliminary isotopic analysis (^{13}C and ^{15}N) of explosives: A likelihood ratio approach to assess the links between Semtex samples. *Forensic Science International* 167: 43–48.

de Rijke, E., J. C. Schoorl, C. Cerli, H. B. Vonhof, S. J. A. Verdegaal, G. Vivó-Truyols, M. Lopatka, et al. 2016. The use of δ^2H and δ^{18}O isotopic analyses combined with chemometrics as a traceability tool for the geographical origin of bell peppers. *Food Chemistry* 204: 122–128.

Rodionova, O. Y., A. V. Titova and A. L. Pomerantsev. 2016. Discriminant analysis is an inappropriate method of authentication. *TrAC Trends in Analytical Chemistry* 78: 17–22.

Royle, A. J. and D. R. Rubenstein. 2004. The role of species abundance in determining breeding origins of migratory birds with stable isotopes. *Ecological Applications* 14: 1780–1788.

Rubenstein, D. R. and K. A. Hobson. 2004. From birds to butterflies: Animal movement patterns and stable isotopes. *Trends in Ecology & Evolution* 19: 256–263.

Simmross, U. 2014. Appraisal of scientific evidence in criminal justice systems: On winds of change and coexisting formats. *Law, Probability and Risk* 13: 105–115.

Skrzypek, G. 2013. Normalization procedures and reference material selection in stable HCNOS isotope analyses: An overview. *Analytical and Bioanalytical Chemistry* 405: 2815–2823.

Skrzypek, G., R. Sadler and D. Paul. 2010. Error propagation in normalization of stable isotope data: A Monte Carlo analysis. *Rapid Communications in Mass Spectrometry* 24: 2697–2705.

Somes, C. J., A. Schmittner, E. D. Galbraith, M. F. Lehmann, M. A. Altabet, J. P. Montoya, R. M. Letelier, A. C. Mix, A. Bourbonnais and M. Eby. 2010. Simulating the global distribution of nitrogen isotopes in the ocean. *Global Biogeochemical Cycles* 24: GB4019.

Still, C. J. and R. L. Powell. 2010. Continental-scale distributions of vegetation stable carbon isotope ratios. pp. 179–193. *In*: J. B. West, G. J. Bowen, T. E. Dawson and K. P. Tu (eds.). *Isoscapes: Understanding Movement, Pattern, and Process on Earth through Isotope Mapping*. Dordrecht: Springer Netherlands.

Vander Zanden, H. B., K. A. Bjorndal, W. Mustin, J. M. Ponciano and A. B. Bolten. 2012. Inherent variation in stable isotope values and discrimination factors in two life stages of green turtles. *Physiological and Biochemical Zoology* 85: 431–441.

Vander Zanden, H. B., A. D. Tucker, K. M. Hart, M. M. Lamont, I. Fujisaki, D. S. Addison, K. L. Mansfield et al. 2015. Determining origin in a migratory marine vertebrate: a novel method to integrate stable isotopes and satellite tracking. *Ecological Applications* 25: 320–335.

Vander Zanden, H. B., M. B. Wunder, K. A. Hobson, S. L. Van Wilgenburg, L. I. Wassenaar, J. M. Welker and G. J. Bowen. 2014. Contrasting assignment of migratory organisms to geographic origins using long-term versus year-specific precipitation isotope maps. *Methods in Ecology and Evolution* 5: 891–900.

Vander Zanden, H. B., M. B. Wunder, K. A. Hobson, S. L. Van Wilgenburg, L. I. Wassenaar, J. M. Welker and G. J. Bowen. 2015. Space-time tradeoffs in the development of precipitation-based isoscape models for determining migratory origin. *Journal of Avian Biology* 46: 658–657.

Van Wilgenburg, S. L. and K. A. Hobson. 2011. Combining stable-isotope (δD) and band recovery data to improve probabilistic assignment of migratory birds to origin. *Ecological Applications* 21: 1340–1351.

West, J. B., G. J. Bowen, T. E. Cerling and J. R. Ehleringer. 2006. Stable isotopes as one of nature's ecological recorders. *Trends in Ecology & Evolution* 21: 408–414.

West, J. B., G. J. Bowen, T. E. Dawson and K. P. Tu. 2010. *Isoscapes: Understanding Movement, Pattern, and Process on Earth through Isotope Mapping*. Dordrecht: Springer Netherlands.

West, J. B., J. R. Ehleringer and T. E. Cerling. 2007. Geography and vintage predicted by a novel GIS model of wine δ^{18}O. *Journal of Agricultural and Food Chemistry* 55: 7075–7083.

West, J. B., A. Sobek and J. R. Ehleringer. 2008. A simplified GIS approach to modeling global leaf water isoscapes. *PLoS ONE* 3: e2447.

Whitledge, G. W., B. M. Johnson and P. J. Martinez. 2006. Stable hydrogen isotopic composition of fishes reflects that of their environment. *Canadian Journal of Fisheries and Aquatic Sciences* 63: 1746–1751.

Własiuk, P., A. Martyna and G. Zadora. 2015. A likelihood ratio model for the determination of the geographical origin of olive oil. *Analytica Chimica Acta* 853: 187–199.

Wunder, M. B. 2010. Using isoscapes to model probability surfaces for determining geographic origins. pp. 251–270. *In*: J. B. West, G. J. Bowen, T. E. Dawson and K. P. Tu (eds.). *Isoscapes: Understanding Movement, Pattern, and Process on Earth through Isotope Mapping*. Dordrecht: Springer Netherlands.

Wunder, M. B., C. L. Kester, F. L. Knopf and R. O. Rye. 2005. A test of geographic assignment using isotope tracers in feathers of known origin. *Oecologia* 144: 607–617.

Zhao, Y., B. Zhang, B. Guo, D. Wang and S. Yang. 2016. Combination of multi-element and stable isotope analysis improved the traceability of chicken from four provinces of China. *CyTA—Journal of Food* 14: 163–168.

CHAPTER 6

Flesh Foods, or What Can Stable Isotope Analysis Reveal About the Meat You Eat?

Lesley A. Chesson

6.1 Introduction

"Dis-moi ce que tu manges, je te dirai ce que tu es."
"Tell me what you eat, and I will tell you what you are."

(*Jean Anthelme Brillat-Savarin,*
French lawyer and politician; 1755–1826)

Most people will have heard the adage: "You are what you eat (and drink)." From an isotopic point-of-view the isotope ratios of your inputs (food and drink) are reflected in the isotope ratios of your body tissues, such as bone, hair, nail, and teeth. Isotope analysis of these tissues can be used to reconstruct aspects of your life history, including dietary choices, changes in consumption, and potentially, changes in location. For an omnivore, dietary inputs encompass three basic groups: animal-based foods, plant-based foods, and beverages. This chapter will discuss animal-derived flesh foods—e.g., *meat*.

How do you know what you're really eating? Food fraud is a serious and pervasive issue with potentially significant economic and consumer safety consequences. When foods marketed with a particular attribute—*organic, grass-fed*, or *protected denomination of origin* (PDO)—command higher market prices, there exists a powerful temptation for deception. An investigative report published in the *Washington Post* (Layton 2010) found that:

IsoForensics, Inc., 421 Wakara Way, Suite 100, Salt Lake City, UT 84108, USA.
Email: lesley@isoforensics.com

"The expensive 'sheep's milk' cheese in a Manhattan [New York, New York, USA] market was really made from cow's milk. And a jar of 'Sturgeon caviar' was, in fact, Mississippi paddlefish.... 'Food fraud' has been documented in fruit juice, olive oil, spices, vinegar, wine, spirits and maple syrup, and appears to pose a significant problem in the seafood industry. Victims range from the shopper at the local supermarket to multimillion dollar companies..."

Three years after this report was published in the USA, a scandal engulfed Europe as products labeled as beef were found to contain a variety of other, undeclared meats, including horse and pork. The scandal highlighted a serious failing in food traceability systems within the affected European countries.

The "real" sources of the isotopes you consume—that are then recorded in your tissues—are largely irrelevant; for example, the isotopic record does not change significantly if you consume beef, horse or pork flesh. From a moralistic, religious, or ethical perspective, however, truth in food labeling has significant relevance. Would isotopic analysis have helped to expose the "Sturgeon caviar" in the *Washington Post* investigation? Or helped to determine where the mislabeled horse meat originated? This chapter considers the *state-of-the-art* of isotope analyses of flesh foods.

6.2 Review of isotopic variation in flesh foods

Isotopic variations of the elements found in flesh foods record a range of information about source, derivation, or history. In general, the isotopes of major elements essential to life—the bio-elements H, C, N, O, and S—provide information on biological and environmental processes affecting animals and their inputs (food and water). The isotopes of radiogenic trace elements such as Sr and Pb more broadly provide information on the underlying geology local to the animal.

The isotopic composition of a flesh food may be impacted by an offset in δ-values from the animal's inputs, due to fractionation processes during metabolism and tissue formation. The phenomenon was first summarized by DeNiro and Epstein as, "You are what you eat (plus or minus a few‰)" (DeNiro and Epstein 1976). The magnitude of this offset depends on the animal, the tissue, and the isotope(s) measured. More detailed information on isotopes and their natural variations is presented in other chapters (particularly Chapters 3 and 4); here, some highlights are reviewed related to meats.

Hydrogen and oxygen: Hydrogen and oxygen isotope variation in flesh foods is primarily related to isotopic variation in the body-water pool of an animal (Kohn 1996; Podlesak et al. 2008; Podlesak et al. 2012; O'Grady et al. 2012). Water can be added directly to this pool as drinking water, or as water contained in the animal's food. Metabolic water is another contributor to the body-water pool; its isotopic composition is related to the isotopic composition of diet but also to inhaled O_2. Animals record the isotopic composition of their body-water pool in the hydrogen and oxygen isotopes of tissues such as muscle that is eaten as a flesh food.

Carbon: The measured $\delta^{13}C$ value of a flesh food is ultimately linked to the photosynthetic pathway of plants at the base of a food web. Plants using C_3 carbon fixation (e.g., wheat, barley, rye, rice, and most fruits) have $\delta^{13}C$ values ranging from approximately −30‰ to approximately −22‰ while plants using C_4 carbon fixation (e.g., maize/corn, sugar cane, millet, and teff) have $\delta^{13}C$ values ranging from approximately −15‰ to approximately −10‰ (Cerling et al. 1997; Tipple and Pagani 2007). The $\delta^{13}C$ values of CAM plants (e.g., pineapples) range between the values of C_3 and C_4 plants, significantly overlapping with the values of C_4 plants (O'Leary 1988). The carbon isotope ratios of an animal's tissues will reflect its diet, with a (mostly) small effect from fractionation (Post 2002). The $\delta^{13}C$ value measured for a bulk tissue—such as a muscle—will represent a variety of different fractionation effects within separate tissue components, such as individual amino acids.

Nitrogen: Most nitrogen isotope variation in flesh foods is driven by food web size, since $\delta^{15}N$ values typically increase with each trophic level (Post 2002). In general, higher nitrogen isotope ratios are seen in aquatic systems than in terrestrial systems due to the many levels of aquatic food webs. The magnitude of the fractionation between trophic levels depends on protein quantity and quality (Robbins et al. 2005; Robbins et al. 2010). As is the case for carbon, the $\delta^{15}N$ value of a bulk tissue represents a combination of different isotope fractionations within separate components.

Sulfur: The major sources for sulfur (as sulfate) to an ecosystem include fertilizers, marine deposits, and the combustion of fossil fuels. Due to variation in sources, there is a wide range of $\delta^{34}S$ values in oceans and across terrestrial landscapes, ranging from approximately −19 to +30‰ (Thode 1991). Marine ecosystems typically have higher $\delta^{34}S$ values than terrestrial ecosystems and measured $\delta^{34}S$ values increase as deposition of sulfur from marine sources increases; this is known as the *sea spray effect* (Richards et al. 2003). Conversely, terrestrial $\delta^{34}S$ values typically decrease as sulfur deposition from anthropogenic sources—such as coal burning power plants—begins to increase. Unlike nitrogen, there is generally little to no change in $\delta^{34}S$ values with change in trophic level of a food web (McCutchan et al. 2003).

6.2.1 A primer on strontium isotope variation

In contrast to the elements H, C, N, O, and S, strontium is not recognized as an element essential for life. It is found in trace amounts, compared to the bio-elements, and has been used less frequently in isotopic investigations of flesh foods since analysis requires clean laboratory facilities. Strontium can, however, be easily substituted for calcium (a major nutritional cation) and is found in significant concentrations in animal bone and tooth enamel, making strontium isotope analysis challenging, but useful. Strontium isotope variation is linked principally to variation in an animal's local geology.

There are four natural stable isotopes of strontium: ^{84}Sr, ^{86}Sr, ^{87}Sr, and ^{88}Sr. The radiogenic isotope ^{87}Sr is formed from the decay of ^{87}Rb. The ratio of ^{87}Sr to ^{86}Sr in a material depends on the initial concentration of $^{87}Sr/^{86}Sr$, the initial relative concentration of Rb/Sr, and the passage of time. Isotope abundances of strontium are typically reported as an absolute abundance ratio, $^{87}Sr/^{86}Sr$, without recourse to δ-notation used for the bio-elements. Significant variations in $^{87}Sr/^{86}Sr$ ratios are seen when materials are older than tens of millions of years. Lower $^{87}Sr/^{86}Sr$ ratios (e.g., 0.702 to 0.706) are associated with younger rocks, such as modern volcanic basalts, while higher $^{87}Sr/^{86}Sr$ ratios (e.g., 0.710 and higher) are associated with older rocks, such as Paleozoic granites. Intermediate $^{87}Sr/^{86}Sr$ ratios are often observed in unaltered limestones that span the Phanerozoic eon (Capo et al. 1998; Beard and Johnson 2000). Researchers have published isoscapes relating bedrock identity and age to predict the $^{87}Sr/^{86}Sr$ ratios of bio-available Sr found in the local environment (Beard and Johnson 2000; Bataille and Bowen 2012).

6.3 Flesh foods

Meat consumption (along with consumption of refined sugar) is a major feature of the so-called *Western diet* (Block 2004) and many studies have used isotope analysis to characterize meat of both aquatic and terrestrial origins. Of the published studies, the majority has focused on terrestrially produced meats, including beef, lamb, poultry, and pork; a single study measured kid goat meat (Longobardi et al. 2012). The earliest publications on isotope analysis of animals reared for human consumption focused on beef cattle (Minson et al. 1975; Jones et al. 1979; Jones et al. 1981); since then, more isotope studies have been published on beef than any other terrestrial meat. One reason for this special attention to bovine meat may be related to the preponderance of fast food restaurants that include *hamburgers* (beef patties) as a staple menu item. Figure 6.1 shows a cartogram (www. worldmapper.org), which resized countries based on the prevalence of McDonald's restaurants. In 2004, there were 30,496 outlets in the fast food chain, approximately 45% of which were located in the USA. In contrast,

Figure 6.1 Distribution of one major brand of fast food outlet (McDonald's) in 2004; countries have been resized based on the prevalence of restaurants per million people living there. Map 364 (International Fast Food), published by www.worldmapper.org; reprinted with permission.

only 150 McDonald's restaurant outlets were located on the African continent, with the majority concentrated in the country of South Africa.

6.3.1 Bovine foodstuffs

"Beef. It's What's For Dinner."

(*A message from the Beef Industry
Council; May 18, 1992*)

Forensic isotope analysis of bovine foodstuffs has been mainly concerned with two facets of cattle production:

What did the beef cow eat? Where was the beef cow raised?

Additionally, researchers have used isotopic analysis to study growth patterns, seasonal variation, dietary turnover rates, and even the effects of cooking (Zhou et al. 2015). Table 6.1 presents a summary of published research on isotope analysis of beef. Investigations into the diet of beef cattle have relied primarily on measurements of carbon and nitrogen isotopes, although sulfur and hydrogen isotopes may also prove useful to reconstruct dietary history. To answer questions of origin, hydrogen and oxygen isotopes have proved more useful; strontium isotopes could also be used to verify beef meat origin.

In a 2004 study, three groups of young bulls were fed diets that varied in the proportion of C_4 plants (e.g., maize) (De Smet et al. 2004). Once the bulls had fattened to a target mass, they were slaughtered and samples of blood, liver, kidney-fat, hair, and muscle collected. Among the samples

Table 6.1 A summary of isotope studies on bovine foodstuffs.

Isotope(s)	Tissue	Citation	Principle finding(s)
C	Hair	(Minson et al. 1975)	Diet
C	Feces	(Jones et al. 1979)	Diet
C	Hair	(Jones et al. 1981)	Diet, dietary turnover
C and N	Hair	(Schwertl et al. 2003)	Diet
H, O, C, N, and S	Extracted water, muscle	(Boner and Förstel 2004)	Organic production
C	Blood, liver, fat, hair, muscle	(De Smet et al. 2004)	Diet
C	Muscle	(Gebbing et al. 2004)	Diet
O	Extracted water	(Renou et al. 2004)	Diet
O	Extracted water	(Thiem et al. 2004)	Origin
C and N	Muscle, lipids	(Bahar et al. 2005)	Diet
C, N, and S	Muscle	(Schmidt et al. 2005)	Origin and diet
C and N	Hair	(Schwertl et al. 2005)	Diet
C	Tooth dentine	(Zazzo et al. 2006)	Diet
C	Hoof	(Harrison et al. 2007)	3D growth
C and N	Hoof	(Harrison et al. 2007)	Diet
C	Hair, hoof	(Zazzo et al. 2007)	Dietary turnover
C, N, and S	Muscle	(Bahar et al. 2008)	Seasonal variation
H, O, C, and N	Muscle	(Chesson et al. 2008)	Spatial variation
O and Sr (via ICP-MS)	Muscle	(Franke et al. 2008a)	Origin
O	Muscle	(Franke et al. 2008b)	Origin
H, O, C, and N	Muscle, lipids	(Heaton et al. 2008)	Origin
O, C, and N	Muscle	(Nakashita et al. 2008)	Origin
C and N	Muscle	(Jahren and Kraft 2008)	Diet
C, N, and S	Muscle	(Bahar et al. 2009)	Dietary turnover
O, C, and N	Muscle	(Bong et al. 2010)	Origin
H and O	Muscle	(Chesson et al. 2010)	Spatial variation
C and N	Hair, muscle	(Guo et al. 2010)	Origin and diet
O	Extracted water	(Horacek et al. 2010)	Origin, cooking effects
H, C, and N	Muscle	(Horacek and Min 2010)	Origin
C, N and Sr (via TIMS)	Muscle	(Baroni et al. 2011)	Origin
C	Muscle	(Martinelli et al. 2011)	Diet
H, C, N, and S	Muscle	(Osorio et al. 2011a)	Origin
H, C, N, and S	Hair, muscle	(Osorio et al. 2011b)	Diet
C (plus ^{14}C via AMS)	Muscle	(Kim et al. 2012)	Diet

Table 6.1 contd. ...

...*Table 6.1 contd.*

Isotope(s)	Tissue	Citation	Principle finding(s)
Sr (via ICP-MS)	Muscle	(Rummel et al. 2012)	Origin
C and N	Hair, muscle	(Yanagi et al. 2012)	Diet
H, C, and N	Hair	(Liu et al. 2013)	Origin
H, C, N, and S	Muscle, lipids	(Osorio et al. 2013)	Diet
C and N	Muscle	(Zhao et al. 2013)	Origin
C and N	Blood, "inner organs"	(Bahar et al. 2014)	Dietary turnover
H, C, and N	Muscle	(Zhou et al. 2015)	Cooking effects

collected within each diet group the isotopic offset between diet and tissue varied depending on tissue type, with the lowest $\delta^{13}C$ values observed in the kidney-fat samples and the highest in the hair samples. The carbon isotopic composition of tissues at slaughter were highly correlated with the carbon isotopic composition of diet, with the highest values seen in the group fed the most maize (35% of whole diet) and the lowest values observed in the group fed no C_4 plants, leading the authors to conclude "the proportion of C_4 material in the diet could be accurately estimated from the $\delta^{13}C$ values in different tissues" (De Smet et al. 2004). However, the authors cautioned "tissues may differ in their usefulness for this purpose" depending on the timing of dietary changes and the turnover rates of different tissue types.

Multiple studies have followed this initial research and confirmed that cattle diets can be estimated using carbon isotopes and, in some cases, nitrogen isotopes (Gebbing et al. 2004; Bahar et al. 2005; Schmidt et al. 2005; Schwertl et al. 2005; Zazzo et al. 2006; Harrison et al. 2007; Jahren and Kraft 2008; Martinelli et al. 2011; Osorio et al. 2011b; Kim et al. 2012; Yanagi et al. 2012; Osorio et al. 2013). A 2005 study found that "there was a strong linear relationship between the level of maize-derived C in the diet and $\delta^{13}C$ of lipid-free muscle" (Bahar et al. 2005) and the authors constructed a simple linear regression relating the $\delta^{13}C$ values of lipid-free muscle to the % maize in the diet:

$$\delta^{13}C_{muscle} = 0.0958 \times \% \text{ maize} - 25.215‰ \qquad \text{Equation 6.1}$$

Measurement of biomarkers—such as fatty acids, lutein, α-tocopherol, and β-carotene—in combination with carbon and nitrogen stable isotope analysis has proved useful to authenticate particular beef production claims, such as *pasture-fed* (Osorio et al. 2013). Several studies have also used carbon and nitrogen isotopes to investigate the dietary turnover rates of different bovine tissues, including hair (Schwertl et al. 2003; Zazzo et al. 2007), hoof (Harrison et al. 2007; Zazzo et al. 2007), muscle (Bahar et al. 2009), and blood plus other organs (Bahar et al. 2014).

Two isotope studies have focused specifically on a staple of the Western diet, the fast food hamburger. Researchers collecting cooked beef patties from restaurants across the USA found an "overwhelming importance of corn [maize] agriculture within virtually every aspect of fast food manufacture" (Jahren and Kraft 2008) [see also (Chesson et al. 2009a; Jahren and Kraft 2009)]. A later study of Big Mac® burger patties from around the world found the proportion of maize used in beef production varied by country and could be indicative of the *glocalization* of a global food commodity, whereby local resources were used to produce a food item that was visually—but not isotopically—the same (Martinelli et al. 2011).

Following these studies, the percentage of maize in the diets of cattle used by the U.S. fast food industry was calculated for a set of 147 hamburger samples purchased between 2007 and 2009 from restaurants across the USA. Samples were prepared as described elsewhere (Chesson et al. 2008) and measurements made on de-fatted dry matter (DDM). Carbon isotope analysis results are shown in Table 6.2, alongside measured $\delta^{15}N$ values; $\delta^{13}C$ values ranged from −20.6‰ to −11.6‰.

Dietary composition calculations, made using Equation 6.1, ranged from 48% maize to 142% maize for the U.S. hamburgers (Table 6.2). Compositions above 100% are likely to be artifacts of the original dataset used to relate % maize and beef muscle $\delta^{13}C$ values. First, the experimental diet used in that feeding study contained some fraction of non-maize concentrates, including barley and soybean; the maize silage diet was thus not entirely C_4 plants. Second, the diet of the animals before the experiment start was unknown, but may well have included C_3 plants (Bahar et al. 2005). Finally, the animals may not have reached equilibrium with the maize silage diet before slaughter—that is, tissue turnover was not complete. In combination, these factors served to decrease the observed carbon isotopic compositions of the animals fed maize silage to values lower than beef cattle raised exclusively on C_4 plants for the entire life span.

The utility of nitrogen isotopes to investigate beef cattle diet has been less clear-cut than that of carbon. A 2005 study found that $\delta^{15}N$ values varied between conventional farms and grassland-based farms; pastured animals generally had lower nitrogen isotope ratios in hair than the conventionally-raised animals (Schwertl et al. 2005). More interestingly, the $\delta^{15}N$ values of cattle hair were positively correlated with stocking rate (livestock units per hectare), which explained 55% of the total variation in nitrogen isotopic composition. One possible explanation for this relationship was the ^{15}N enrichment in animal excrement, which was present at much higher concentrations on farms with higher stocking rates (Schwertl et al. 2005). A 2008 study found that $\delta^{15}N$ values of organic and conventional Irish beef differed, with organic samples having lower values at each sampling time point (Bahar et al. 2008). This difference may have been due to the diet of cattle raised organically, which included clover and legumes in specific

Table 6.2 Measurement results from the isotope analysis of hamburgers (beef patties) collected at fast food restaurants across the USA between 2007 and 2009. Proportion of maize in the diet has been calculated following Bahar et al. 2005.

Sample ID	Restaurant	Type	City	State	$\delta^{15}N$, ‰	$\delta^{13}C$, ‰	$\delta^{2}H$, ‰	$\delta^{18}O$, ‰	% maize
1615	McDonald's		Montgomery	AL	6.7	-17.6	-112	16.5	79
1071	Burger King		Nogales	AZ	7.3	-20.6	-137	13.7	48
1099	Burger King		Yuma	AZ	6.9	-18.8	-117	15.8	67
1083	In-N-Out Burger		Chandler	AZ	6.6	-17.3	-140	11.8	83
1093	Jack in the Box		San Luis	AZ	7.7	-19.3	-120	16.0	62
1096	McDonald's		Yuma	AZ	5.1	-19.1	-114	15.2	64
1080	McDonald's		Tempe	AZ	6.5	-15.9	-118	14.9	97
1065	McDonald's		Tucson	AZ	7.1	-14.8	-119	14.5	109
1077	McDonald's		Green Valley	AZ	6.2	-13.8	-113	14.8	119
1090	McDonald's		San Luis	AZ	6.6	-13.7	-115	15.0	120
1074	McDonald's		Nogales	AZ	6.3	-13.7	-119	13.7	121
1114	Burger King		Chula Vista	CA	6.6	-17.4	-138	12.0	82
1566	McDonald's		Riverside	CA	6.1	-19.0	-147	11.2	65
1117	McDonald's		Riverside	CA	6.3	-18.2	-148	10.4	74
1102	McDonald's		El Centro	CA	6.4	-16.1	-132	12.9	95
1105	McDonald's		Chula Vista	CA	6.3	-14.8	-130	13.0	108
925	McDonald's		Riverside	CA	5.8	-14.7	-130	13.7	109
763	McDonald's		San Diego	CA	6.4	-14.5	-122	14.1	112
252	McDonald's		Berkeley	CA	6.2	-14.1	-128	13.4	116
1184	Good Times		Denver	CO	5.7	-13.9	-130	12.5	118

Table 6.2 contd. ....

...Table 6.2 contd.

Sample ID	Restaurant	Type	City	State	$\delta^{15}N$, ‰	$\delta^{13}C$, ‰	δ^2H, ‰	$\delta^{18}O$, ‰	% maize
1187	McDonald's		Grand Junction	CO	6.7	−16.1	−112	16.2	95
834	McDonald's		Fredrick	CO	6.1	−15.4	−111	17.9	102
1181	McDonald's		Denver	CO	6.8	−14.3	−111	15.9	114
207	Wendy's		Grand Junction	CO	6.7	−12.1	−116	15.6	137
1607	McDonald's		Shalimar	FL	6.5	−14.8	−112	15.0	109
1623	McDonald's		Honolulu	HI	5.6	−14.2	−146	11.8	115
1207	McDonald's		Iowa City	IA	7.6	−13.4	−113	15.0	123
946	Burger King		Mountain Home	ID	6.2	−17.3	−121	14.5	83
665	McDonald's		Idaho Falls	ID	6.0	−15.3	−137	11.3	104
819	McDonald's		Pocatello	ID	6.3	−14.0	−115	16.1	117
949	McDonald's		Twin Falls	ID	6.1	−13.8	−109	16.9	119
1575	McDonald's		Moscow	ID	6.3	−18.8	−136	12.1	67
1195	McDonald's		Frankfort	IL	5.7	−13.6	−118	14.6	121
240	McDonald's		Indianapolis	IN	6.3	−14.5	−119	15.0	112
868	McDonald's		Lawrence	KS	6.4	−16.3	−117	16.0	93
1172	McDonald's		Colby	KS	6.6	−16.1	−114	15.5	95
1166	McDonald's		Wichita	KS	6.2	−15.7	−116	15.4	99
1175	McDonald's		Salina	KS	5.8	−15.0	−126	13.1	107
1178	Spangles		Salina	KS	7.3	−15.2	−111	15.3	104
1129	McDonald's		Lake Charles	LA	6.4	−15.1	−128	18.0	105
1591	McDonald's		Hammond	LA	6.4	−14.9	−116	15.3	108

871	McDonald's	Boston	MA	6.2	-14.7	-117	14.5	110
838	Wendy's	Boston	MA	6.6	-11.6	-120	13.0	142
1651	McDonald's	South Portland	ME	5.8	-12.5	-130		132
1203	McDonald's	Fenton	MI	5.8	-18.7	-114	13.3	68
658	McDonald's	Detroit	MI	6.2	-13.9	-123	13.2	118
1557	McDonald's	Minneapolis	MN	5.7	-18.2	-120	13.6	73
1599	McDonald's	Gulfport	MS	6.0	-16.2	-126	14.9	94
700	McDonald's	Livingstone	MT	6.2	-15.2	-125	13.7	105
217	McDonald's	Troy	NC	6.3	-15.8	-120	13.6	98
1199	McDonald's	Omaha	NE	5.5	-16.1	-115	14.6	96
1197	McDonald's	North Platte	NE	5.9	-14.2	-111	15.4	115
266	McDonald's	Lincoln	NE	6.8	-11.8	-124	17.1	140
874	McDonald's	Lancaster	NH	6.3	-16.3	-113	14.0	93
886	McDonald's	West Lebanon	NH	5.9	-14.8	-113	16.3	109
880	McDonald's	Lincoln	NH	6.9	-13.9	-107	16.4	118
877	McDonald's	Plymouth	NH	6.8	-13.3	-118	17.8	124
1059	McDonald's	Las Cruces	NM	6.5	-17.3	-117	15.0	82
1062	McDonald's	Lordsburg	NM	6.2	-17.0	-123	15.1	86
952	McDonald's	Santa Fe	NM	6.0	-15.4	-130	14.0	103
1583	McDonald's	Albuquerque	NM	6.0	-12.8	-106	12.5	130
1056	Whataburger	Las Cruces	NM	6.9	-14.7	-142	16.2	110
1123	In-N-Out Burger	Las Vegas	NV	6.3	-16.4	-112	11.2	92
866	McDonald's	Mesquite	NV	6.1	-15.8	-112	17.4	98

Table 6.2 contd. ...

...*Table 6.2 contd.*

Sample ID	Restaurant	Type	City	State	$\delta^{15}N$, ‰	$\delta^{13}C$, ‰	$\delta^{2}H$, ‰	$\delta^{18}O$, ‰	% maize
1120	McDonald's		Las Vegas	NV	6.4	−14.3	−132	12.1	114
928	McDonald's		North Las Vegas	NV	6.5	−18.2	−145	11.6	73
1542	McDonald's		Searchlight	NV	6.5	−17.3	−136	11.9	83
1213	McDonald's		Parma Heights	OH	6.5	−18.0	−136	11.7	76
1160	McDonald's		Perry	OK	5.9	−18.7	−139	12.3	68
1169	McDonald's		Oklahoma City	OK	6.5	−12.0	−109	16.0	137
895	McDonald's		Philomath	OR	6.2	−16.1	−129	14.0	95
943	McDonald's		Pendleton	OR	7.0	−16.0	−109	17.5	96
1193	McDonald's		Monroeville	PA	5.6	−16.1	−127	13.5	95
1205	McDonald's		Camp Hill	PA	5.9	−15.6	−122	14.0	100
810	McDonald's		Wakefield	RI	6.6	−16.1	−127	12.8	95
804	Bessinger's		Mt. Pleasant	SC	6.4	−15.0	−115	17.4	107
801	Boulevard Diner		Mt. Pleasant	SC	6.3	−19.9	−135	13.1	55
780	Burger King		Mt. Pleasant	SC	6.7	−20.5	−119	15.1	49
807	Five Guys		Mt. Pleasant	SC	6.2	−18.3	−111	16.0	73
789	Hardee's		Mt. Pleasant	SC	8.2	−18.7	−103	19.0	68
783	Jim 'n Nick's		Charleston	SC	6.3	−12.4	−133	14.1	134
792	Majestic Grill		Charleston	SC	6.8	−16.1	−110	14.9	95
680	McDonald's		Mt. Pleasant	SC	6.4	−13.7	−111	14.4	120
786	Sonic		Mt. Pleasant	SC	7.0	−14.0	−105	16.5	117
798	Wendy's		Mt. Pleasant	SC	5.7	−12.7	−126	14.7	130
795	Ye Ole Fashioned		Mt. Pleasant	SC	6.6	−17.5	−143	12.1	81

898	McDonald's		Oak Ridge	TN	6.4	-16.2	-137	11.8	94
1163	Braum's		Gainesville	TX	7.0	-20.4	-111	16.8	50
1050	Burger King		El Paso	TX	7.1	-19.7	-120	15.2	58
760	Burger King		La Porte	TX	6.5	-14.8	-111	15.3	109
1135	McDonald's		Houston	TX	6.8	-17.9	-115	14.7	76
1132	McDonald's		Beaumont	TX	7.1	-17.7	-121	14.0	78
1570	McDonald's		San Antonio	TX	6.2	-16.7	-127	12.5	89
1041	McDonald's		El Paso	TX	6.7	-16.1	-130	12.9	95
863	McDonald's		Houston	TX	6.0	-14.9	-123	15.1	108
892	McDonald's		Austin	TX	6.1	-14.5	-119	14.8	111
757	McDonald's		La Porte	TX	6.0	-14.5	-119	14.1	112
1151	McDonald's		Robinson	TX	6.3	-13.1	-108	15.5	127
1157	McDonald's		Gainesville	TX	6.4	-12.8	-110	15.7	129
1145	McDonald's		Austin	TX	6.6	-12.7	-108	15.8	131
1154	McDonald's		Dallas	TX	7.0	-12.5	-112	15.7	132
1141	McDonald's		San Antonio	TX	6.3	-12.5	-108	15.6	133
1044	Weinerschnitzel		El Paso	TX	7.6	-17.1	-109	18.2	85
1148	Whataburger		Austin	TX	6.3	-15.5	-106	16.6	102
1053	Whataburger		El Paso	TX	8.2	-12.6	-106	16.4	132
1138	Whataburger		Houston	TX	8.1	-12.3	-98	17.1	135
227	Apollo Burgers	local chain	West Valley City	UT	6.3	-18.9	-161	10.8	66
576	Apollo Burgers	local chain	West Jordan	UT	6.2	-17.0	-156	9.7	86

Table 6.2 contd. ....

...Table 6.2 contd.

Sample ID	Restaurant	Type	City	State	$\delta^{15}N$, ‰	$\delta^{13}C$, ‰	$\delta^{2}H$, ‰	$\delta^{18}O$, ‰	% maize
573	Arctic Circle	local chain	Riverton	UT	6.2	−12.8	−124	14.9	130
201	Astro Burgers	local chain	South Jordan	UT	6.2	−18.5	−162	10.1	70
889	B&D Burgers	local chain	Salt Lake City	UT	6.4	−18.2	−163	8.9	73
854	B&D Burgers	local chain	Salt Lake City	UT	6.3	−17.9	−156	10.5	77
198	B&D Burgers	local chain	Salt Lake City	UT	6.0	−17.0	−146	11.7	86
844	Big Ed's	local chain	Salt Lake City	UT	6.3	−15.7	−132	13.2	99
258	Burger King	national chain	Salt Lake City	UT	6.3	−12.2	−130	14.4	136
255	Carl's, Jr.	national chain	Salt Lake City	UT	6.0	−12.2	−124	15.5	136
195	Crown Burgers	local chain	Salt Lake City	UT	6.6	−20.3	−169	9.5	51
678	Crown Burgers	local chain	Salt Lake City	UT	6.6	−19.6	−163	9.3	58
1030	Crown Burgers	local chain	Salt Lake City	UT	6.1	−18.4	−158	10.0	72
847	Crown Burgers	local chain	Salt Lake City	UT	6.5	−18.2	−150	10.8	73
243	Crown Burgers	local chain	Salt Lake City	UT	6.0	−18.2	−159	10.6	74
766	Crown Burgers	local chain	Salt Lake City	UT	6.2	−17.6	−156	10.3	80
916	Crown Burgers	local chain	Salt Lake City	UT	5.9	−17.2	−153	10.3	84
813	Crown Burgers	local chain	Salt Lake City	UT	6.3	−17.1	−157	9.9	85
901	Crown Burgers	local chain	Salt Lake City	UT	6.7	−16.7	−153	10.3	89
754	Hires Big H	local chain	Salt Lake City	UT	6.1	−14.9	−153	10.1	107
230	KFC/A&W	national chain	Salt Lake City	UT	6.3	−16.4	−131	14.0	92
922	Larry's Drive In	local chain	Fillmore	UT	6.3	−17.1	−156	10.8	84
1190	McDonald's	national chain	Price	UT	6.1	−19.0	−155	10.3	65

675	McDonald's	national chain	Delta	UT	6.3	-19.0	-136	11.6	65
939	McDonald's	national chain	Salt Lake City	UT	6.6	-18.4	-140	11.8	72
827	McDonald's	national chain	Salt Lake City	UT	6.2	-18.4	-129	14.1	72
224	McDonald's	national chain	Salt Lake City	UT	6.5	-17.3	-141	12.9	83
1014	McDonald's	national chain	Kaysville	UT	5.9	-17.1	-116	14.7	84
1126	McDonald's	national chain	St. George	UT	6.8	-15.8	-130	13.0	98
860	McDonald's	national chain	Salt Lake City	UT	6.2	-15.3	-132	13.1	103
743	McDonald's	national chain	Centerville	UT	6.0	-14.7	-123	13.8	109
767	McDonald's	national chain	Salt Lake City	UT	6.0	-13.7	-115	14.7	121
204	Wendy's	national chain	Draper	UT	7.0	-11.7	-123	15.8	141
1549	Winger's	national chain	Salt Lake City	UT	5.7	-16.6	-114	15.9	90
483	McDonald's		Fairfax	VA	6.2	-13.2	-112	14.9	125
1201	McDonald's		Alexandria	VA	6.2	-12.5	-116	14.1	133
883	McDonald's		White River Junction	VT	6.1	-13.5	-117	16.0	123
1033	McDonald's		Bothell	WA	5.7	-13.0	-116	15.4	128
1211	McDonald's		Evanston	WY	6.3	-15.4	-130	13.4	103
831	McDonald's		Rock Springs	WY	6.1	-15.3	-116	17.7	104
1209	McDonald's		Laramie	WY	6.0	-12.6	-107	15.7	132
				min	5.1	-20.6	-169	8.9	48
				max	8.2	-11.6	-98	19.0	142

seasons. Most often, however, $\delta^{15}N$ values of beef have been measured with other isotope ratios (e.g., H, C, O, and/or S) to differentiate sample groups, feeding regimes, or even food source using multivariate data (Heaton et al. 2008; Nakashita et al. 2008; Bong et al. 2010; Guo et al. 2010; Horacek and Min 2010; Osorio et al. 2011b; Yanagi et al. 2012; Liu et al. 2013; Osorio et al. 2013; Zhao et al. 2013).

To verify cattle origin, hydrogen, oxygen, and, more recently, strontium isotopes have proven useful (Table 6.1). In 2004, researchers demonstrated that German and Argentinian beef could be differentiated using ^2H and ^{18}O of extracted meat-water, due to variations in the isotopic composition of the animals' drinking water (Boner and Förstel 2004). That same year, another group of researchers measured the ^{18}O composition of meat-water in an attempt to determine the origin of bovine (as well as porcine) meat, but found several storage and handling factors impacted oxygen isotopic composition; the authors concluded that origin assignment using "$\delta^{18}O$-values of meat-water is hence bound to be impossible" (Thiem et al. 2004). A 2010 study, however, found a relatively small impact from commercial processing, concluding "meat (e.g., beef and pork) kept in industrial cold storages... preserves its primary isotopic $\delta^{18}O$ meat juice value" (Horacek et al. 2010). Interestingly, one study on meat-water found that oxygen isotopes were more useful for discriminating diet than production site of beef from France (Renou et al. 2004). Additional studies have used ^2H and/or ^{18}O of protein (i.e., "muscle")—and, in one case, hair (Liu et al. 2013)—to discriminate beef provenance in various important cattle-producing regions of the globe (e.g., China, Korea, Japan) (Franke et al. 2008b; Heaton et al. 2008; Nakashita et al. 2008; Bong et al. 2010; Horacek and Min 2010; Osorio et al. 2011a; Osorio et al. 2011b; Osorio et al. 2013).

Returning to the set of 147 fast food hamburgers purchased between 2007 and 2009, there are possible geographic patterns to be seen in the measured δ^2H and $\delta^{18}O$ signatures of DDM, when samples are grouped by purchase location and restaurant. Hydrogen and oxygen isotope ratios span large ranges when the entire dataset is considered: from −179‰ to −98‰ for H and from +8.9‰ to +19.0‰ for O, as shown in Table 6.2. [Note that ^2H results represent the calculated "intrinsic" fraction; for more details, see Chapter 2 and (Chesson et al. 2008; Chesson et al. 2009b; Chesson et al. 2010).] When only samples purchased from the state of Utah are considered, δ^2H and $\delta^{18}O$ values still span large ranges, from −179‰ to −114‰ and from +8.9‰ to +15.9‰, respectively, as seen in Figure 6.2 (top). Further grouping the Utah hamburgers by restaurant type—local *vs.* national chain—reveals that samples from local chain restaurants, or those with only a few outlets in Utah and surrounding states, generally have the lowest δ^2H and $\delta^{18}O$ values of the 147 hamburgers purchased across the USA (Figure 6.2, bottom). Considering the low δ^2H and $\delta^{18}O$ values expected for water in Utah and

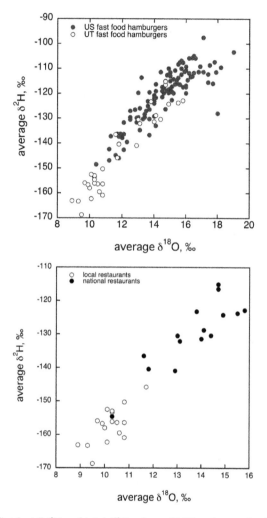

Figure 6.2 The "intrinsic" δ^2H and total $\delta^{18}O$ values of 147 hamburger (cooked beef patties) purchased across the USA between 2007 and 2009 (top), with those purchased in the state of Utah identified. Samples were analyzed as de-fatted dry matter (DDM). Utah hamburger samples are further grouped by restaurant type (local *vs.* national chain, bottom), revealing that hamburgers from local chain restaurants, or those with only a few outlets in Utah and surrounding states, generally had the lowest δ^2H and total $\delta^{18}O$ values of the dataset.

surrounding states (Bowen and Revenaugh 2003; Bowen et al. 2007), these data suggest the local-chain restaurants in Utah may be sourcing beef more locally than outlets of national-chain fast food restaurants also located in Utah. This possibility is investigated in detail in Chapter 5.

Relatively few studies have used Sr isotope analysis for beef origin verification (Table 6.1). The first study, published in 2008, measured samples of dried beef from around the globe and found that $^{87}Sr/^{86}Sr$ ratios spanned a narrow range and the effect of sample origin was only weakly significant (Franke et al. 2008a). In contrast, a study in 2011 of beef from three major cattle-producing regions of Argentina found that $^{87}Sr/^{86}Sr$ ratios—with other analytes, including ^{13}C and ^{15}N—were useful for differentiating origin (Baroni et al. 2011). A year later, strontium isotope results from a large collection of 206 beef samples from across Europe were published (Rummel et al. 2012). The authors found that sample processing, such as de-fatting, could have a significant impact on $^{87}Sr/^{86}Sr$ ratios. With careful and consistent preparation, however, significant differences could be observed in the Sr isotopic composition of beef meat originating from regions with distinct geological features (e.g., limestone bedrock).

6.3.2 Ovine foodstuffs

"The real fact is that I could no longer stand their eternal cold mutton."

(Cecil Rhodes, British businessman and
South African politician, purportedly said on
why he left England for South Africa; 1853–1902)

Similar to beef, forensic isotope analyses of ovine foodstuffs are generally concerned with questions of sheep production—how and where were the animals raised? The isotope ratios used to help answer these questions include hydrogen, carbon, nitrogen, oxygen, and sulfur; studies utilizing these isotopes are listed in Table 6.3. Sheep are raised not only for their flesh (lamb, hogget, or mutton) but also their fleece and isotope studies have focused on measurements of both meat and wool.

To date, there have been no studies of strontium isotope ratio variation in sheep-meat, although Sr isotope analysis has been applied in some studies of wool. While these investigations focused primarily on the provenance of ancient textiles [e.g., (Frei et al. 2009; Frei 2014)], results have shown that $^{87}Sr/^{86}Sr$ ratios in sheep wool "agree well with the compositions of biologically available (soluble) strontium fractions from the respective feeding ground soils" (Frei et al. 2009). This suggests that Sr isotope analysis could be a useful tool for investing sheep meat origin as well as sheep wool origin.

Much of the forensic investigative work performed using hydrogen, carbon, nitrogen, and oxygen isotopes within sheep tissues is similar to that performed in studies of cattle tissues [i.e., investigations of diet and origin (Piasentier et al. 2003; Sacco et al. 2005; Moreno-Rojas et al. 2008a; Zazzo et al. 2008; Prache et al. 2009; Harrison et al. 2010; Zazzo et al. 2010; Sun et al. 2012; Devincenzi et al. 2014; Erasmus et al. 2016)], and this section, therefore, focuses on sulfur isotopes. In 2007, ^{34}S—along with ^{15}N—was measured in de-fatted lamb-meat samples collected across Europe; both

Table 6.3 A summary of isotope studies on ovine foodstuffs.

Isotope(s)	Tissue	Citation	Principle finding(s)
C and N	Muscle, lipids	(Piasentier et al. 2003)	Origin and diet
C and N (plus ¹H via NMR)	Muscle	(Sacco et al. 2005)	Origin
H, C, N, and S	Muscle	(Camin et al. 2007)	Origin
C	Muscle, "fat", wool	(Moreno-Rojas et al. 2008a)	Diet
C	Wool	(Zazzo et al. 2008)	Dietary turnover
H, O, C, N, and S	Muscle, "fat"	(Perini et al. 2009)	Origin and diet
N	Muscle	(Prache et al. 2009)	Diet
C	Muscle	(Harrison et al. 2010)	Dietary turnover
H, O, C, and S	Muscle, lipids	(Harrison et al. 2011)	Dietary turnover
C	Tooth enamel	(Zazzo et al. 2010)	Dietary turnover
S	Wool	(Zazzo et al. 2011)	Origin
C and N (via NIRS)	Muscle	(Sun et al. 2012)	Origin
H, O, C, and S	Muscle, blood plasma	(Biondi et al. 2013)	Diet
N	Muscle	(Devincenzi et al. 2014)	Diet
H, O, C, and N	Wool	(Zazzo et al. 2015)	Seasonal variation
O, C, N, and S	Blood plasma, erythrocytes	(Bontempo et al. 2016)	Diet, dietary turnover
C and N	Muscle	(Erasmus et al. 2016)	Origin and diet

isotopes could delineate samples from different regions, with a total range of +1.6‰ to +12.8‰ for $\delta^{34}S$ (and +3.8‰ to +9.2‰ for $\delta^{15}N$) (Camin et al. 2007). When combined with measurements of ^2H and ^{13}C, via linear discriminant analysis, sulfur and nitrogen isotope ratios helped to correctly classify approximately 78% of the samples. Two years later, a study of de-fatted dry lamb-meat, collected exclusively in Italy, found that $\delta^{34}S$ values ranged from a minimum of −9.0‰ to a maximum of +9.5‰ (Perini et al. 2009), an overall larger interval than that seen previously in the study of European lamb meat samples (Camin et al. 2007). This variation was linked to the environments inhabited by the sheep, which included areas near coastlines—and thus impacted by sea spray—and regions with distinctive geological features, such as volcanic sulfur sources.

A diet-switch experiment found no offset or "diet-tissue fractionation between the de-fatted muscle and the control diet," suggesting that $\delta^{34}S$ values measured for lamb-meat directly reflect the sulfur isotopic composition of diet (Harrison et al. 2011). In 2014, researchers showed that diet affected the δ^{34}S values of DDM of lambs switched from pasture-based diet to a concentrate diet, indicating the sulfur isotope analysis could be used to trace changes in feed (Biondi et al. 2013). A more recent study compared the δ^{34}S (as well as δ^{13}C and δ^{15}N) signatures of lamb blood plasma and erythrocytes (red blood cells) following a brief 14-day switch from pasture diet to concentrate diet, equivalent to a *finishing period* (Bontempo et al. 2016). The authors found that the isotopic composition of the red blood cells was not affected by a diet change during the finishing period, while "analysis of plasma could detect very short finishing periods with concentrate and hay" and provide a possible tool for authenticating sheep meat production claims.

In a departure from meat studies, researchers measured δ^{34}S signatures of sheep's wool to assess the impact of sea spray on samples collected from across Ireland (Zazzo et al. 2011). Measured δ^{34}S values ranged from +5.3 to +17.0‰, with higher values found nearer the coasts. Values were also generally higher in wool samples collected from the west coast of Ireland than those collected from eastern portions of the country, which the authors expected based on the direction of prevailing winds—south and west. It was possible to construct a sulfur wool isoscape based on the results, "describing inland regions accurately in terms of their distance (tens to a few hundred km) to the coasts" (Zazzo et al. 2011). More recently, researchers have used isotope analysis (H, C, N, and O) to investigate seasonal variations in sheep wool, finding that measurements may provide useful information on short-term changes in sheep diet and local climatic conditions (Zazzo et al. 2015).

6.3.3 Poultry

"EAT MOR CHIKIN."

(The Cow Campaign *from Chick-fil-A, 1995*)

Poultry, domesticated birds kept by human for food, include chicken, quail, turkey, and waterfowl (ducks and geese). The majority of forensic isotope studies on poultry have focused on carbon and nitrogen isotopic compositions of chicken meat to understand poultry diet, as shown in Table 6.4. This work has primarily taken place in Brazil, with researchers attempting to authenticate production methods:

- How were nutrients from the diet routed into different tissues?
- Was the bird fed maize (corn)?
- Were any animal byproducts included in the feed?

Only three publications have used isotopic analysis specifically to investigate poultry origin, using oxygen and/or strontium (Franke et al. 2008a; Franke et al. 2008b), or a combination of hydrogen, carbon, nitrogen, oxygen, sulfur, and strontium isotopes (Rees et al. 2016). By and large,

Table 6.4 A summary of isotope studies on poultry.

Isotope(s)	Tissue	Citation	Principle finding(s)
C	Muscle	(Sakamoto et al. 2002)	Diet
C	Muscle, liver	(Cruz et al. 2004)	Metabolic routing of diet
C	Muscle, liver	(Cruz et al. 2005)	Dietary turnover
C and N	Muscle	(Carrijo et al. 2006)	Diet (animal byproduct)
C and N	Muscle, bone*	(Móri et al. 2007)	Diet (animal byproduct)
O and Sr (via ICP-MS)	Muscle	(Franke et al. 2008a)	Origin
O	Muscle	(Franke et al. 2008b)	Origin
C and N	Muscle, bone	(Oliveira et al. 2010)	Diet (animal byproduct)
C	Muscle	(Rhodes et al. 2010)	Diet (maize-fed)
C and N	Muscle	(Coletta et al. 2012)	Diet (barn *vs.* free-range)
C and N	Muscle	(Cruz et al. 2012)	Diet (animal byproduct)
C and N	Muscle*	(Sernagiotto et al. 2013)	Diet (animal byproduct)
C and N	Muscle	(Zhao et al. 2016a)	Origin
H, C, N, O, S and Sr (via TIMS)	Muscle[a]	(Rees et al. 2016)	Origin

*From quail.
[a]Some turkey samples were included in this study.

however, hydrogen, oxygen, sulfur, and strontium isotopes have been under-utilized in studies of poultry meat.

In 2004 and 2005, researchers used chicken meat ^{13}C composition to trace the flow of energy from diet to different tissues (Cruz et al. 2004; Cruz et al. 2005). As noted by the authors, "diet constitutes one of the main factors of the efficient and healthy development of animals" (Cruz et al. 2004) and understanding the impact of different diets on chicken metabolism is important for the meat industry. Of equal importance to consumers is information on the safety of chicken diets—specifically, whether or not the diets contain bovine meat and/or bone meal (MBM), which has been linked to bovine spongiform encephalopathy (BSE or "mad cow disease"), or animal byproducts from the same species (e.g., poultry offal meal, POM; or feather meal). In 2006, researchers fed chickens a variety of diets containing between 0 and 8% MBM and measured breast muscle $\delta^{13}C$ and $\delta^{15}N$ signatures, demonstrating that carbon and nitrogen isotopes could discriminate diets containing as little as 2% bovine byproducts from diets containing no byproducts (Carrijo et al. 2006). A year after this study, researchers fed quails a variety of diets containing MBM, POM, feather meal, and mixtures thereof, then measured the carbon and nitrogen isotopic compositions of breast muscle, keel (an extension of the breastbone), and tibia (shinbone) (Móri et al. 2007). The authors found that it was possible to discriminate birds fed diets containing animal byproducts from those eating byproduct-free diets using ^{13}C and ^{15}N. It was not possible, however, to discriminate between the different byproduct diets with carbon and nitrogen isotopes alone. A 2010 study used carbon and nitrogen isotopes to test for the inclusion of POM in the diets of chickens demonstrated that the δ^{13}C and δ^{15}N values of breast muscle from birds fed 8% and 16% POM were significantly different from values in birds fed 0, 2, and 4% POM (Oliveira et al. 2010). When keel and tibia were analyzed (in place of breast muscle) the birds fed 4% POM were also found to be significantly different from animals fed 0 and 2% POM. A 2012 study demonstrated the sensitivity of carbon and nitrogen isotopes to detect POM in chicken diets by switching birds from vegetarian diets to diets containing animal byproducts for varying lengths of time, concluding that "the animal ingredient has to be a part of the feeding for 21 d[ays] or longer to be detected by this method" (Cruz et al. 2012). In a further study, researchers switched quails from a vegetable diet to a diet of 8% POM and found that it was "possible to trace poultry offal meal inclusion in a strictly vegetable diet after the diet was changed for a least 14 days" (Sernagiotto et al. 2013).

In addition to providing information on the safety of poultry diets—by verifying the absence of animal byproducts in the feed—carbon and/or nitrogen isotopes can potentially be used to authenticate diet claims. A study of chickens fed varying amounts of maize demonstrated that "the ^{13}C content of the protein was a reliable marker for the dietary status of the chickens" and could be used to distinguish birds fed maize from those with no maize in the diet (Rhodes et al. 2010). In 2012, researchers measured the carbon and nitrogen isotopic composition of breast muscle from chickens allowed to range freely and eat maize as well as birds kept in a barn and fed diets of maize or maize plus soybean (Coletta et al. 2012). The δ^{15}N signatures were higher for free-range chickens than for barn-raised birds, most likely due to the inclusion of animal proteins (earthworms, insects, etc.), in addition to maize, in the diet.

A few isotopic studies of poultry have focused on questions of origin (Table 6.4). A pioneering study in 2002 measured ^{13}C in the "body component" (neck, breast, back, wing, and leg) of chicken from different countries (Japan, China, and USA) and observed variations, which were attributed to "consumption of corn [maize] fed to chicken" (Sakamoto et al. 2002) although no information about diet was available to the researchers at the time. Several years later, researchers used a combination of oxygen and strontium isotope ratios in the analysis of frozen chicken breast samples from Brazil, France, Germany, Hungary, and Switzerland (Franke et al. 2008a). Although differentiation of sample origin was possible using oxygen isotope ratios, "no clear discrimination between individual countries of origin was possible using $^{87}Sr/^{86}Sr$." A following study combined elemental concentration measurements with $\delta^{18}O$ signatures and found that "combining data did not clearly reduce the percentage of incorrectly classified individual samples compared to the two approaches applied separately" (Franke et al. 2008b).

In 2015, carbon and nitrogen isotopes—along with elemental concentrations—were measured for samples of de-fatted chicken breast muscle from four Chinese provenances (Zhao et al. 2016a). There were differences among the $\delta^{13}C$ and $\delta^{15}N$ values of samples, which, combined with concentrations of K, Zn, and Rb using discriminant analysis, provided correct classification for 100% of the samples. More recently, a combination of $\delta^{2}H$, $\delta^{13}C$, and $\delta^{15}N$ signatures and concentrations of Mg, Tl, Rb, and Mo were used in discriminant analysis to correctly classify a total 88.3% of 384 known-origin poultry samples, with the authors noting that "92.0% of cross-validated European cases were correctly classified and 100% of cross-validated Chinese cases were correctly classified" which demonstrated "the reliability of the multivariate approach in classifying the origin of poultry" (Rees et al. 2016).

6.3.4 Porcine foodstuffs

"Pork. The Other White Meat."

> (*A message from the National Pork Board; March 2, 1987—Despite this commercial branding, pork is not a white meat according to the U.S. Department of Agriculture.*)

Although porcine foodstuffs have been less studied than other terrestrial meats, some of the earliest published studies using forensic isotope analysis to investigate flesh foods focused on pork, as summarized in Table 6.5. The majority of pork meat investigations have attempted to answer questions of production method with some additional work on origin using hydrogen, carbon, nitrogen, oxygen, and sulfur isotopes. To date, no studies of strontium isotopic variation in pork meat have been published.

The first published studies of pork meat used isotope analysis to differentiate Iberian swine raised under distinct "feeding regimes"—e.g., fattened on acorns *vs.* commercial feed during the finishing period (Gonzalez-Martin et al. 1999; Gonzalez-Martin et al. 2001). Tissues analyzed included adipose, muscle, and liver (Gonzalez-Martin et al. 1999), and liver alone (Gonzalez-Martin et al. 2001); both ^{13}C and ^{34}S were found to be useful for discriminating feed type through measurement of liver tissue. A 2006 study investigated the offset in ^{13}C, as well as ^{15}N, between swine diet and tissue, measuring liver, muscle, cartilage, and fat tissues plus hair and nail keratin (Nardoto et al. 2006). In general, tissues had similar or lower $\delta^{13}C$

Table 6.5 A summary of isotope studies on porcine foodstuffs.

Isotope(s)	Tissue	Citation	Principle finding(s)
C	Muscle, "fat", and liver	(Gonzalez-Martin et al. 1999)	Diet
C and S	Liver	(Gonzalez-Martin et al. 2001)	Diet
O	Extracted water	(Thiem et al. 2004)	Origin
C and N	Muscle, "fat", liver, hair, nail, cartilage	(Nardoto et al. 2006)	Diet-tissue fractionation
H and O	Extracted water	(Horacek et al. 2010)	Origin
H, O, C, N, and S	Muscle, "fat"	(Perini et al. 2013)	Origin, diet, and processing
C and N	Muscle	(Kim et al. 2013)	Origin
C (via GC-C-IRMS)	Fatty acids	(Recio et al. 2013)	Diet
C and N	Muscle	(Zhao et al. 2016b)	Organic production

signatures when compared to diet, ranging from −2.4‰ (liver) to +0.9‰ (cartilage). In contrast, all tissues had higher $\delta^{15}N$ signatures than diet, ranging from +2.2‰ (liver) to +3.0‰ (nail).

A study in 2013 used carbon, nitrogen, and sulfur along with hydrogen and oxygen isotopes to investigate 86 samples of dry-cured ham reared in three geographical areas—two in Italy and one in Spain (Perini et al. 2013). Isotope differences were observed between the hams from the different areas. For $\delta^{13}C$ signatures, the authors noted that variability was "related to the dietary abundance of the C_4 plants." Isotopic variation in meteoric waters available to the pigs impacted δ^2H and $\delta^{18}O$ signatures of hams from different areas, which "showed a geographic and climatic gradient caused by systematic global variations in the isotope composition of precipitated water, transferred to some extent to the isotopic values of pork fractions" (Perini et al. 2013). Recent studies have used carbon and nitrogen isotopes to investigate differences in pork meat from different countries and raised under different systems (conventional *vs.* organic). When samples of de-fatted muscle from South Korea, Europe, and the USA were compared, there was little overlap between the groups in ^{13}C and ^{15}N (Kim et al. 2013). Similarly, there was little overlap in ^{13}C and ^{15}N for samples of de-fatted muscle from pigs raised under conventional and organic systems (Zhao et al. 2016b).

In 2013, researchers published a study on the use of compound-specific isotope analysis to investigate swine diet (Recio et al. 2013). The carbon isotope signatures of 159 samples of fat from European pigs were analyzed via gas-chromatography combustion IRMS (GC-C-IRMS) following derivatization of samples to fatty acid methyl esters (FAMEs). The authors found that the $\delta^{13}C$ values of oleic acid (C18:1) in particular were good discriminators of feeding regime, separating *Bellota* swine (raised on a diet of grass and winter acorns) from *Cebo* and *Recebo* swine (raised on diets that included all or some formulated animal feed, respectively). As noted by Recio and colleagues, "no single sample classed as *Bellota* has $\delta^{13}C_{C18:1} > -25.9‰$" (Recio et al. 2013).

Notwithstanding this study of pig meat, no other studies of flesh foods appear to have used compound specific isotope analysis techniques, although specific meat compounds—such as lipids or fats—have been the focus of research. A 2008 study measured the hydrogen and oxygen isotope ratios of "beef lipid" (Heaton et al. 2008), but gas chromatography was not used for analysis and solvent-extracted lipids were simply weighed into silver capsules for measurement via TC/EA-IRMS. Similarly, a 2011 study collected different lipids from lamb muscle using a variety of solvent and solid phase extraction techniques; lipids in solution were injected into capsules and the solvent evaporated prior to analysis via EA-IRMS (Harrison et al. 2011). This general approach has been used in other studies that measured the isotopic composition of meat lipids [e.g., (Perini et al. 2009; Perini et al. 2013)]. Investigations into the isotopic compositions of animal "fat" have focused on samples collected directly as adipose tissue [e.g., (Gonzalez-Martin et al. 1999; Piasentier et al. 2003; De Smet et al. 2004; Nardoto et al. 2006; Moreno-Rojas et al. 2008a; Osorio et al. 2013; Perini et al. 2013)].

6.3.5 Aquatic foodstuffs

"It has always been my private conviction that any man who puts his intelligence up against a fish and loses had it coming."

(John Steinbeck, American author; 1902–1968)

The final section of this chapter focuses on aquatic flesh foods, derived from animals living most or all of life in freshwater or saltwater environments. While not as extensively studied as terrestrial animals, there are nevertheless many stable isotopes studies of aquatic meats, both invertebrates (e.g., shrimp or prawns) and vertebrates (e.g., bony fish), which are listed in Table 6.6. In aquatic settings, hydrogen and oxygen isotopes have been generally less studied than carbon, nitrogen, and strontium isotopes, which have been used to investigate questions of food origin and authenticity.

Isotope ratios of fish *otoliths*—also known as "earstones"—have been measured to reconstruct the history of migratory fish like salmon. Otoliths are comprised of calcium carbonate, a new ring of which is added each year, and measuring the isotopic composition of these yearly annuli can

Table 6.6 A summary of isotope studies on aquatic foodstuffs.

Animal	Isotope(s)	Citation	Principle finding(s)
Atlantic salmon	Sr (of otoliths)	(Kennedy et al. 1997)	Origin
Chinook (Pacific) salmon	Sr (of otoliths, via TIMS)	(Ingram and Weber 1999)	Origin
Atlantic salmon	H and C	(Aursand et al. 2000)	Authentication – wild *vs.* farmed
Atlantic salmon	C and N	(Dempson and Power 2004)	Authentication – wild *vs.* farmed
Freshwater fish (variety)	H (of otoliths and muscle)	(Whitledge et al. 2006)	Origin
Sea bass	O, C, and N	(Bell et al. 2007)	Authentication – wild *vs.* farmed
Atlantic salmon	C and N	(Molkentin et al. 2007)	Organic production
Sea bream	O, C, and N	(Morrison et al. 2007)	Authentication – wild *vs.* farmed
Sea bream	C and N	(Serrano et al. 2007)	Authentication – wild *vs.* farmed
Chinook (Pacific) salmon	Sr (of otoliths, via ICP-MS)	(Barnett-Johnson et al. 2008)	Origin
Turbot	C and N	(Busetto et al. 2008)	Authentication – wild *vs.* farmed
Rainbow trout	C and N	(Moreno-Rojas et al. 2008b)	Diet of farmed fish
Atlantic salmon	O and N	(Thomas et al. 2008)	Authentication – wild *vs.* farmed
Rainbow trout	Sr (of otoliths, via ICP-MS)	(Gibson-Reinemer et al. 2009)	Origin
Murray cod	O, C, and N	(Turchini et al. 2009)	Authentication – wild *vs.* farmed, origin
Freshwater fish (variety)	O and C (of otoliths)	(Whitledge 2009)	Origin
Salmon (variety)	C and N	(Anderson et al. 2010)	Authentication – wild *vs.* farmed
Chinook (Pacific) salmon	Sr (of otoliths, via ICP-MS)	(Barnett-Johnson et al. 2010)	Origin
Sea bass	C and N	(Fasolato et al. 2010)	Authentication – wild *vs.* farmed
Brazilian freshwater fish	C and N	(Sant'Ana et al. 2010)	Authentication – wild *vs.* farmed

Table 6.6 contd. ...

...Table 6.6 contd.

Animal	Isotope(s)	Citation	Principle finding(s)
Freshwater fish (variety)	O and C (of otoliths)	(Zeigler and Whitledge 2010)	Origin
Cod, saithe	C and N	(Oliveira et al. 2011)	Discrimination of salted fish
Trout (variety)	O and S	(Trembaczowski 2011)	Origin
Cutthroat trout	Sr (of otoliths, via ICP-MS)	(Muhlfeld et al. 2012)	Origin
Piscivores (pike, bass, walleye)	Sr (of otoliths, via ICP-MS)	(Wolff et al. 2012)	Origin
Seatrout	O and C (of otoliths)	(Curtis et al. 2014)	Authentication – wild *vs.* farmed
Mackerel, haddock	C and N	(Fernandes et al. 2014)	Cooking effects
Shrimp	C and N	(Gamboa-Delgado et al. 2014)	Authentication – wild *vs.* farmed
Shrimp	C and N	(Ostermeyer et al. 2014)	Authentication – wild *vs.* farmed
Freshwater fish (variety)	O	(Rossier et al. 2014)	Origin
Chinook (Pacific) salmon	Sr (of otoliths, via ICP-MS)	(Brennan et al. 2015)	Origin
Prawn (shrimp)	H, O, C, and N	(Carter et al. 2015)	Authentication – local *vs.* imported
Mackerel, croaker, pollock	C and N	(Kim et al. 2015a)	Country of origin
Hairtail fish, shrimp	C and N	(Kim et al. 2015b)	Country of origin
Trout, salmon (variety)	C and N	(Molkentin et al. 2015)	Organic production
Shrimp	C and N	(Ortea and Gallardo 2015)	Authentication – wild *vs.* farmed
Bering cisco	Sr (of otoliths, via ICP-MS)	(Padilla et al. 2015)	Origin
Shrimp	H, C, N, O, and S	(Fry et al. 2016)	Pollution

thus provide information on the waters a fish inhabited when each ring was formed. The first work on strontium isotopes in Atlantic salmon otoliths was published in 1997 and used $^{87}Sr/^{86}Sr$ ratios to discriminate fish reared in various Vermont (USA) rivers (Kennedy et al. 1997). This was followed in 1999 by a study of otoliths from Chinook (Pacific) salmon, in which the strontium isotope variation in river water of the Sacramento-San Joaquin

(California, USA) system was mapped and related to $^{87}Sr/^{86}Sr$ ratios in the fish reared there (Ingram and Weber 1999).

The pioneering studies on strontium isotopes in salmon otoliths have been followed by additional studies on Chinook salmon (Barnett-Johnson et al. 2008; Barnett-Johnson et al. 2010; Brennan et al. 2015) as well as trout [rainbow (Gibson-Reinemer et al. 2009) and cutthroat (Muhlfeld et al. 2012)], a variety of freshwater piscivores [e.g., bass, pike, walleye (Wolff et al. 2012)], and Bering cisco (Padilla et al. 2015). Results based on strontium isotope analysis of otoliths have been important to identify key fish habitats for conservation and management. As noted in a study on fish from the Upper Colorado River Basin (USA) (Wolff et al. 2012), strontium isotopes have also been useful "as a deterrent against illicit transfer of aquatic organisms" such as the illegal stocking of bass, pike and walleye in reservoirs. In addition to strontium, oxygen and carbon isotope ratios of otoliths have been measured to determine the usefulness of other isotopes for origin investigations. In 2009, $\delta^{13}C$ signatures together with Sr:Ca and Ba:Ca ratios were used to distinguish samples from the fresh waters of Lake Michigan and the upper Illinois River plus three of its tributaries (Whitledge 2009). A year later, researchers found that ^{13}C and ^{18}O in otoliths were useful to distinguish samples collected from the lower Illinois River, its tributaries, or floodplain lakes (Zeigler and Whitledge 2010). More recently, otolith chemistry—including carbon and oxygen isotope ratios with element concentrations—was successfully used to classify spotted seatrout from the bay waters of Texas (USA) as wild or hatchery-raised (Curtis et al. 2014).

Hydrogen, carbon, nitrogen, and oxygen isotopes have been measured in other aquatic animal tissues to investigate origin. A 2006 study measured hydrogen isotope ratios of fish collected from a variety of environments in Canada and demonstrated that "significant linear relationships ($r^2 \geq 0.97$) exist between fish otolith and muscle δ^2H and δ^2H of waters that fish inhabit" (Whitledge et al. 2006). Similarly, a study on fish from Swiss lakes found that there was a "rather good linear correlation (0.986) between the $\delta^{18}O$ values of fish water and of water they grew in" (Rossier et al. 2014). Recently, researchers used the $\delta^{13}C$ and $\delta^{15}N$ signatures of known-origin fish fillets to distinguish samples from different countries (Kim et al. 2015a), while a combination of ^2H, ^{13}C, and elemental content measurements could be used to distinguish Australian prawns (shrimp) from prawns imported from China, Malaysia, and Vietnam (Carter et al. 2015). A single study has investigated sulfur (along with carbon) isotopes of fish for determining origin, but the work was limited to samples of two fish (Trembaczowski 2011). In 2016 sulfur isotopic compositions—with additional isotope ratios and element contents—were measured for prawns collected in several Asia-Pacific countries to investigate the extent of coastal pollution and the potential contamination of a marine food resource (Fry et al. 2016). Despite these publications, it is likely too soon to determine the overall usefulness of sulfur isotopes in forensic examinations of aquatic foodstuffs.

Measurements of carbon and/or nitrogen isotope ratios—along with measurements of hydrogen (Aursand et al. 2000) and oxygen (Bell et al. 2007; Morrison et al. 2007; Thomas et al. 2008) isotope ratios—have also been used to authenticate production claims of aquatic foods. In some cases, isotope signatures alone were sufficient to discriminate wild *vs.* farmed fish (Dempson and Power 2004; Serrano et al. 2007; Sant'Ana et al. 2010; Molkentin et al. 2015) and shrimp (Gamboa-Delgado et al. 2014; Ostermeyer et al. 2014; Kim et al. 2015b; Ortea and Gallardo 2015). In other cases, discrimination of wild *vs.* farmed fish was more powerful with additional analytes, such as fatty acids (Aursand et al. 2000; Bell et al. 2007; Molkentin et al. 2007; Morrison et al. 2007; Busetto et al. 2008; Thomas et al. 2008; Fasolato et al. 2010; Muhlfeld et al. 2012) or elemental concentrations (Anderson et al. 2010). A 2008 study used carbon and nitrogen isotopes to identify fish fed on plant-based diets from those fed diets containing fish protein (Moreno-Rojas et al. 2008b). Subsequently, researchers found that carbon and nitrogen isotopes were useful to identify fish on commercial diets; oxygen isotopes in the fish were linked to specific water source and could provide additional data to distinguish samples from different farms (Turchini et al. 2009). A 2011 study measured ^{13}C and ^{15}N in salted fish and found it was possible to discriminate two different species (cod and saithe) based on isotopic composition (Oliveira et al. 2011).

6.4 Conclusion

This chapter has reviewed the use of hydrogen, carbon, nitrogen, oxygen, sulfur, and strontium isotopes to investigate the authenticity and origin of flesh foods, or meat, both terrestrial and aquatic. Application of forensic isotope analysis to flesh foods has generally focused on understanding what an animal ate and drank, thereby providing information on meat production—for example, to verify that swine were finished on a diet of acorns or raised in a particular region. In addition to the research studies cited in this chapter, several earlier reviews, dissertations, and data compilations have discussed the isotope analysis of meat [e.g., (Franke et al. 2005; Kelly et al. 2005; Prache et al. 2005; Prache 2007; Schmidt et al. 2008; Gonzalvez et al. 2009; Chesson et al. 2011a; Chesson et al. 2011b; Drivelos and Georgiou 2012; Monahan et al. 2012; Montowska and Pospiech 2012; Vinci et al. 2012; Perini 2013; Zhao et al. 2014; Krivachy (Tanz) et al. 2015; Li et al. 2016)]. These summaries include a 2015 study of the potential usefulness of Pb isotopes for meat authenticity and traceability (Evans et al. 2015), the first of its kind.

Returning again to the *Washington Post* exposé on food fraud in that Manhattan market (Layton 2010), it appears forensic isotope analysis could have helped discriminate traditionally-harvested, expensive caviar from cheaper paddlefish roe; the H, C, N, O, S, and Sr isotope ratios of a wild sturgeon from the Caspian or Black Seas should be significantly different from a paddlefish living in the fresh waters of the Mississippi River. While isotope analysis alone may not be able to discriminate horse meat from beef meat, the hydrogen, oxygen, and strontium isotopic compositions of the mislabeled horse meat discovered in European markets in 2013 could have guided investigation of the source of the adulterated meat. In short, forensic isotope analysis of flesh foods has the potential to provide valuable information to food inspectors and investigators, information on the history, production, and origin of an animal-derived meat product that cannot be easily gathered using any other analytical test.

References

Anderson, K. A., K. A. Hobbie and B. W. Smith. 2010. Chemical profiling with modeling differentiates wild and farm-raised salmon. *Journal of Agricultural and Food Chemistry* 58: 11768–11774.

Aursand, M., F. Mabon and G. J. Martin. 2000. Characterization of farmed and wild salmon (*Salmo salar*) by a combined use of compositional and isotopic analyses. *Journal of the American Oil Chemists' Society* 77: 659–666.

Bahar, B., S. M. Harrison, A. P. Moloney, F. J. Monahan and O. Schmidt. 2014. Isotopic turnover of carbon and nitrogen in bovine blood fractions and inner organs. *Rapid Communications in Mass Spectrometry* 28: 1011–1018.

Bahar, B., A. P. Moloney, F. J. Monahan, S. M. Harrison, A. Zazzo, C. M. Scrimgeour, I. S. Begley and O. Schmidt. 2009. Turnover of carbon, nitrogen, and sulfur in bovine longissimus dorsi and psoas major muscles: Implications for isotopic authentication of meat. *Journal of Animal Science* 87: 905–913.

Bahar, B., F. J. Monahan, A. P. Moloney, P. O'Kiely, C. M. Scrimgeour and O. Schmidt. 2005. Alteration of the carbon and nitrogen stable isotope composition of beef by substitution of grass silage with maize silage. *Rapid Communications in Mass Spectrometry* 19: 1937–1942.

Bahar, B., O. Schmidt, A. Moloney, C. Scrimgeour, I. Begley and F. Monahan. 2008. Seasonal variation in the C, N and S stable isotope composition of retail organic and conventional Irish beef. *Food Chemistry* 106: 1299–1305.

Barnett-Johnson, R., T. E. Pearson, F. C. Ramos, C. B. Grimes and R. B. MacFarlane. 2008. Tracking natal origins of salmon using isotopes, otoliths, and landscape geology. *Limnology and Oceanography* 53: 1633–1642.

Barnett-Johnson, R., D. J. Teel and E. Casillas. 2010. Genetic and otolith isotopic markers identify salmon populations in the Columbia River at broad and fine geographic scales. *Environmental Biology of Fishes* 89: 533–546.

Baroni, M. V., N. S. Podio, R. G. Badini, M. Inga, H. A. Ostera, M. Cagnoni, E. Gallegos et al. 2011. How much do soil and water contribute to the composition of meat? A case study: Meat from three areas of Argentina. *Journal of Agricultural and Food Chemistry* 59: 11117–11128.

Bataille, C. P. and G. J. Bowen. 2012. Mapping $^{87}Sr/^{86}Sr$ variations in bedrock and water for large scale provenance studies. *Chemical Geology* 304-305: 39–52.

Beard, B. L. and C. M. Johnson. 2000. Strontium isotope composition of skeletal material can determine the birth place and geographic mobility of humans and animals. *Journal of Forensic Sciences* 45: 1049–1061.

Bell, J. G., T. Preston, R. J. Henderson, F. Strachan, J. E. Bron, K. Cooper and D. J. Morrison. 2007. Discrimination of wild and cultured European Sea Bass (*Dicentrarchus labrax*) using chemical and isotopic analyses. *Journal of Agricultural and Food Chemistry* 55: 5934–5941.

Biondi, L., M. G. D'Urso, V. Vasta, G. Luciano, M. Scerra, A. Priolo, L. Ziller, L. Bontempo, P. Caparra and F. Camin. 2013. Stable isotope ratios of blood components and muscle to trace dietary changes in lambs. *Animal* 7: 1559–1566.

Block, G. 2004. Foods contributing to energy intake in the US: Data from NHANES III and NHANES 1999–2000. *Journal of Food Composition and Analysis* 17: 439–447.

Boner, M. and H. Förstel. 2004. Stable isotope variation as a tool to trace the authenticity of beef. *Analytical and Bioanalytical Chemistry* 378: 301–310.

Bong, Y. -S., W. -J. Shin, A. -R. Lee, Y. -S. Kim, K. Kim and K. -S. Lee. 2010. Tracking the geographical origin of beefs being circulated in Korean markets based on stable isotopes. *Rapid Communications in Mass Spectrometry* 24: 155–159.

Bontempo, L., F. Camin, L. Ziller, L. Biondi, M. G. D'Urso, V. Vasta and G. Luciano. 2016. Variations in stable isotope ratios in lamb blood fractions following dietary changes: A preliminary study. *Rapid Communications in Mass Spectrometry* 30: 170–174.

Bowen, G. J., J. R. Ehleringer, L. A. Chesson, E. Stange and T. E. Cerling. 2007. Stable isotope ratios of tap water in the contiguous United States. *Water Resources Research* 43: W03419.

Bowen, G. J. and J. Revenaugh. 2003. Interpolating the isotopic composition of modern meteoric precipitation. *Water Resources Research* 39: 1299–1311.

Brennan, S. R., C. E. Zimmerman, D. P. Fernandez, T. E. Cerling, M. V. McPhee and M. J. Wooller. 2015. Strontium isotopes delineate fine-scale natal origins and migration histories of Pacific salmon. *Science Advances* 1: e1400124.

Busetto, M. L., V. M. Moretti, J. M. Moreno-Rojas, F. Caprino, I. Giani, R. Malandra, F. Bellagamba and C. Guillou. 2008. Authentication of farmed and wild turbot (*Psetta maxima*) by fatty acid and isotopic analyses combined with chemometrics. *Journal of Agricultural and Food Chemistry* 56: 2742–2750.

Camin, F., L. Bontempo, K. Heinrich, M. Horacek, S. D. Kelly, C. Schlicht, F. Thomas, F. Monahan, J. Hoogewerff and A. Rossmann. 2007. Multi-element (H, C, N, S) stable isotope characteristics of lamb meat from different European regions. *Analytical and Bioanalytical Chemistry* 389: 309–320.

Capo, R. C., B. W. Stewart and O. A. Chadwick. 1998. Strontium isotopes as tracers of ecosystem processes: Theory and methods. *Geoderma* 82: 197–225.

Carrijo, A. S., A. C. Pezzato, C. Ducatti, J. R. Sartori, L. Trinca and E. T. Silva. 2006. Traceability of bovine meat and bone meal in poultry by stable isotope analysis. *Brazilian Journal of Poultry Science* 8: 63–68.

Carter, J. F., U. Tinggi, X. Yang and B. Fry. 2015. Stable isotope and trace metal compositions of Australian prawns as a guide to authenticity and wholesomeness. *Food Chemistry* 170: 241–248.

Cerling, T. E., J. M. Harris, B. J. MacFadden, M. G. Leakey, J. Quade, V. Eisenmann and J. R. Ehleringer. 1997. Global vegetation change through the Miocene/Pliocene boundary. *Nature* 389: 153–158.

Chesson, L. A., J. Ehleringer and T. Cerling. 2009a. American fast food isn't all corn-based. *Proceedings of the National Academy of Sciences* doi:10.1073/pnas.0811787106.

Chesson, L. A., J. R. Ehleringer and T. E. Cerling. 2011a. Chapter 33. Light-element isotopes (H, C, N, and O) as tracers of human diet: A case study on fast food meals. pp. 707–723. *In*: M. Baskaran (ed.). *Handbook of Environmental Isotope Geochemistry*. Berlin: Springer-Verlag.

Chesson, L. A., D. W. Podlesak, T. E. Cerling and J. R. Ehleringer. 2009b. Evaluating uncertainty in the calculation of non-exchangeable hydrogen fractions within organic materials. *Rapid Communications in Mass Spectrometry* 23: 1275–1280.

Chesson, L. A., D. W. Podlesak, B. R. Erkkila, T. E. Cerling and J. R. Ehleringer. 2010. Isotopic consequences of consumer food choice: Hydrogen and oxygen stable isotope ratios in foods from fast food restaurants versus supermarkets. *Food Chemistry* 119: 1250–1256.

Chesson, L. A., D. W. Podlesak, A. H. Thompson, T. E. Cerling and J. R. Ehleringer. 2008. Variation of hydrogen, carbon, nitrogen, and oxygen stable isotope ratios in an American diet: Fast food meals. *Journal of Agricultural and Food Chemistry* 56: 4084–4091.

Chesson, L. A., L. O. Valenzuela, G. J. Bowen, T. E. Cerling and J. R. Ehleringer. 2011b. Consistent predictable patterns in the hydrogen and oxygen stable isotope ratios of animal proteins consumed by modern humans in the USA. *Rapid Communications in Mass Spectrometry* 25: 3713–3722.

Coletta, L. D., A. L. Pereira, A. A. D. Coelho, V. J. M. Savino, J. F. M. Menten, E. Correr, L. C. França and L. A. Martinelli. 2012. Barn vs. free-range chickens: Differences in their diets determined by stable isotopes. *Food Chemistry* 131: 155–160.

Cruz, V. C., P. C. Araujo, J. R. Sartori, A. C. Pezzato, J. C. Denadai, G. V. Polycarpo, L. H. Zanetti and C. Ducatti. 2012. Poultry offal meal in chicken: Traceability using the technique of carbon ($^{13}C/^{12}C$)- and nitrogen ($^{15}N/^{14}N$)-stable isotopes. *Poultry Science* 91: 478–486.

Cruz, V. C., C. Ducatti, A. C. Pezzato, D. F. Pinheiro, J. R. Sartori, J. C. Gonçalves and A. S. Carrijo. 2005. Influence of diet on assimilation and turnover of ^{13}C in the tissues of broiler chickens. *British Poultry Science* 46: 382–389.

Cruz, V. C., A. C. Pezzato, C. Ducatti, D. F. Pinheiro, J. R. Sartori and J. C. Gonçalves. 2004. Tracing metabolic routes of feed ingredients in tissues of broiler chickens using stable isotopes. *Poultry Science* 83: 1376–1381.

Curtis, J. M., G. W. Stunz, R. D. Overath and R. R. Vega. 2014. Otolith chemistry can discriminate signatures of hatchery-reared and wild spotted seatrout. *Fisheries Research* 153: 31–40.

Dempson, J. B. and M. Power. 2004. Use of stable isotopes to distinguish farmed from wild Atlantic salmon, *Salmo salar*. *Ecology of Freshwater Fish* 13: 176–184.

DeNiro, M. J. and S. Epstein. 1976. You are what you eat (plus a few permil): The carbon isotope cycle in food chains. *In: Geological Society of American Abstracts with Programs* 8: 834–835.

De Smet, S., A. Balcaen, E. Claeys, P. Boeckx and O. van Cleemput. 2004. Stable carbon isotope analysis of different tissues of beef animals in relation to their diet. *Rapid Communications in Mass Spectrometry* 18: 1227–1232.

Devincenzi, T., O. Delfosse, D. Andueza, C. Nabinger and S. Prache. 2014. Dose-dependent response of nitrogen stable isotope ratio to proportion of legumes in diet to authenticate lamb meat produced from legume-rich diets. *Food Chemistry* 152: 456–461.

Drivelos, S. A. and C. A. Georgiou. 2012. Multi-element and multi-isotope-ratio analysis to determine the geographical origin of foods in the European Union. *Trends in Analytical Chemistry* 40: 38–51.

Erasmus, S. W., M. Muller, M. van der Rijst and L. C. Hoffman. 2016. Stable isotope ratio analysis: A potential analytical tool for the authentication of South African lamb meat. *Food Chemistry* 192: 997–1005.

Evans, J. A., V. Pashley, G. J. Richards, N. Brereton and T. G. Knowles. 2015. Geogenic lead isotope signatures from meat products in Great Britain: Potential for use in food authentication and supply chain traceability. *Science of the Total Environment* 537: 447–452.

Fasolato, L., E. Novelli, L. Salmaso, L. Corain, F. Camin, M. Perini, P. Antonetti and S. Balzan. 2010. Application of nonparametric multivariate analyses to the authentication of wild and farmed sea bass (*Dicentrarchus labrax*). Results of a survey on fish sampled in the retail trade. *Journal of Agricultural and Food Chemistry* 58: 10979–10988.

Fernandes, R., J. Meadows, A. Dreves, M. -J. Nadeau and P. Grootes. 2014. A preliminary study on the influence of cooking on the C and N isotopic composition of multiple organic fractions of fish (mackerel and haddock). *Journal of Archaeological Science* 50: 153–159.

Franke, B. M., G. Germaud, R. Hadorn and M. Kreuzer. 2005. Geographic origin of meat —elements of an analytical approach to its authentication. *European Food Research and Technology* 221: 493–503.

Franke, B. M., S. Koslitz, F. Micaux, U. Piantini, V. Maury, E. Pfammatter, S. Wunderli et al. 2008a. Tracing the geographic origin of poultry meat and dried beef with oxygen and strontium isotope ratios. *European Food Research and Technology* 226: 761–769.

Franke, B. M., R. Hadorn, J. O. Bosset, G. Gremaud and M. Kreuzer. 2008b. Is authentication of the geographic origin of poultry meat and dried beef improved by combining multiple trace element and oxygen isotope analysis? *Meat Science* 80: 944–947.

Frei, K. M. 2014. Provenance of archaeological wool textiles: New case studies. *Open Journal of Archaeometry* 2: 1–5.

Frei, K. M., R. Frei, U. Mannering, M. Gelba, M. L. Nosch and H. Lyngstrøm. 2009. Provenance of ancient textiles—A pilot study evaluating the strontium isotope system in wool. *Archaeometry* 51: 252–276.

Fry, B., J. F. Carter, U. Tinggi, A. Arman, M. Kamal, M. Metian, V. A. Waduge and R. B. Yaccup. 2016. Prawn biomonitors of nutrient and trace metal pollution along Asia-Pacific coastlines. *Isotopes in Environmental and Health Studies* 52: 619–632.

Gamboa-Delgado, J., C. Molina-Poveda, D. E. Godínez-Siordia, D. Villarreal-Cavazos, D. Ricque-Marie, L. E. Cruz-Suárez and B. Gillanders. 2014. Application of stable isotope analysis to differentiate shrimp extracted by industrial fishing or produced through aquaculture practices. *Canadian Journal of Fisheries and Aquatic Sciences* 71: 1520–1528.

Gebbing, T., J. Schellberg and W. Kühbauch. 2004. Switching from grass to maize diet changes the C isotope signature of meat and fat during fattening of steers. *Proceedings of the 20th General Meeting of the European Grassland Federation* 9: 1130–1132.

Gibson-Reinemer, D. K., B. M. Johnson, P. J. Martinez, D. L. Winkelman, A. E. Koenig and J. D. Woodhead. 2009. Elemental signatures in otoliths of hatchery rainbow trout (*Oncorhynchus mykiss*): Distinctiveness and utility for detecting origins and movement. *Canadian Journal of Fisheries and Aquatic Sciences* 66: 513–524.

Gonzalez-Martin, I., C. Gonzalez-Perez, J. Henandez Mendez, E. Marques-Macias and F. Sanz Poveda. 1999. Use of isotope analysis to characterize meat from Iberian-breed swine. *Meat Science* 52: 437–441.

Gonzalez-Martin, I., C. Gonzalez-Perez, J. Hernandez Mendez and C. Sanchez Gonzalez. 2001. Differentiation of dietary regimen of Iberian swine by means of isotopic analysis of carbon and sulphur in hepatic tissue. *Meat Science* 58: 25–30.

Gonzalvez, A., S. Armenta and M. de la Guardia. 2009. Trace-element composition and stable-isotope ratio for discrimination of foods with protected designation of origin. *Trends in Analytical Chemistry* 28: 1295–1311.

Guo, B. L., Y. M. Wei, J. R. Pan and Y. Li. 2010. Stable C and N isotope ratio analysis for regional geographical traceability of cattle in China. *Food Chemistry* 118: 915–920.

Harrison, S. M., F. J. Monahan, A. P. Moloney, S. D. Kelly, J. Hoogewerff and O. Schmidt. 2010. Intra-muscular and inter-muscular variation in carbon turnover of ovine muscles as recorded by stable isotope ratios. *Food Chemistry* 123: 203–209.

Harrison, S. M., F. J. Monahan, A. Zazzo, B. Bahar, A. P. Moloney, C. M. Scrimgeour and O. Schmidt. 2007. Three-dimensional growth of bovine hoof as recorded by carbon stable isotope ratios. *Rapid Communications in Mass Spectrometry* 21: 39171–3976.

Harrison, S. M., O. Schmidt, A. P. Moloney, S. D. Kelly, A. Rossmann, A. Schellenberg, F. Camin, M. Perini, J. Hoogewerff and F. J. Monahan. 2011. Tissue turnover in ovine muscles and lipids as recorded by multiple (H, C, O, S) stable isotope ratios. *Food Chemistry* 124: 291–297.

Harrison, S. M., A. Zazzo, B. Bahar, F. J. Monahan, A. P. Moloney, C. M. Scrimgeour and O. Schmidt. 2007. Using hooves for high-resolution isotopic reconstruction of bovine dietary history. *Rapid Communications in Mass Spectrometry* 21: 479–486.

Heaton, K., S. D. Kelly, J. Hoogewerff and M. Woolfe. 2008. Verifying the geographical origin of beef: The application of multi-element isotope and trace element analysis. *Food Chemistry* 107: 506–515.

Horacek, M., E. Eisinger and W. Papesch. 2010. Reliability of stable isotope values from meat juice for the determination of the meat origin. *Food Chemistry* 118: 910–914.

Horacek, M. and J. -S. Min. 2010. Discrimination of Korean beef from beef of other origin by stable isotope measurements. *Food Chemistry* 121: 517–520.

Ingram, B. L. and P. K. Weber. 1999. Salmon origin in California's Sacramento-San Joaquin river system as determined by otolith strontium isotopic composition. *Geology* 27: 851–854.

Jahren, A. H. and R. Kraft. 2009. Reply to Chesson et al.: Carbon stable isotopes in beef differ distinctly between corporations. *Proceedings of the National Academy of Sciences* doi: 10.1073/pnas.0812302106.

Jahren, A. H. and R. A. Kraft. 2008. Carbon and nitrogen stable isotopes in fast food: Signatures of corn and confinement. *Proceedings of the National Academy of Sciences* 105: 17855–17860.

Jones, R. J., M. M. Ludlow, J. H. Troughton and C. G. Blunt. 1979. Estimation of the proportion of C_3 and C_4 plant species in the diet of animals from the ratio of natural ^{12}C and ^{13}C isotopes in the faeces. *The Journal of Agricultural Science* 92: 91–100.

Jones, R. J., M. M. Ludlow, J. H. Troughton and C. G. Blunt. 1981. Changes in the natural carbon isotope ratios of the hair from steers fed diets in C_4, C_3 and C_4 species in sequence. *SEARCH* 12: 85–87.

Kelly, S., K. Heaton and J. Hoogewerff. 2005. Tracing the geographical origin of food: The application of multi-element and multi-isotope analysis. *Trends in Food Science and Technology* 16: 555–567.

Kennedy, B. P., C. L. Folt, J. D. Blum and C. P. Chamberlain. 1997. Natural isotope markers in salmon. *Nature* 387: 766–767.

Kim, H., K. S. Kumar and K. -H. Shin. 2015a. Applicability of stable C and N isotope analysis in inferring the geographical origin and authentication of commercial fish (Mackerel, Yellow Croaker and Pollock). *Food Chemistry* 172: 523–527.

Kim, H., K. S. Kumar, S. Y. Hwang, B. -C. Kang, H. -B. Moon and K. -H. Shin. 2015b. Utility of stable isotope and cytochrome oxidase I gene sequencing analyses in inferring origin and authentication of Hairtail fish and shrimp. *Journal of Agricultural and Food Chemistry* 63: 5548–5556.

Kim, K. S., J. S. Kim, I. M. Hwang, I. S. Jeong, N. Khan, S. I. Lee, D. B. Jeon, Y. H. Song and K. S. Kim. 2013. Application of stable isotope ratio analysis for origin authentication of pork. *Korean Journal for Food Science of Animal Resources* 33: 39–44.

Kim, S. -H., G. D. Cruz, J. G. Fadel and A. J. Clifford. 2012. Food authenticity using natural carbon isotopes (^{12}C, ^{13}C, ^{14}C) in grass-fed and grain-fed beef. *Food Science and Biotechnology* 21: 295–298.

Kohn, M. J. 1996. Predicting animal $\delta^{18}O$: Accounting for diet and physiological adaptation. *Geochimica et Cosmochimica Acta* 60: 4811–4829.

Krivachy (Tanz), N., A. Rossmann and H. -L. Schmidt. 2015. Potentials and caveats with oxygen and sulfur stable isotope analyses in authenticity and origin checks of food and food commodities. *Food Control* 48: 143–150.

Layton, L. 2010. FDA pressured to combat rising "food fraud." *The Washington Post*. www.washingtonpost.com/wp-dyn/content/article/2010/03/29/AR2010032903824.html.

Li, L., C. E. Boyd and Z. Sun. 2016. Authentication of fishery and aquaculture products by multi-element and stable isotope analysis. *Food Chemistry* 194: 1238–1244.

Liu, X., B. Guo, Y. Wei, J. Shi and S. Sun. 2013. Stable isotope analysis of cattle tail hair: A potential tool for verifying the geographical origin of beef. *Food Chemistry* 140: 135–140.

Longobardi, F., D. Sacco, G. Casiello, A. Ventrella, A. Contessa and A. Sacco. 2012. Garganica kid goat meat: Physico-chemical characterization and nutritional impacts. *Journal of Food Composition and Analysis* 28: 107–113.

Martinelli, L. A., G. B. Nardoto, L. A. Chesson, F. D. Rinaldi, J. P. H. B. Ometto, T. E. Cerling and J. R. Ehleringer. 2011. Worldwide stable carbon and nitrogen isotopes of Big Mac® patties: An example of a truly "glocal" food. *Food Chemistry* 127: 1712–1718.

McCutchan, J. H., W. M. Lewis, C. Kendall and C. C. McGrath. 2003. Variation in trophic shift for stable isotope ratios of carbon, nitrogen, and sulfur. *Oikos* 102: 378–390.

Minson, D. J., M. M. Ludlow and J. H. Troughton. 1975. Differences in natural carbon isotope ratios of milk and hair from cattle grazing in tropical and temperate pastures. *Nature* 256: 602.

Molkentin, J., I. Lehmann, U. Ostermeyer and H. Rehbein. 2015. Traceability of organic fish— Authenticating the production origin of salmonids by chemical and isotopic analyses. *Food Control* 53: 55–66.

Molkentin, J., H. Meisel, I. Lehmann and H. Rehbein. 2007. Identification of organically farmed Atlantic salmon by analysis of stable isotopes and fatty acids. *European Food Research and Technology* 224: 535–543.

Monahan, F. J., A. P. Moloney, M. T. Osorio, F. T. Röhrle, O. Schmidt and L. Brennan. 2012. Authentication of grass-fed beef using bovine muscle, hair or urine. *Trends in Food Science & Technology* 28: 69–76.

Montowska, M. and E. Pospiech. 2012. Is authentication of regional and traditional food made of meat possible? *Critical Reviews in Food Science and Nutrition* 52: 475–487.

Moreno-Rojas, J. M., V. Vasta, A. Lanza, G. Luciano, V. Ladroue, C. Guillou and A. Priolo. 2008a. Stable isotopes to discriminate lambs fed herbage or concentrate both obtained from C_3 plants. *Rapid Communications in Mass Spectrometry* 22: 3701–3705.

Moreno-Rojas, J. M., F. Tulli, M. Messina, E. Tibaldi and C. Guillou. 2008b. Stable isotope ratio analysis as a tool to discriminate between rainbow trout (*O. mykiss*) fed diets based on plant and fish-meal proteins. *Rapid Communications in Mass Spectrometry* 22: 3706–3710.

Móri, C., E. A. Garcia, C. Ducatti, J. C. Denadai, K. Pelícia, R. Gottmann, A. O. M. Mituo and A. M. Bordinhon. 2007. Traceability of animal byproducts in quail (*Coturnix coturnix japonica*) tissues using carbon ($^{13}C/^{12}C$) and nitrogen ($^{15}N/^{14}N$) stable isotopes. *Brazilian Journal of Poultry Science* 9: 263–269.

Morrison, D. J., T. Preston, J. E. Bron, R. J. Hemderson, K. Cooper, F. Strachan and J. G. Bell. 2007. Authenticating production origin of gilthead sea bream (*Sparus aurata*) by chemical and isotopic fingerprinting. *Lipids* 42: 537–545.

Muhlfeld, C. C., S. R. Thorrold, T. E. McMahon and B. Marotz. 2012. Estimating westslope cutthroat trout (*Oncorhynchus clarkii lewisi*) movements in a river network using strontium isotopes. *Canadian Journal of Fisheries and Aquatic Sciences* 69: 906–915.

Nakashita, R., Y. Suzuki, F. Akamatsu, Y. Iizumi, T. Korenaga and Y. Chikaraishi. 2008. Stable carbon, nitrogen, and oxygen isotope analysis as a potential tool for verifying geographical origin of beef. *Analytica Chimica Acta* 617: 148–152.

Nardoto, G. B., P. B. de Godoy, E. S. de B. Ferraz, J. P. H. B. Ometto and L. A. Martinelli. 2006. Stable carbon and nitrogen isotopic fractionation between diet and swine tissues. *Scientia Agricola* 63: 579–582.

O'Grady, S. P., L. O. Valenzuela, C. H. Remien, L. E. Enright, M. J. Jorgensen, J. R. Kaplan, J. D. Wagner, T. E. Cerling and J. R. Ehleringer. 2012. Hydrogen and oxygen isotope ratios in body water and hair: Modeling isotope dynamics in nonhuman primates. *American Journal of Primatology* 74: 651–660.

O'Leary, M. H. 1988. Carbon isotopes in photosynthesis. *Bioscience* 38: 328–336.

Oliveira, E. J. V. M., L. S. Sant'Ana, C. Ducatti, J. C. Denadai and C. R. de S. Kruliski. 2011. The use of stable isotopes for authentication of gadoid fish species. *European Food Research and Technology* 232: 97–101.

Oliveira, R. P., C. Ducatti, A. C. Pezzato, J. R. Denadai, V. C. Cruz, J. R. Sartori, A. S. Carrijo and F. R. Caldara. 2010. Traceability of poultry offal meal in broiler feeding using isotopic analysis ($\delta^{13}C$ and $\delta^{15}N$) of different tissues. *Revista Brasileira de Ciência Avícola* 12: 13–20.

Ortea, I. and J. M. Gallardo. 2015. Investigation of production method, geographical origin and species authentication in commercially relevant shrimps using stable isotope ratio and/or multi-element analyses combined with chemometrics: An exploratory analysis. *Food Chemistry* 170: 145–153.

Osorio, M. T., G. Downey, A. P. Moloney, F. T. Röhrle, G. Luciano, O. Schmidt and F. J. Monahan. 2013. Beef authentication using dietary markers: Chemometric selection and modelling of significant beef biomarkers using concatenated data from multiple analytical methods. *Food Chemistry* 141: 2795–2801.

Osorio, M. T., A. P. Moloney, O. Schmidt and F. J. Monahan. 2011a. Multielement isotope analysis of bovine muscle for determination of international geographical origin of meat. *Journal of Agricultural and Food Chemistry* 59: 3285–3294.

Osorio, M. T., A. P. Moloney, O. Schmidt and F. J. Monahan. 2011b. Beef authentication and retrospective dietary verification using stable isotope ratio analysis of bovine muscle and tail hair. *Journal of Agricultural and Food Chemistry* 59: 3295–3305.

Ostermeyer, U., J. Molkentin, I. Lehmann, H. Rehbein and H. -G. Walte. 2014. Suitability of instrumental analysis for the discrimination between wild-caught and conventionally and organically farmed shrimps. *European Food Research and Technology* 239: 1015–1029.

Padilla, A. J., R. J. Brown and M. J. Wooller. 2015. Strontium isotope analyses ($^{87}Sr/^{86}Sr$) of otoliths from anadromous Bering cisco (*Coregonus laurettae*) to determine stock composition. *ICES Journal of Marine Science: Journal Du Conseil* 72: 2110–2117.

Perini, M. 2013. Stable isotope ratio analysis for the traceability of raw and cured meat. Ph.D., Universita Degli Studi di Undine.

Perini, M., F. Camin, L. Bontempo, A. Rossmann and E. Piasentier. 2009. Multielement (H, C, N, O, S) stable isotope characteristics of lamb meat from different Italian regions. *Rapid Communications in Mass Spectrometry* 23: 2573–2585.

Perini, M., F. Camin, J. Sánchez del Pulgar and E. Piasentier. 2013. Effects of origin, breeding and processing conditions on the isotope ratios of bioelements in dry-cured ham. *Food Chemistry* 136: 1543–1550.

Piasentier, E., R. Valusso, F. Camin and G. Versini. 2003. Stable isotope ratio analysis for authentication of lamb meat. *Meat Science* 64: 239–247.

Podlesak, D. W., G. J. Bowen, S. P. O'Grady, T. E. Cerling and J. R. Ehleringer. 2012. δ^2H and $\delta^{18}O$ of human body water: A GIS model to distinguish residents from non-residents in the contiguous USA. *Isotopes in Environmental and Health Studies* 48: 259–279.

Podlesak, D. W., A. -M. Torregrossa, J. R. Ehleringer, M. D. Dearing, B. H. Passey and T. E. Cerling. 2008. Turnover of oxygen and hydrogen isotopes in the body water, CO_2, hair, and enamel of a small mammal. *Geochimica et Cosmochimica Acta* 72: 19–35.

Post, D. M. 2002. Using stable isotopes to estimate trophic position: Models, methods and assumptions. *Ecology* 83: 702–718.

Prache, S. 2007. Developing a diet authentication system from the composition of milk and meat in sheep: A review. *The Journal of Agricultural Science* 145: 435–444.

Prache, S., A. Cornu, J. L. Berdagué and A. Priolo. 2005. Traceability of animal feeding diet in the meat and milk of small ruminants. *Small Ruminant Research* 59: 157–168.

Prache, S., N. Kondjoyan, O. Delfosse, B. Chauveau-Duriot, D. Andueza and A. Cornu. 2009. Discrimination of pasture-fed lambs from lambs fed dehydrated alfalfa indoors using different compounds measured in the fat, meat and plasma. *Animal* 3: 598–605.

Recio, C., Q. Martín and C. Raposo. 2013. GC-C-IRMS analysis of FAMEs as a tool to ascertain the diet of Iberian pigs used for the production of pork products with high added value. *Grasas y Aceites* 64: 181–190.

Rees, G., S. D. Kelly, P. Cairns, H. Ueckermann, S. Hoelzl, A. Rossmann and M. J. Scotter. 2016. Verifying the geographical origin of poultry: The application of stable isotope and trace element (SITE) analysis. *Food Control* 67: 144–154.

Renou, J. -P., G. Bielicki, C. Deponge, P. Gachon, D. Micol and P. Ritz. 2004. Characterization of animal products according to geographic origin and feeding diet using nuclear magnetic resonance and isotope ratio mass spectrometry. Part II: Beef meat. *Food Chemistry* 86: 251–256.

Rhodes, C. N., J. H. Lofthouse, S. Hird, P. Rose, P. Reece, J. Christy, R. Macarthur and P. A. Bereton. 2010. The use of stable carbon isotopes to authenticate claims that poultry have been corn-fed. *Food Chemistry* 118: 927–932.

Richards, M. P., B. T. Fuller, M. Sponheimer, T. Robinson and L. Ayliffe. 2003. Sulphur isotopes in palaeodietary studies: A review and results from a controlled feeding experiment. *International Journal of Osteology* 13: 37–45.

Robbins, C. T., L. A. Felicetti and S. T. Florin. 2010. The impact of protein quality on stable nitrogen isotope ratio discrimination and assimilated diet estimation. *Oecologia* 162: 571–579.

Robbins, C. T., L. A. Felicetti and M. Sponheimer. 2005. The effect of dietary protein quality on nitrogen isotope discrimination in mammals and birds. *Oecologia* 144: 534–540.

Rossier, J. S., V. Maury, B. de Voogd and E. Pfammatter. 2014. Use of isotope ratio mass spectrometry (IRMS) determination ($^{18}O/^{16}O$) to assess the local origin of fish and asparagus in Western Switzerland. *CHIMIA International Journal for Chemistry* 68: 696–700.

Rummel, S., C. H. Dekant, S. Hölzl, S. D. Kelly, M. Baxter, N. Marigheto, C. R. Quetel et al. 2012. Sr isotope measurements in beef—analytical challenge and first results. *Analytical and Bioanalytical Chemistry* 402: 2837–2848.

Sacco, D., M. A. Brescia, A. Buccolieri and A. Caputi Jambrenghi. 2005. Geographical origin and breed discrimination of Apulian lamb meat samples by means of analytical and spectroscopic determinations. *Meat Science* 71: 542–548.

Sakamoto, N., T. Ishida, T. Arima, K. Idemitsu, Y. Inagaki, H. Furuya, H. Kawamura, N. Matsuoka and S. Tawaki. 2002. Concentrations of radiocarbon and isotope compositions of stable carbon in food. *Journal of Nuclear Science and Technology* 39: 323–328.

Sant'Ana, L. S., C. Ducatti and D. G. Ramires. 2010. Seasonal variations in chemical composition and stable isotopes of farmed and wild Brazilian freshwater fish. *Food Chemistry* 122: 74–77.

Schmidt, H. -L., A. Rossmann, S. Rummel and N. Tanz. 2008. Stable Isotope Analysis for Meat authenticity and origin check. pp. 767–787. *In*: L. Nollet and F. Toldrá (eds.). *Handbook of Muscle Foods Analysis*. CRC Press.

Schmidt, O., J. Quilter, B. Bahar, A. Moloney, C. Scrimgeour, I. Begley and F. Monahan. 2005. Inferring the origin and dietary history of beef from C, N and S stable isotope ratio analysis. *Food Chemistry* 91: 545–549.

Schwertl, M., K. Auerswald, R. Schaufele and H. Schnyder. 2005. Carbon and nitrogen stable isotope composition of cattle hair: Ecological fingerprints of production systems? *Agriculture, Ecosystems & Environment* 109: 153–165.

Schwertl, M., K. Auerswald and H. Schnyder. 2003. Reconstruction of the isotopic history of animal diets by hair segmental analysis. *Rapid Communications in Mass Spectrometry* 17: 1312–1318.

Sernagiotto, E., C. Ducatti, J. Sartori, A. Stradiotti, M. Maruno, P. Araujo, F. Carvalho and A. Pezzato. 2013. The use of carbon and nitrogen stable isotopes for the detection of poultry offal meal in meat-type quail feeds. *Revista Brasileira de Ciência Avícola* 15: 65–70.

Serrano, R., M. A. Blanes and L. Orero. 2007. Stable isotope determination in wild and farmed gilthead sea bream (*Sparus aurata*) tissues from the western Mediterranean. *Chemosphere* 69: 1075–1080.

Sun, S., B. Guo, Y. Wei and M. Fan. 2012. Classification of geographical origins and prediction of $\delta^{13}C$ and $\delta^{15}N$ values of lamb meat by near infrared reflectance spectroscopy. *Food Chemistry* 135: 508–514.

Thiem, I., M. Lüpke and H. Seifert. 2004. Factors influencing the $^{18}O/^{16}O$-ratio in meat juices. *Isotopes in Environmental and Health Studies* 40: 191–197.

Thode, H. G. 1991. Sulphur isotopes in nature and the environment: An overview. *In*: H. R. Krouse and V. A. Grinenko (eds.). *Stable Isotopes in the Assessment of Natural and Anthropogenic Sulphur in the Environment*. John Wiley & Sons, Ltd.

Thomas, F., E. Jamin, K. Wietzerbin, R. Guérin, M. Lees, E. Morvan, I. Billault et al. 2008. Determination of origin of Atlantic salmon (*Salmo salar*): The use of multiprobe and multielement isotopic analyses in combination with fatty acid composition to assess wild or farmed origin. *Journal of Agricultural and Food Chemistry* 56: 989–997.

Tipple, B. J. and M. Pagani. 2007. The early origins of terrestrial C_4 photosynthesis. *Annual Review of Earth and Planetary Sciences* 35: 435–461.

Trembaczowski, A. 2011. Use of sulphur and carbon stable-isotope composition of fish scales and muscles to identify the origin of fish. *Mineralogia* 42: 33–37.

Turchini, G. M., G. P. Quinn, P. L. Jones, G. Palmeri and G. Gooley. 2009. Traceability and discrimination among differently farmed fish: A case study on Australian Murray cod. *Journal of Agricultural and Food Chemistry* 57: 274–281.

Vinci, G., R. Preti, A. Tieri and S. Vieri. 2012. Authenticity and quality of animal origin food investigated by stable-isotope ratio analysis. *Journal of the Science of Food and Agriculture* 93: 439–448.

Whitledge, G. W. 2009. Otolith microchemistry and isotopic composition as potential indicators of fish movement between the Illinois River drainage and Lake Michigan. *Journal of Great Lakes Research* 35: 101–106.

Whitledge, G. W., B. M. Johnson and P. J. Martinez. 2006. Stable hydrogen isotopic composition of fishes reflects that of their environment. *Canadian Journal of Fisheries and Aquatic Sciences* 63: 1746–1751.

Wolff, B. A., B. M. Johnson, A. R. Breton, P. J. Martinez, D. L. Winkelman and B. Gillanders. 2012. Origins of invasive piscivores determined from the strontium isotope ratio ($^{87}Sr/^{86}Sr$) of otoliths. *Canadian Journal of Fisheries and Aquatic Sciences* 69: 724–739.

Yanagi, Y., H. Hirooka, K. Oishi, Y. Choumei, H. Hata, M. Arai, M. Kitagawa, T. Gotoh, S. Inada and H. Kumagai. 2012. Stable carbon and nitrogen isotope analysis as a tool for inferring beef cattle feeding systems in Japan. *Food Chemistry* 134: 502–506.

Zazzo, A., M. Balasse, B. H. Passey, A. P. Moloney, F. J. Monahan and O. Schmidt. 2010. The isotope record of short- and long-term dietary changes in sheep tooth enamel: Implications for quantitative reconstruction of paleodiets. *Geochimica et Cosmochimica Acta* 74: 3571–3586.

Zazzo, A., M. Balasse and W. P. Patterson. 2006. The reconstruction of mammal individual history: Refining high-resolution isotope record in bovine tooth dentine. *Journal of Archaeological Science* 33: 1177–1187.

Zazzo, A., T. E. Cerling, J. R. Ehleringer, A. P. Moloney, F. J. Monahan and O. Schmidt. 2015. Isotopic composition of sheep wool records seasonality of climate and diet. *Rapid Communications in Mass Spectrometry* 29: 1357–1369.

Zazzo, A., S. M. Harrison, B. Bahar, A. P. Moloney, F. J. Monahan, C. M. Scrimgeour and O. Schmidt. 2007. Experimental determination of dietary carbon turnover in bovine hair and hoof. *Canadian Journal of Zoology* 85: 1239–1248.

Zazzo, A., A. P. Moloney, F. J. Monahan, C. M. Scrimgeour and O. Schmidt. 2008. Effect of age and food intake on dietary carbon turnover recorded in sheep wool. *Rapid Communications in Mass Spectrometry* 22: 2937–2945.

Zazzo, A., F. J. Monahan, A. P. Moloney, S. Green and O. Schmidt. 2011. Sulphur isotopes in animal hair track distance to the sea. *Rapid Communications in Mass Spectrometry* 25: 2371–2378.

Zeigler, J. M. and G. W. Whitledge. 2010. Assessment of otolith chemistry for identifying source environment of fishes in the lower Illinois River, Illinois. *Hydrobiologia* 638: 109–119.

Zhao, Y., B. Zhang, B. Guo, D. Wang and S. Yang. 2016a. Combination of multi-element and stable isotope analysis improved the traceability of chicken from four provinces of China. *CyTA—Journal of Food* 14: 163–168.

Zhao, Y., S. Yang and D. Wang. 2016b. Stable carbon and nitrogen isotopes as a potential tool to differentiate pork from organic and conventional systems. *Journal of the Science of Food and Agriculture* 96: 3950–3955.

Zhao, Y., B. Zhang, G. Chen, A. Chen, S. Yang and Z. Ye. 2013. Tracing the geographic origin of beef in China on the basis of the combination of stable isotopes and multielement analysis. *Journal of Agricultural and Food Chemistry* 61: 7055–7060.

Zhao, Y., B. Zhang, G. Chen, A. Chen, S. Yang and Z. Ye. 2014. Recent developments in application of stable isotope analysis on agro-product authenticity and traceability. *Food Chemistry* 145: 300–305.

Zhou, J., B. Guo, Y. Wei, G. Zhang, S. Wei and Y. Ma. 2015. The effect of different cooking processes on stable C, N and H isotopic compositions of beef. *Food Chemistry* 182: 23–26.

CHAPTER 7

Fruits and Vegetables

James F. Carter

7.1 Introduction

"I'm President of the United States and I'm not going to eat any more broccoli."

George H. W. Bush, U.S. President, 1990

Of all the foodstuffs discussed in this volume, fruits and vegetables have perhaps the lowest commercial values—with a few exceptions, such as coffee—and very likely for this reason comparatively little research exists into the authentication of fruits and vegetables. Aside from economic reasons, anyone who has bought fruits or vegetables at the store will know that the nature and quality are easily judged by appearance, feel, and aroma. At a more technical level, DNA technology now provides a ready (and increasingly cheaper) means to identify species, varieties, and cultivars.

Despite a low unit price, the vast scale of some commodity markets makes the extension of products such as orange juice a temping proposition and many early publications sought to develop methods to detect the adulteration of fruit juices. More recent publications have focused largely on fruits and vegetables at the premium price point of the market, such as basmati rice, coffee, pistachios, etc., and applied stable isotope, elemental, and chemical analysis techniques in attempts to link these characteristics to countries or regions of origin. Due to increased consumer awareness and choice, a significant body of research also exists concerning the authentication of produce branded *organic*, as described in detail in Chapter 12.

Forensic and Scientific Services, Health Support Queensland, 39 Kessels Road, Coopers Plains QLD 4108, AUSTRALIA.
Email: Jim.Carter@health.qld.gov.au
With contributions from Karyne M. Rogers

7.2 Fruits, vegetables, and isotopes

Chapter 3 provides a comprehensive review of the processes that bring about the characteristic stable isotope ratios within plants. The following is a brief summary relevant to the authentication of fruits and vegetables and especially fruit juices.

Water is the most abundant component of all fruits and vegetables, varying from about 82% in grapes to 92% in strawberries. The water in most foods is ultimately derived from groundwater and variations in groundwater δ^2H and $\delta^{18}O$ values will be reflected in plant water and tissues. Early studies, however, showed that the waters in fruits and vegetables are enriched in both ^2H and ^{18}O with respect to the water in which the plants were grown (e.g., Dunbar and Wilson 1983), as illustrated in Table 7.1. The source of the enrichment is universally accepted as being a process of evapotranspiration—migration of water across the leaf/air boundary—producing a characteristic plot of δ^2H *vs.* $\delta^{18}O$ with a slope of approximately 2.5, whereas meteoric waters have a slope of approximately 8. In contrast, plots of δ^2H *vs.* $\delta^{18}O$ for orange and grape juices were found to have a slope of approximately 4, which was attributed to a combination of evapotranspiration and equilibration between fruit-water and atmospheric moisture. Research has demonstrated a strong correlation between the

Table 7.1 The carbon, oxygen, and hydrogen isotopic compositions of commercial sugars and fruits. Based on (a) Krueger 1998, (b) Jamin et al. 1998, (c) Dunbar and Wilson 1983, and (d) Martin et al. 1991.

Source	Sugar $\delta^{13}C$, ‰[a]	Protein $\delta^{13}C$, ‰[b]	Sugar $\delta^{18}O$, ‰[c]	(D/H)$_I$[d]	(D/H)$_{II}$[d]
corn (maize) syrup	−12.4 to −9.5			110.6	122.9
cane sugar	−12.2 to −10.3			112.0	127.7
beet sugar	−28.0 to −26.7			92.7	124.1
apple	−27.9 to −22.5		−0.4		
cranberry	−26.3 to −24.1				
grape	−30.5 to −23.5		−2.4		
grapefruit	−28.0 to −24.6	−28.5 to −26.9	+1.8		
lemon	−26.3 to −24.1	−27.2 to −24.6	−2.5	97.2 [d]	125.2 [d]
orange	−28.1 to −23.5	−27.4 to −24.6	−1.5		
pineapple	−14.3 to −11.2	−13.9 to −11.1			
raspberry	−26.1 to −21.5				
strawberry	−25.7 to −23.1	−26.0 to −23.4	−0.5		
local water			−7.5		

a: Values from the original reference are cited *vs.* PDB, not VPDB.
c: Values from the original reference are cited *vs.* SMOW, not VSMOW.
d: The source of these values in the original reference is given as "fruit."

$\delta^{18}O$ composition of water and organic matter within orange juice and that neither is affected by changes in precipitation in the few days preceding harvest (Jamin et al. 2003). It has also been shown that similar varieties of fruit, grown in the same region, have similar δ^2H and $\delta^{18}O$ compositions, whereas geographical effects were much larger—i.e., similar varieties grown in different areas had different δ^2H and $\delta^{18}O$ compositions, primarily driven by latitude and available precipitation (Dunbar and Wilson 1983). The same authors proposed a general trend that fast-growing vegetables (such as zucchini, cucumbers, and watermelons) did not exhibit as much ^{18}O enrichment as slow-growing fruits (such as apples, pears, and plums). A more recent study of fast-growing vegetables and slow-growing fruit from the same, small orchard recorded the opposite effect (Bong et al. 2008), whereby fast-growing vegetables were subject to greater evaporation than slow-growing fruits.

Carbohydrates are the next most abundant component of fruits and vegetables and comprise approximately 10% of juice, typically expressed in degrees Brix ($1\,^{\circ}Bx = 1$ g of sugar per 100 g of solution). The carbohydrates present are mostly polysaccharides such as starch, cellulose, and pectin, although simple saccharides such as sucrose, fructose, and dextrose occur in all fruits and become more abundant when the fruit is mature. The polyalcohol sorbitol is also abundant in fruits such as pears and plums, accounting for their laxative properties!

Proteins and lipids each comprise less than 1% of most of fruits and vegetables; probably for this reason these components have been studied less as a means of authentication. Minerals in fruits and vegetables (mostly potassium, calcium, and magnesium) typically occur in combination with organic acids and, because these are readily water soluble, juices will contain most of the minerals originally present in the fruit or vegetable.

7.3 Fruit juices

Although the term *fruit juice* may appear obvious, legal definitions exist, such as Council Directive 2001/112/EC:

> "The fermentable but unfermented product obtained from the edible part of fruit which is sound and ripe, fresh or preserved by chilling or freezing of one or more kinds mixed together having the characteristic colour, flavour and taste typical of the juice of the fruit from which it comes."

The process of converting fruit into juice can be summarized as: (1) removing foreign matter and diseased fruit, (2) washing, and (3) pressing (Ashurst 2009), although this process is modified in many ways to accommodate the physical characteristics of different fruits. For example, apples, pears, and red/black berries are milled and treated with pectolytic enzyme to break down cell walls and release the juice; citrus fruits are processed to release essential oils before the fruit is pressed; and stone fruits are heated to soften the flesh and facilitate removal of the stone. After pressing, the pulp may be washed and pressed several times to produce *pulp-wash*, which can be added back to the juice. Historically, the use of pulp-wash has been controversial and the subject of many authentication studies, but using an in-line process to add back about 5% of the pulp-wash to what is still regarded as an authentic juice is now widely accepted and legal:

> "Flavour, pulp, and cells obtained by suitable physical means from the same species of fruit may be restored to the juice. In the case of citrus fruits, the fruit juice must come from the endocarp. Lime juice, however, may be obtained from the whole fruit" (2001/112/EC).

Most orange and grapefruit juices have solid contents between 8–15°Bx and in order to prepare a product that is economically viable to freeze and transport, these juices are concentrated to between 60–66°Bx. Apple juice is typically concentrated to 70°Bx but is shelf-stable without being frozen. During the concentration process essential oils, which contribute to the character of the juice, are removed along with water and are frequently recovered and added back to the juice.

Although much of the early food forensic detective work aimed to identify the illegal dilution of fruit juices with groundwater and/or sugar, legislation now allows for the dilution of many fruit juices to produce *fruit nectars*, which may be "...obtained by adding water with or without the addition of sugars and/or honey ... to fruit purée and/or to concentrated fruit purée" (2001/112/EC):

> "Fruits with acidic juice unpalatable in the natural state," such as cranberries, passion fruit, lemons, and limes, may be diluted to a final juice content of between 25 to 50%,
>
> "Low-acid, pulpy or highly flavoured fruits with juice unpalatable in the natural state," such as bananas, mangoes, and pomegranates, may be diluted to a final juice content of 25%, and
>
> "Fruits with juice palatable in the natural state," such as apples, pineapples, and tomatoes, may be diluted to a final juice content of 50%.

7.3.1 Detecting the adulteration of fruits juices

Today, packaged fruit juices are available to retail customers in virtually every country of the world thanks to plentiful production and modern packaging technology. The production of fruit juices has increased steadily since the mid 1900s, dominated by the production of concentrates, which are reconstituted by the addition of water immediately prior to final packaging. The majority of the products available on supermarket shelves are *long-life* juices—typically packaged in aseptic cartons, which negate the need for heat treatment. An alternative but rapidly growing sector of the market relies on chill-chain distribution to supply *short-life* juices, prepared with minimum processing and often specifically designated *not-from-concentrate* (NFC).

The price of orange juice concentrate is typically two to three times the price of sugar and, on a commercial scale, there is significant profit to be made from substituting even a few percent of the juice solids with sugar, especially cane sugar and High Fructose Corn Syrup (HFCS), both of which are cheap and readily available. A UK government survey in 1991 concluded that there was widespread and systematic adulteration of the orange juice sold in the UK (MAFF 1991). Other fraudulent activities include the adulteration of citrus juices with pulp-wash or the extension of expensive red/black berry juices (raspberry, black current, etc.) with less expensive juices such as apple, pear, or grape.

Typically the adulteration of fruit juices occurs at the packaging facility, and once detected, there are often records that can substantiate whether the amounts of packaged product leaving the factory were consistent with the amounts of raw material entering.

7.3.2 Chemical profiling of fruit juices

Many analytical and statistical methods have been proposed to detect the various methods used to adulterate fruit juices. Historically, the concentrations of various components were measured and compared against a database derived from authentic, pure juices. The European Fruit Juice Association (AIJN) Code of Practice consists of a set of reference guidelines describing, in detail, the characteristic natural components of industrially processed fruits, including stable isotopic composition. Unfortunately, the natural concentration ranges for all of these analytes can vary widely, meaning that adulterated juice can be identified as genuine and *vice versa*; for example, it can be impossible to distinguish juice with added pulp-wash from juice that has been extracted at too high a pressure.

In an ideal world, there would be a single *marker compound* and the concentration of this compound would determine the purity of a foodstuff. In practice this will never be true, but the search for such a compound illustrates the cat-and-mouse game played by food forensic detectives and fraudsters. For example, the concentration of malic acid in fruit juices varies considerably but the acid, naturally present, exists exclusively as the L-enantiomer. If D-malic acid is detected in a juice it is evidence for the addition of industrially manufactured D,L-malic acid; the method was successfully used to detect the adulteration of apple, grapefruit, orange, and pineapple juices in the early 1990s. Unfortunately, once fraudsters became aware of this technique, they simply switched to adding commercially available, high purity L-malic acid, albeit more expensive than the racemic mix. Tartaric acid is another organic acid that provides a marker for the addition of grape juice as it is largely absent from other fruit juices. Unfortunately, commercially available *rectified grape juice* has the tartaric acid removed.

Most juices have characteristic distributions of glucose, fructose, and sucrose, and these profiles can be used as a basis to detect the addition of sugars or other juices. For example, raspberry juice naturally contains very low levels of sucrose and, therefore any sucrose present must come from an external source. Again, a knowledgeable fraudster can use a mixture of sugars to disguise the addition—for example, Beet Medium Invert Syrup (BMIS), made by hydrolysis of beet sucrose, contains one part glucose, one part fructose, and two parts sucrose, a very similar profile to the sugars naturally present in orange juice. Although the addition of these types of syrups can sometimes be detected by the presence of oligosaccharides, these complex molecules can also form during heat treatment of juices. One promising compound is sorbitol, which is present in apple and pear juice but largely absent from red/black berry juices and, therefore, provides a marker for the extension of the more expensive juices.

Because of the increasing skill of the fraudster it is often considered good practice to perform analyses of multiple components in order to detect potential fraud—e.g., measuring the profiles of sugars, organic acids, cations (K, Na, Mg, Ca), anions (Cl, SO_4, PO_4, NO_3), amino acids, polyphenols, polymethoxyflavones, carotenoids, anthrocyanins, etc. (Hammond 1996). As noted above, the natural concentration range for all of these compounds can vary widely but a systematic change across a large number of analytes must arouse suspicion.

7.3.3 Detecting the addition of cane/corn sugars

Carbon isotope ratio measurements are frequently used to identify the source of sugars in food products and, for almost four decades, have been used to detect the addition of cane/corn (maize) sugar (C_4 plants) to juice from C_3 fruits such as oranges and grapefruits (e.g., Bricout and Fontes 1974; Smith 1975). Table 7.1 shows the typical ranges for $\delta^{13}C$ compositions of commercial sugars and fruit juices. It is apparent that fruit juices (with the exception of pineapple juice) are significantly depleted in ^{13}C with respect to cane/corn sugars—the upper range for fruits being −21.5‰ and the lower range for commercial sugars being −12.4‰. This difference forms the basis of the AOAC Official Methods for the detection of corn syrup, including HFCS, in apple juice (AOAC 981.09) and orange juice (AOAC 982.21). These methods define the average $\delta^{13}C$ composition of apple and orange juices as −25.3 and −24.5‰, respectively, and the average composition of corn syrup as −9.7‰. [Unfortunately, even in the latest edition (20th Ed., 2016) these methods reference carbon isotope ratios *vs.* PDB and not VPDB.]

From these defined values it is possible to estimate corn syrup addition by simple mass balance, based on Equation 7.1. Samples for which $\%C_{CORN}$ is greater than 7% would be deemed suspect.

$$\%C_{CORN} = \frac{(\delta^{13}C_{JUICE} - \delta^{13}C_3)}{(\delta^{13}C_4 - \delta^{13}C_3)} \times 100 \qquad \text{Equation 7.1}$$

The detection of sugar addition to pineapple juice is less straightforward, as this plant fixes carbon via Crassulacean Acid Metabolism (CAM) and produces sugars with a carbon isotopic composition of approximately −12‰ (Jamin et al. 1997). One group of researchers (González et al. 1999) has separated the organic acids from pineapple juice by anion exchange and isolated the individual sugars (fructose, glucose, and sucrose) by preparative HPLC for $\delta^{13}C$ analysis. This research concluded that pineapple juice was characterized by the differences between the isotopic compositions of the individual sugars.

In efforts to improve the detection of exogenous sugars in fruit juices a number of researchers have proposed methods in which the isotopic composition of the sugars are compared to other components of the juice—an approach coined *internal standardization* or *intermolecular reference*—rather than to a different, albeit authentic, juice. Early work compared the carbon isotopic composition of sugars from oranges to that of the lipid-free dried pulp (Bricout and Koziet 1987). In this study $\Delta^{13}C_{SUGAR-PULP}$ varied between −0.5 and +1.4‰, and the authors concluded that differences greater than 2‰ were indicative of adulteration with cane/corn sugar.

Subsequent analysis of the data presented by Bricout and Koziet (1987) using Least Absolute Deviation (LAD) Regression returns R^2 (observed *vs.* predicted) of 0.235, which does not appear to support a clear relationship between $\delta^{13}C_{SUGAR}$ and $\delta^{13}C_{PULP}$.

A subsequent inter-laboratory comparison, organized by the European Commission, tested the potential of the $\Delta^{13}C_{SUGAR-PULP}$ method to determine the addition of equal amounts of cane and beet sugars to orange, grapefruit, and pineapple juices (Rossmann et al. 1997). Based on results from 19 laboratories, the study concluded that this was an acceptable method for the detection of sugar addition to fruit juices and is still used by commercial laboratories. Proteins have also been proposed as an intermolecular reference for adulteration (Jamin et al. 1998). The authors isolated protein from juices by bentonite precipitation and found a good correlation between the $\delta^{13}C$ composition of the protein and that of the sugar ($R^2 = 0.995$), Table 7.1. The authors noted that the $\delta^{13}C$ composition of the protein was less sensitive to industrial processes than the $\delta^{13}C$ composition of the pulp.

Research has also examined the $\delta^{13}C$ compositions of heterosides (mainly benzyl and 2-phethyl glycosides) present in concentrated rectified (grape) musts as an internal standard (Versini et al. 2006). The authors found that $\Delta^{13}C_{\text{SUGAR-HETEROSIDE}}$ ranged from 0.66 to 5.68‰, which was attributed to the aromatic part of the molecule. The authors then isolated the sugar fraction of the heterosides (mostly glucose) by chemical hydrolysis and found that this had a strong correlation ($R^2 = 0.765$) to the composition of the free sugar, albeit depleted by 1.7 ± 0.4‰.

7.3.4 Detecting the addition of beet sugars

Although the methods above provide a means to detect the addition of cane/corn sweeteners to fruit juice, it is not possible to distinguish fruit sugars naturally found in juice from those derived from sugar beet, as these are photosynthesized by the same C_3 mechanism. To avoid detection, the knowledgeable fraudster will use beet sugar for fruit juice extension, even though beet sugar is more expensive than cane/corn sweeteners.

Researchers have determined the δ^2H composition of cane, corn, and beet sugars as well as orange juice from North America, Israel, and France (Bricout and Koziet 1987). Prior to analysis, sugars were converted to nitrate esters, by treatment with nitric acid, to remove oxygen bound hydrogen (see Chapter 2 for a discussion of intrinsic *vs.* extrinsic hydrogen). For North American beet sugars the authors observed a gradient of ^2H concentration with latitude from $\delta^2H = -160$‰ for Canada to -109‰ for Texas; values for French and Israeli beet sugars fell within these extremes. In contrast, cane/corn sugars from the same regions were significantly enriched in ^2H with compositions ranging from -63 to -31‰. The authors concluded that consideration of both the $\delta^{13}C$ and δ^2H composition of sugars from orange juice provided a means to detect cane, corn, and beet sugars—the main commercial sources of adulterants.

The partitioning of ^2H (deuterium, D) within organic molecules is not a simple statistical distribution but is governed by botanical origin and possibly geographical location. The (D/H) ratios of the methyl [$(D/H)_I$] and methylene groups [$(D/H)_{II}$] of ethanol derived from fermentation are characteristic of the sugar substrate and can distinguish fruit sugars from beet sugar and BMIS (Martin et al. 1991). For the method to be effective, $>95\%$ of the sugars must be fermented to ethanol, which is then isolated by distillation and, again $>95\%$ must be recovered. Table 7.1 illustrates the differences in the D distribution of ethanol produced from cane, corn, and beet sugars plus sugars derived from fruit juices; the addition of beet sugar depresses the deuterium content at the methyl site whereas the addition of cane sugar has the opposite effect.

The AOAC has issued an Official Method of analysis (AOAC 995.17) for the measurement of (D/H) ratios in ethanol using Stable Natural Isotope Fractionation-NMR (SNIF-NMR; see Chapter 8). Inter-laboratory comparability of the method relies on the use of an internal standard of tetramethylurea (TMU), which is distributed by the European Commission

Institute for Reference Materials and Measurements (IRMM) as IRMM-425. An authentic sample of fruit juice is used to determine the natural ratio, $(D/H)_I$min.

If the ratio $(D/H)_I$ of a questioned juice (sample, or "samp") is greater than the authentic juice, no addition of beet sugar is assumed. If the ratio $(D/H)_I$ of a questioned juice is less than the authentic juice, the percentage of added beet sugar can be calculated according to Equation 7.2.

$$\%\text{Beet sugar} = \frac{(D/H)_I\text{min} - (D/H)_I\text{samp}}{(D/H)_I\text{min} - 92} \times 100 \qquad \text{Equation 7.2}$$

If added cane or corn sugar has been detected by $\delta^{13}C$ analysis, the value of $(D/H)_I$samp must be corrected for the percentage of C_4 sugar ($\%C_4$) according to Equation 7.3.

$$\text{corrected } (D/H)_I = (D/H)_I - (\%C_4 \times 100) \times [110 - (D/H)_I\text{min}] \qquad \text{Equation 7.3}$$

In practice these calculation can give a significant underestimation of the proportion of added sugar, especially when mixtures of C_3 and C_4 sugars are used; this will favor the defendant in any legal proceeding. Although the technique is effective in determining the addition of beet sugar to citrus juices, detection is reported to be less sensitive for juices grown in cooler climates such as apples and pears.

It has also been proposed that the internal D distribution can be used to identify citric acid manufactured from beet sugar or other C_3 plants (González et al. 1998). In this study, citric acid was isolated from lemon juices by preparative HPLC followed by conversion to triethyl citrate (TEC) and analysis by SNIF-NMR. Subsequently, a simplified method was proposed for the detection of exogenous citric acid measuring D/H ratio by IRMS (Jamin et al. 2005) after precipitation as the calcium salt. This study found that citric acid from citrus, pineapple, and red fruit juices was systematically enriched in ^2H compared to citric acid produced by fermentation of various sugar sources.

7.3.5 A case study

A recent court case was brought against Company J, who supplies "fresh orange juice" to the Australian market. Several suspect juices were tested for sugar adulteration and found to contain up to 50% added sugar based on their $\delta^{13}C$ values plotted along the carbon isotope mixing line of pure juice and cane sugar (Figure 7.1). The pure juice $\delta^{13}C$ end-member was determined by analyzing two composite samples of six freshly squeezed oranges from oranges grown near Company J's orchards, and the C_4 sugar end-member was taken as $-10.3‰$, consistent with cane sugar available in the Australian market. Furthermore, on testing the $\delta^{18}O$ values of the local tap water and the composite orange juice samples, it was evident that there had been some considerable addition of tap water or pulp-wash to the juice. Estimates through a two-point mixing model suggested up to 60% of the juice was not from local oranges, although it is possible that juices from other countries were combined to change the isotopic composition of the juice. Nonetheless, the labelling of the product was still misleading, as it was claimed to be "100% pure orange juice from Australian oranges." As a result, Company J was fined more than $125,000 (AUD) for misleading and deceptive conduct.

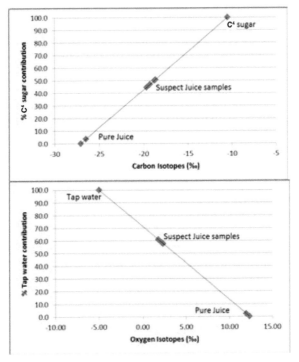

Figure 7.1 The carbon and oxygen isotopic compositions of pure orange juice (from freshly squeezed oranges), suspect orange juice, and cane sugar (top panel) or tap water (bottom panel) from southern Australia.

7.3.6 Authentication of NFC juices

As noted above, fruit-waters are naturally enriched in both ^{2}H and ^{18}O with respect to the plant growth water as a result of evapotranspiration and any addition of meteoric waters, which are almost universally depleted relative to fruit-water, will be evident in lower-than-expected δ^2H and δ^{18}O signatures (Nissenbaum et al. 1974). As a complication, the water in juice concentrate is further enriched relative to the starting juice, typically by +10 to +15‰ in δ^{18}O (Brause 1992), and, as a further complication, concentrates are often reconstituted in production facilities remote from the region in which the fruit was grown, possibly with groundwater with very different isotopic composition. Despite these obstacles, reconstituted juices will still be significantly depleted in ^{18}O compared to not-from-concentrate juices made purely of fruit-water. $\delta^{18}O$ composition has recently been proposed as a means to differentiate tomato passata from diluted tomato paste (Bontempo et al. 2014). The deuterium content of fruit juice has also been proposed as a marker for the addition of groundwater (Bricout and Koziet 1987). As a cautionary note (and possible further complication), discharge waters from large-scale juice concentration facilities would normally be considered as wastewater but can have significant commercial value if sold for the adulteration of juice concentrates.

Researchers claim to have improved the detection of reconstituted orange juice by separating sugars and citric acid from the water and measuring the $\delta^{18}O$ composition of the three components (Houerou et al. 1999). This study of single strength orange juice noted a correlation between $\delta^{18}O$ values of sugar, citric acid, and water; the isotopic composition of the sugars ranged from +29.1 to +38.8‰, the citric acid from +18.9 to +25.4‰, and the water from −2.1 to +7.8‰. Other researchers have reported a correlation between the $\delta^{13}C$ compositions of sugars, citric acid, and 1-malic acid (González et al. 1998).

More recent research has used HPLC-IRMS technology to measure the carbon isotopic composition of citric acid, glucose, and fructose in lemon and lime juices (Guyon et al. 2014). This study reported the average $\delta^{13}C$ value for authentic juices as −25.40 ± 1.62‰, −23.83 ± 1.82‰, and −25.67 ± 1.72‰ for citric acid, glucose, and fructose, respectively. The authors compared 30 commercial juices against these ranges and found that 10 samples showed indications of the addition of undeclared C_4 acids or sugars. Another recent innovation proposes the analysis of strontium isotope ratios to detect the blending of single strength orange juice with concentrates by comparing the $^{87}Sr/^{86}Sr$ ratio of the soluble and insoluble components (Rummel et al. 2010).

The δ^2H and $\delta^{18}O$ compositions of orange juice have also been proposed as a means to detect adulteration with beet sugar since this would typically be added as a syrup prepared using groundwater (Doner et al. 1987), and may also be indicative of the introduction of pulp-wash. The analysis of $\delta^{18}O$ composition provides the basis for an AOAC Official Method for the detection of sugar beet-derived syrups to frozen concentrated orange juice (AOAC 992.09). The method offers a very straightforward interpretation: "A $\delta^{18}O$ value < +8.9‰ for frozen concentrated orange juice indicates the presence of groundwater-prepared products (typically sugar-beet-derived syrups)." Unfortunately the method appears very much in need of updating as it requires $\delta^{18}O$ measurements to be expressed *vs*. PDB and recommends an unorthodox conversion factor of: $\delta^{18}O_{SMOW} = \delta^{18}O_{PDB} - 0.26‰$ (cf. Coplen et al. 1983).

An untargeted study of more than 300 orange juice samples by ^{1}H-NMR (Le Gall et al. 2001) identified the concentration of dimethylproline as a potential marker compound for the addition of pulp-wash.

7.4 Determining countries or regions of origin

7.4.1 Fruit juice and whole fruit

From the previous sections it will be apparent that considerable research effort has been expended to develop methods to identify the illicit extension of fruit juices; however, far less research appears to have been targeted at identifying the geographical origins of juices.

In 1987 Doner and Phillips reported a correlation between the growth location of sugar beets and both the δ^2H and δ^{18}O compositions of nitrated sugars (Doner and Phillips 1987) and in the same year Bricout and Koziet (1987) noted the relationship between latitude and the δ^2H composition of orange juice sugars. Although these trends doubtless exist, a simple plot of the data presented in the latter paper (Figure 7.2a) shows significant overlap between U.S. orange juices and Israeli orange juices prepared from two varieties of orange. Data derived from U.S. orange juices also overlap data from Brazilian orange juices. It will be evident from many chapters of this book that isotopic composition is governed by climate and precipitation more so than simple geography and it is common sense that oranges will be grown in regions with a suitable (and therefore similar) climate. A decade later Kornexl et al. (1996) proposed that a consideration of both the δ^{13}C and δ^{15}N compositions of citrus pulp allowed an improvement in geographical assignment accuracy. Again, a simple plot of these data (Figure 7.2b) reveals that although juices from southern and south America have somewhat characteristic isotopic signatures, there are no significant differences between juices from countries as disparate as Australia, Israel, Italy, and Spain.

Moving forward another decade, a report of results from the EU "Pure Juice" project (Rummel et al. 2010) claimed that a "multi-element approach" (δ^2H, δ^{13}C, δ^{15}N, δ^{34}S, ^{87}Sr/^{86}Sr) was most successful in a regional assessment of 150 authentic orange juices from North and South America, Africa, and Europe. The caveat to this report is that the juices were collected from processing factories, which potentially received fruit from 100 km^2 areas. The isotopic compositions for all of the bio-elements showed significant variation and overlap between cultivation regions, although δ^{34}S values were very well defined for some sites. Strontium isotope ratios were found to reflect known, underlying geology. When the data were combined using discriminant analysis, the discriminant functions were dominated by δ^2H and ^{87}Sr/^{86}Sr (Figure 7.3)—the parameters most closely associated with geographical location.

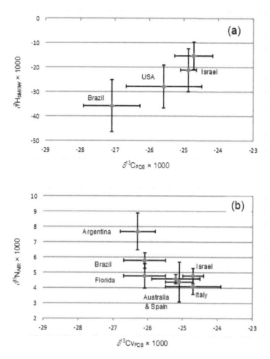

Figure 7.2 Stable isotopic compositions of orange juices (a) hydrogen *vs.* carbon (data from Bricout and Koziet 1987) and (b) nitrogen *vs.* carbon (data from Kornexl et al. 1996).

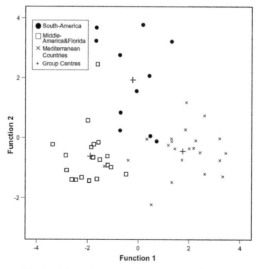

Figure 7.3 The geographical origin assignment of orange juices based on discriminant analysis of δ^2H, $\delta^{13}C$, $\delta^{15}N$, $\delta^{34}S$, and $^{87}Sr/^{86}Sr$ measurements of orange juice pulp.

Reproduced from Rummel et al. 2010 with permission. Copyright Elsevier.

In contrast to fruit juices, the few studies investigating the geographic origins of whole fruit mostly focused on variations over small regions. A study of the chemical and isotopic compositions of Slovenian apples (Bizjak Bat et al. 2012) found that cultivar was the major influence on both the $\delta^{13}C$ and $\delta^{15}N$ compositions of proteins as well as the $\delta^{2}H$ and $\delta^{18}O$ compositions of waters. Geographical regions were found to be well defined by $\delta^{2}H$ and $\delta^{18}O$ values of water together with the concentrations of S and Rb. A similar study of four apple varieties, grown in northern Italy (Mimmo et al. 2015), found that the $\delta^{13}C$ and $\delta^{15}N$ compositions of peel, pulp, and pips could distinguish between four growing regions but could only distinguish between varieties when these were grown at the same location.

7.4.2 Vegetables

In common with many other areas of food forensics, much of the published research to determine the regions of origin of fruits and vegetables has relied on a comparison of stable isotopic compositions with existing geospatial databases. Typically, the ratios of radiogenic isotopes (strontium and lead) are compared to maps of underlying geology, while $\delta^{2}H$ and $\delta^{18}O$ compositions are compared to meteoric/groundwater—e.g., the Global Network of Isotopes in Precipitation (GNIP) database.

Researchers have demonstrated that the $^{87}Sr/^{86}Sr$ composition of the bio-available strontium fraction of local soils corresponds well with the composition measured in fruits and vegetables, including asparagus (Swoboda et al. 2008), hot peppers and rice (Song et al. 2014), and Argentinean wheat (Podio et al. 2013). The first of these Sr isotope analysis applications was successfully used to distinguish Austrian *Marchfelder Spargel*, which has a protected geographical identification, from asparagus imported from neighboring Hungary and Slovakia. Strontium isotope ratios have been used to determine the origins of tomatoes (Trincherini et al. 2014) and have been shown to remain constant throughout the production chain involving tomatoes, tinned tomatoes, passata, tomato sauce, and tomato concentrates (Bontempo et al. 2011). The data from this study were used to distinguish between products from China and Italy and between different production areas within Italy. Strontium isotope ratios, in combination with concentrations of other Group II metals, have also been used to distinguish Korean grown *Chinese cabbage* from Chinese imports (Bong et al. 2012).

The $^{87}Sr/^{86}Sr$ of brown rice (*Oryza sativa* L.) has been reported for Japan (0.706–0.709), China and Vietnam (0.710–0.711), Australia (0.715–0.717), and California (0.706) (Kawasaki et al. 2002) and was considered "key information" for the estimation of rice provenance. A study combining $^{87}Sr/^{86}Sr$, $^{204}Pb/^{206}Pb$, and $^{207}Pb/^{208}Pb$ measurements with the concentrations of eight elements of 350 rice samples from Japan, USA, China, and Thailand found that this method gave 97% correct classification (Ariyama et al. 2012). Other studies of rice have relied on the isotope ratios of bio-elements and elemental concentrations. Kelly et al. (2002) studied premium long grain rice and found that American rice was characterized by high concentrations of boron, European rice by high concentrations of magnesium, and Indian/Pakistani rice by relatively low $\delta^{18}O$ values. Chung et al. (2016) concluded that a combination of $\delta^{18}O$ and $\delta^{34}S$ measurements was effective in discriminating rice from the Philippines, China, and Korea. Gonzálvez et al. (2011) measured the concentration of 32 elements in 107 glutinous rice samples from Spain, Brazil, India, and Japan and found that these measurements provided 91% correct classification. The elemental concentrations of rice samples from four provinces of China were found to correlate with the concentrations of bio-available elements in soil (Shen et al. 2013) and the concentrations of Mg, Cu, K, Ni, Be, Mn, and Ca could be used for approximately 94% correct geographical prediction.

In addition to comparing the $^{87}Sr/^{86}Sr$ composition of asparagus to local soils, researchers have also compared the $\delta^{18}O$ compositions of asparagus to local groundwater as a means of authentication (Rossier et al. 2014). The $\delta^{18}O$ composition of waters from bell peppers has been proposed as a means of authentication by comparison with the GNIP database (de Rijke et al. 2016). These authors also determined the δ^2H composition of four hydrocarbons present in the bell pepper water by GC-IRMS, and suggested that these measurements could additionally be used to verify geographical origin. In a similar study, the δ^2H composition of the intrinsic (non-exchangeable) component of potato tubers was found to have a consistent offset of $-161 \pm 11‰$ with respect to modelled meteoric waters, ranging from $-95‰$ in Sweden to $+25‰$ in Egypt (Keppler and Hamilton 2008). The authors proposed this comparison as an effective tool to constrain possible geographical origins of potato tubers and other foods.

A pilot study in 2003 claimed that $\delta^{13}C$ measurements alone were sufficient to distinguish wheat grown in Europe, Canada, and the USA (Branch et al. 2003). This finding may be supported by a recent study into the effects of region, genotype, and harvest year on the δ^2H, $\delta^{13}C$, and $\delta^{15}N$ compositions of 270 wheat kernels, which concluded that region accounted for the largest proportion of the total variation observed (Liu et al. 2015). To date the most comprehensive study of wheat involved 500 samples from across Europe and determined $^{87}Sr/^{86}Sr$, $\delta^{13}C$, $\delta^{15}N$, $\delta^{18}O$, and $\delta^{34}S$, plus the concentrations of Na, K, Ca, Cu, and Rb. Analysis of these data was able to distinguish 15 of the 17 sampling sites (Asfaha et al. 2011).

Even without recourse to geospatial mapping, stable isotope ratio measurements of the bio-elements are often able to distinguish between different countries of origin. Measurements of the $\delta^{13}C$ and $\delta^{15}N$ compositions of pistachios were able to distinguish between the three main growing regions of USA, Iran, and Turkey (Anderson and Smith 2006).

These authors found no statistical differences between pistachios from two growing seasons or between two varieties grown at the same location. More recently researchers have reported differences in the δ^2H, $\delta^{13}C$, $\delta^{18}O$, and $\delta^{34}S$ compositions of 93 lentil samples grown in Canada or Italy, but no difference in $\delta^{15}N$ composition was observed (Longobardia et al. 2015).

7.5 Tea

Tea is brewed from the leaves of the evergreen shrub *Camellia sinensis*, which are dried/cured to varying degrees to produce white, green, Oolong, black, etc. varieties. Second only to water, tea is the most consumed drink in the world (MacFarlane and MacFarlane 2009) and yet there have been surprisingly few authentication studies—also surprising given the high value that can be associated with certain tea growing regions, such as Darjeeling.

As with vegetables (see above), it has been shown that the $^{87}Sr/^{86}Sr$ ratios in tea leaves are indistinguishable from those in corresponding soils (Lagad et al. 2013) and that this ratio provided "excellent" distinction between teas grown in four geographically distinct regions of India: Assam, Darjeeling, Kangra, and Munnar. In addition, the $\delta^{13}C$ composition of these teas provided clear distinction between Kangra and Munnar regions. In Taiwan, measurements of both $^{87}Sr/^{86}Sr$ and $\delta^{11}B$ compositions were able to distinguish between tea leaves from the major tea gardens with the exception of Hualien (Chang et al. 2016). Large variation in $\delta^{11}B$ (+0.38 to +23.73‰) were attributed to different fertilizers.

GC-IRMS has been used to measure the $\delta^{13}C$ compositions of three characteristic flavor compounds in green teas: 2-phenylethanol, 3-hexenol, and benzyl alcohol (Murata et al. 2013). The isotopic compositions of these compounds were found to have much larger variations than bulk $\delta^{13}C$ measurements of the tea leaves and the authors speculated that this was a potential strategy to establish geographical location.

The concentrations of eight metallic elements have been shown to distinguish between three classes of tea—green, Oolong, and black—based on 48 samples of commercial tea (Herrador and González 2001). A later work claimed that δ^2H and $\delta^{13}C$ measurements in combination with the concentrations of 20 metallic elements could distinguish teas from Sri Lanka, China, India, and Taiwan (Pilgrim et al. 2010) using discriminant analysis (DA).

CAUTION—when appraising regional authentication based on DA (see Chapter 5):

1) the number of observations must be at least five times the number of independent variables, and

2) the number of observations in the smallest group must be greater than the number of independent variables.

It is a moot point as to whether multiple measurements of the same physical sample count as "observations."

7.6 Coffee

Coffee is brewed from the roasted beans of the plants *Coffea arabica* and/or *Coffea canephor* (*robusta*) and is widely consumed across most continents and cultures. At the premium end of the coffee bean market, coffees from Jamaica or Panama can command prices between US$ 128 and 260 per kg whereas Brazilian coffee can be purchased for as little as US$ 15 per kg (www.ico. org; accessed July 2016). As with many other commodities discussed in this volume, this price differential may well lead unscrupulous vendors to substitute or adulterate premium brands with cheaper coffee.

Researchers have considered various chemical components of coffee beans as proxies for both cultivar and region of origin (Anderson and Smith 2002; Bertrand et al. 2008; Alonso-Salces et al. 2009; Choi et al. 2010). Others have applied spectroscopic techniques such as ^{1}H-NMR (Charlton et al. 2002; Arana et al. 2015; Monakhova et al. 2015), ^{13}C-NMR (Wei et al. 2012), and infrared spectroscopy (Briandet et al. 1996; Downey et al. 1997) to the same ends. As with most other areas of food forensics, research has increasingly focused on the stable isotopic and elemental compositions of coffee beans as a means to determine the region of origin—mostly green coffee beans.

Surprisingly, one of the earliest applications of stable isotopes to determine geographical origin did not analyze whole beans (which comprise structural carbohydrates, simple sugars, proteins, lipids, alkaloids, minerals, etc.) but the caffeine contained within the beans (Weckerle et al. 2002). Analyzing the δ^2H, $\delta^{13}C$, and $\delta^{18}O$ compositions of caffeine from 45 coffee samples it was possible to distinguish samples from Central America, South America, and Africa with correct classification better than 98%.

In contrast, analysis of the $\delta^{13}C$, $\delta^{15}N$, and $\delta^{18}O$ compositions of 68 samples of whole coffee beans, from 20 difference geographical locations, was reported to provide "some discrimination" for "some of the coffees" (Rodrigues et al. 2009). The limited degree of discrimination may reflect a mixed isotopic signal from a complex matrix such as coffee beans as opposed to the analysis of an isolated component like caffeine. The authors still recorded that the observed differences were mainly explained by altitude and precipitation associated with different geographical locations. An earlier study, which measured $\delta^{11}B$ together with $\delta^{13}C$ and $\delta^{15}N$ compositions for a similar range of samples (46 samples, 19 different countries, three continents), found 88% overall correct classification for continental origin—Americas, Asia, and Africa (Serra et al. 2005).

Despite the cautionary note presented above on the use of DA, subsequent investigations into the authentication of coffee resemble something of an *arms race* in terms of the number of analytes measured.

A study of 47 samples of green coffee beans from five Hawaiian islands meaured the $\delta^{13}C$, $\delta^{15}N$ $\delta^{18}O$, and $\delta^{34}S$ compositions, the $^{87}Sr/^{86}Sr$ ratio, and the concentrations of 30 elements; the goal was to examine the relationship between the measured parameters, altitude, annual $\delta^{18}O$ precipitation data, and volcanic activity, despite a note from the authors that "with all multivariate analyses, the ratio of samples to variables should be high" (Rodrigues et al. 2011). The authors concluded that carbon and nitrogen isotopic compositions did not contribute to region discrimination whereas $\delta^{18}O$ values showed a positive correlation with altitude. Higher $\delta^{34}S$ values were observed at altitudes less than 200 m and at locations close to the sea. $^{87}Sr/^{86}Sr$ ratios were found to be characteristic of regions of volcanic activity and those close to the sea; the Hawaiian islands have relatively few sources of strontium. In the same year, a study found that $^{87}Sr/^{86}Sr$ ratio could be used to characterize coffee produced on Réunion Island (Bourbon Pointu) (Techer et al. 2011). This study also noted that the contribution from Sr-rich fertilizers was low compared to the cultivation soils.

A study that measured the elemental profile of 54 elements plus δ^2H, $\delta^{13}C$, $\delta^{15}N$, and $\delta^{18}O$ compositions of 62 samples of green coffee beans from four regions claimed 95% correct classification using a combination of δ^2H plus the concentrations of Mg, Fe, Co, and Ni and 98% correct classification using a combination of $\delta^{13}C$, $\delta^{15}N$, $\delta^{18}O$ plus the concentrations of 13 elements (Santato et al. 2012).

A study that measured the concentrations of 59 elements in 42 green coffee bean samples from five regions and 15 countries claimed approximately 97% correct classification with respect to continent and 100% correct classification with respect to country within each continent (Valentin and Watling 2013). An incidental, but useful, finding of this paper was that both the isotopic and elemental concentrations of green coffee beans remained largely unchanged following roasting. A study of 21 coffee samples from 14 countries in Africa, America, and Asia, measured the Sr and B isotope ratios and the concentrations of B, Rb, Sr, Ba, Fe, Mn, and Zn (Liu et al. 2014); the authors concluded that Sr and B isotope ratios were more sensitive to growth location and provided "excellent indicators of the origin of coffee beans."

A recent study measured the $\delta^{13}C$, $\delta^{15}N$, $\delta^{18}O$ compositions and the concentrations of 13 elements in 54 roasted coffees from 20 countries in five continents; samples were purchased from retail outlets (Carter et al. 2015). Stepwise DA provided up to 77% correct classification of regions of production with the isotopic data being the major contributors to the discriminant functions. Samples from Africa and India were

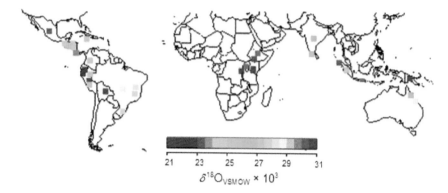

Figure 7.4 A geo-spatial map of the $\delta^{18}O$ compositions of roasted coffee beans. Reproduced from Carter et al. 2015 with permission. Copyright ACS Publications.

readily classified whereas the wide range in both isotopic and elemental compositions of samples from other regions, especially central/south America, resulted in poorer discrimination. A simple geo-spatial plot of the isotopic data (Figure 7.4) confirmed significant variation in the distribution of the $\delta^{18}O$ compositions; for example, the variation in $\delta^{18}O$ values across South America was equivalent to the variation observed from eastern Australia to India. The authors reasoned that using DA to assign samples to broad, man-made groupings such as "Asia" was not appropriate. The use of geo-spatial mapping, however, provided a means to consider whether the isotopic composition of a sample was consistent with the proposed region of origin. As with every field of food forensics, the value of such comparisons will require large datasets and/or models linking sample data to existing databases such as GNIP.

References

Alonso-Salces, R. M., F. Serra, F. Reniero and K. Héberger. 2009. Botanical and geographical characterization of green coffee (*Coffea arabica* and *Coffea canephora*): Chemometric evaluation of phenolic and methylxanthine contents. *Journal of Agricultural and Food Chemistry* 57: 4224–4235.

Anderson, K. A. amd B. W. Smith. 2002. Chemical profiling to differentiate geographic growing origins of coffee. *Journal of Agricultural and Food Chemistry* 50: 2068–2075.

Anderson, K. A. and B. W. Smith. 2006. Effect of season and variety on the differentiation of geographic growing origin of pistachios by stable isotope profiling. *Journal of Agricultural and Food Chemistry* 54: 1747–1752.

Arana, V. A., J. Medina, R. Alarcon, E. Moreno, L. Heintz, H. Schäfer and J. Wist. 2015. Coffee's country of origin determined by NMR: The Colombian case. *Food Chemistry* 175: 500–506.

Ariyama, K., M. Shinozaki and A. Kawasaki. 2012. Determination of the geographic origin of rice by chemometrics with strontium and lead isotope ratios and multielement concentrations. *Journal of Agricultural and Food Chemistry* 60: 1628–1634.

Asfaha, D. G., C. R. Quétel, F. Thomas, M. Horacek, B. Wimmer, G. Heiss, C. Dekant et al. 2011. Combining isotopic signatures of $^{87}Sr/^{86}Sr$ and light stable elements (C, N, O, S) with multi-elemental profiling for the authentication of provenance of European cereal samples. *Journal of Cereal Science* 53: 170–177.

Ashurst, P. 2009. New directions in fruit juice processing. pp. 299–317. *In*: P. Paquin (ed.). *Functional & Speciality Beverage Technology*, CRC.

Bertrand, B., D. Villarreal, A. Laffargue, H. Posada, P. Lashermes and S. Dussert. 2008. Comparison of the effectiveness of fatty acids, chlorogenic acids, and elements for the chemometric discrimination of coffee (*Coffea arabica* L.) varieties and growing origins. *Journal of Agricultural and Food Chemistry* 56: 2273–2280.

Bizjak Bat, K., R. Vidrih, M. Nečemer, B. Mozetič Vodopivec, I. Mulič, P. Kump and N. Ogrinc. 2012. Characterization of Slovenian apples with respect to their botanical and geographical origin and agricultural production practice. *Food Technology & Biotechnology* 50: 107–116.

Bong, Y.-S., K.-S. Lee, W.-J. Shin and J.-S. Ryu. 2008. Comparison of the oxygen and hydrogen isotopes in the juices of fast-growing and slow-growing fruits. *Rapid Communications in Mass Spectrometry* 22: 2809–2812.

Bong, Y.-S., W.-J. Shin, M. K. Gautam, Y.-J. Jeong, A.-R. Lee, C.-S. Jang, Y.-P. Lim, G.-S. Chung and K.-S. Lee. 2012. Determining the geographical origin of Chinese cabbages using multielement composition and strontium isotope ratio analyses. *Food Chemistry* 135: 2666–2674.

Bontempo, L., F. Camin, L. Manzocco, G. Nicolini, R. Wehrens, L. Ziller and R. Larcher. 2011. Traceability along the production chain of Italian tomato products on the basis of stable isotopes and mineral composition. *Rapid Communications in Mass Spectrometry* 25: 899–909.

Bontempo, L., F. A. Ceppa, M. Perini, A. Tonon, G. Gagliano, R. M. Marianella, M. Marega, A. Trifirò and F. Camin. 2014. Use of $\delta^{18}O$ authenticity thresholds to differentiate tomato passata from diluted tomato paste. *Food Control* 35: 413–418.

Branch, S., S. Burke, P. Evans, B. Fairman and C. Wolff Briche. 2003. A preliminary study in determining the geographical origin of wheat using isotope ratio inductively coupled plasma mass spectrometry with ^{13}C, ^{15}N mass spectrometry. *Journal of Analytical Atomic Spectrometry* 18: 17–22.

Brause, A. R. 1992. Detection of juice adulteration 1991 perspective. *International Food Ingredients* 1: 4–11.

Briandet, R., E. K. Kemsley and R. H. Wilson. 1996. Discrimination of Arabica and Robusta in instant coffee by Fourier transform infrared spectroscopy and chemometrics. *Journal of Agricultural and Food Chemistry* 44: 170–174.

Bricout, J. and J. Koziet. 1987. Control of the authentication of orange juice by isotopic analysis. *Journal of Agricultural and Food Chemistry* 35: 758–760.

Bricout, J. and J. C. Fontes. 1974. Analytical distinction between cane and beet sugars. *Annales des Falsifications et de L'expertise Chimique et Toxicologique* 67: 211–215.

Carter, J. F., H. S. A. Yates and U. Tinggi. 2015. Isotopic and elemental composition of roasted coffee as a guide to authenticity and origin. *Journal of Agricultural and Food Chemistry* 63: 5771–5779.

Chang, C.-T., C.-F. You, S. K. Aggarwal, C.-H. Chung, H.-C. Chao and H.-C. Liu. 2016. Boron and strontium isotope ratios and major/trace elements concentrations in tea leaves at four major tea growing gardens in Taiwan. *Environmental Geochemistry and Health* 38: 737–748.

Charlton, A. J., W. H. H. Farrington and P. Brereton. 2002. Application of ^1H NMR and multivariate statistics for screening complex mixtures: Quality control and authenticity of instant coffee. *Journal of Agricultural and Food Chemistry* 50: 3098–3103.

Choi, M.-Y., W. Choi, J. H. Park, J. Lim and S. W. Kwon. 2010. Determination of coffee origins by integrated metabolomic approach of combining multiple analytical data. *Food Chemistry* 121: 1260–1268.

Chung, I.-M., J.-K. Kim, M. Prabakaran, J.-H. Yang and S.-H. Kim. 2016. Authenticity of rice (*Oryza sativa* L.) geographical origin based on analysis of C, N, O and S stable isotope ratios: A preliminary case report in Korea, China and Philippine. *Journal of the Science of Food and Agriculture* 96: 2433–2439.

Coplen, T. B., C. Kendall and J. Hopple. 1983. Comparison of stable isotope reference samples. *Nature* 302: 236–238.

de Rijke, E., J. C. Schoorl, C. Cerli, H. B. Vonhof, S. J. A. Verdegaal, G. Vivó-Truyols, M. Lopatka et al. 2016. The use of δ^2H and $\delta^{18}O$ isotopic analyses combined with chemometrics as a traceability tool for the geographical origin of bell peppers. *Food Chemistry* 204: 122–128.

Doner, L. W., H. O. Ajie and L. Sternberg. 1987. Detecting sugar beet syrups in orange juice by D/H and $^{18}O/^{16}O$ analysis of sucrose. *Journal of Agricultural and Food Chemistry* 35: 610–612.

Doner, L. W. and J. P. Phillips. 1987. Correlation of growth location of sugar beets and ratios of the $^{18}O/^{16}O$ and carbon bound D/H in derived sugars. *Abstracts of the 193rd National Meeting of the American Chemical Society*, Denver, April 5–10.

Downey, G., R. Briandet, R. H. Wilson and E. K. Kemsley. 1997. Near- and mid-infrared spectroscopies in food authentication: Coffee varietal identification. *Journal of Agricultural and Food Chemistry* 45: 4357–4361.

Dunbar, J. and A. T. Wilson. 1983. Oxygen and hydrogen isotopes in fruit and vegetable juices. *Plant Physiology* 72: 725–727.

González, J., E. Jamin, G. Remaud, Y. -L. Martin, G. G. Martin and M. L. Marin. 1998. Authentication of lemon juices and concentrates by a combined multi-isotope approach using SNIF-NMR and IRMS. *Journal of Agricultural and Food Chemistry* 46: 2200–2205.

González, J., G. Remaud, E. Jamin, N. Naulet and G. G. Martin. 1999. Specific natural isotope profile studies by isotope ratio mass spectrometry (SNIP-NMR): $^{13}C/^{12}C$ ratios of fructose, glucose, and sucrose for improved detection of sugar addition to pineapple juices and concentrates. *Journal of Agricultural and Food Chemistry* 47: 2316–2321.

Gonzálvez, A., S. Armenta and M. de la Guardia. 2011. Geographical traceability of "Arròs de Valencia" rice grain based on mineral element composition. *Food Chemistry* 126: 1254–1260.

Guyon, F., P. Auberger, L. Gaillard, C. Loublanches, M. Viateau, N. Sabathié, M. H. Salagoïty and B. Médina. 2014. $^{13}C/^{12}C$ isotope ratios of organic acids, glucose and fructose determined by HPLC-co-IRMS for lemon juice authentication. *Food Chemistry* 146: 36–40.

Hammond, D. A. 1996. Authenticity of fruit juices, jams and preserves. pp. 15–59. *In:* P. R. Ashurst and M. J. Dennis (eds.). *Food Authenticity*. Springer.

Herrador, M. A. and A. G. González. 2001. Pattern recognition procedures for differentiation of green, black and Oolong teas according to their metal content from inductively coupled plasma atomic emission spectrometry. *Talanta* 53: 1249–12.

Houerou, G., S. D. Kelly and M. J. Dennis. 1999. Determination of the oxygen-18/oxygen-16 isotope ratios of sugar, citric acid and water from single strength orange juice. *Rapid Communications in Mass Spectrometry* 13: 1257–1262.

Jamin, E., J. González, G. Remaud, N. Naulet and G. G. Martin. 1997. Detection of exogenous sugars or organic acids addition in pineapple juices and concentrates by ^{13}C IRMS analysis. *Journal of Agricultural and Food Chemistry* 45: 3961–3967.

Jamin, E., J. González, I. Bengoechea, G. Kerneur, G. Remaud, C. Iriondo and G. G. Martin. 1998. Proteins as intermolecular isotope reference for detection of adulteration of fruit juices. *Journal of Agricultural and Food Chemistry* 46: 5118–5123.

Jamin, E., R. Guerin, M. Retif, M. Lee and G. J. Martin. 2003. Improved detection of added water in orange juice by simultaneous determination of the oxygen-18/oxygen-16 isotope ratios of water and ethanol derived from sugars. *Journal of Agricultural and Food Chemistry* 51: 5202–5206.

Jamin, E., F. Martin, R. Santamaria-Fernandez and M. Lees. 2005. Detection of exogenous citric acid in fruit juices by stable isotope ratio analysis. *Journal of Agricultural and Food Chemistry* 53: 5130–5133.

Kawasaki, A., H. Oda and T. Hirata. 2002. Determination of strontium isotope ratio of brown rice for estimating its provenance. *Soil Science and Plant Nutrition* 48: 635–640.

Kelly, S., M. Baxter, S. Chapman, C. N. Rhodes, J. Dennis and P. A. Bereton. 2002. The application of isotopic and elemental analysis to determine the geographical origin of premium long grain rice. *European Food Research and Technology* 214: 72–78.

Keppler, F. and J. T. G. Hamilton. 2008. Tracing the geographical origin of early potato tubers using stable hydrogen isotope ratios of methoxyl groups. *Isotopes in Environmental and Health Studies* 44: 337–347.

Kornexl, B. E., A. Rossman and H. -L. Schmidt. 1996. Improving fruit juice origin assignment by combined carbon and nitrogen isotope ratio. *Zeitschrift für Lebensmittel-Untersuchung und Forschung* 202: 55–59.

Krueger, D. A. 1998. Stable isotope analysis by mass spectrometry. pp. 14–34. *In*: P. R. Ashurst and M. J. Dennis (eds.). *Analytical Methods of Food Authentication*. Springer.

Lagad, R. A., D. Alamelu, A. H. Laskar, V. K. Rai, S. K. Singh and S. K. Aggarwal. 2013. Isotope signature study of the tea samples produced at four different regions in India. *Analytical Methods* 5: 1604–1611.

Le Gall, G., M. Puaud and I. J. Calquhoun. 2001. Discrimination between orange juice and pulp-wash by ¹H nuclear magnetic resonance spectroscopy: Identification of marker compounds. *Journal of Agricultural and Food Chemistry* 49: 580–588.

Liu, H. -C., C. -F. You, C. -F., C. -Y. Chen, Y. -C. Liu and M. -T. Chung. 2014. Geographic determination of coffee beans using multi-element analysis and isotope ratios of boron and strontium. *Food Chemistry* 142: 439–445.

Liu, H., B. Guo, Y. Wei, S. Wei, Y. Ma and W. Zhang. 2015. Effects of region, genotype, harvest year and their interactions on $\delta^{13}C$, $\delta^{15}N$ and δD in wheat kernels. *Food Chemistry* 171: 56–61.

Longobardi, F., G. Casiello, M. Cortese, M. Perini, F. Camin, L. Catuccia and A. Agostiano. 2015. Discrimination of geographical origin of lentils (*Lens culinaris* Medik.) using isotope ratio mass spectrometry combined with chemometrics. *Food Chemistry* 188: 343–349.

Macfarlane, A. and I. Macfarlane. 2009. *The Empire of Tea*. The Overlook Press, New York.

Mimmo, T., F. Camin, L. Bontempo, C. Capici, M. Tagliavini, S. Cesco and M. Scampicchio. 2015. Traceability of different apple varieties by multivariate analysis of isotope ratio mass spectrometry data. *Rapid Communications in Mass Spectrometry* 29: 1984–1990.

Ministry of Agriculture Fisheries and Food, UK. 1991. Comparison of methods for the detection of added substances in orange juice. HMSO, London, UK.

Martin, G. J., D. Danho and C. Vallet. 1991. Natural isotopic fractionation in the discrimination of sugar origins. *Journal of the Science of Food and Agriculture* 56: 419–434.

Monakhova, Y. B., W. Ruge, T. Kuballa, M. Ilse, O. Winkelmann, B. Diehl, F. Thomas and D. W. Lachenmeier. 2015. Rapid approach to identify the presence of Arabica and Robusta species in coffee using ¹H NMR spectroscopy. *Food Chemistry* 182: 178–184.

Murata, A., U. H. Engelhardt, P. Fleischmann, K. Yamada, N. Yoshida, D. Juchelka, A. Hilkert et al. 2013. Purification and gas chromatography–combustion–isotope ratio mass spectrometry of aroma compounds from green tea products and comparison to bulk analysis. *Journal of Agricultural and Food Chemistry* 61: 11321–11325.

Nissenbaum, A., A. Lifshitz and Y. Stepek. 1974. Detection of citrus juice adulteration using the distribution of natural ¹⁸O and D analysis. *Lebensmittel-Wissenschaft & Technologie* 7: 152–154.

Pilgrim, T. S., R. J. Watling and K. Grice. 2010. Application of trace element and stable isotope signatures to determine the provenance of tea (*Camellia Sinensis*) samples. *Food Chemistry* 118: 921–926.

Podio, N. S., M. V. Baroni, R. G. Badini, M. Inga, H. A. Ostera, M. Cagnoni, E. A. Gautier, P. Peral-García, J. Hoogewerff and D. A. Wunderlin. 2013. Elemental and isotopic fingerprint of Argentinean wheat. Matching soil, water, and crop composition to differentiate provenance. *Journal of Agricultural and Food Chemistry* 61: 3763–3773.

Rodrigues, C. I., R. Maia, M. Miranda, M. Ribeirinho, J. M. F. Nogueira and C. Máguas. 2009. Stable isotope analysis for green coffee bean: A possible method for geographic origin discrimination. *Journal of Food Composition and Analysis* 22: 463–471.

Rodrigues, C., M. Brunner, S. Steiman, S. G. J. Bowen, J. M. F. Nogueira, L. Gautz, T. Prohaska and C. Máguas. 2011. Isotopes as tracers of the Hawaiian coffee-producing regions. *Journal of Agricultural and Food Chemistry* 59: 10239–10246.

Rossier, J. S., V. Maury, B. de Voogd and E. Pfammatter. 2014. Use of isotope ratio mass spectrometry (IRMS) determination ($^{18}O/^{16}O$) to assess the local origin of fish and asparagus in Western Switzerland. *CHIMIA International Journal for Chemistry* 68: 696–700.

Rossmann, A., J. Koziet, G. J. Martin and M. J. Dennis. 1997. Determination of the carbon-13 content of sugars and pulp from fruit juices by isotope-ratio mass spectrometry (internal reference method). A European interlaboratory comparison. *Analytica Chimica Acta* 340: 21–29.

Rummel, S., S. Hoelz, P. Horn, A. Rossman and C. Schlicht. 2010. The combination of stable isotope abundance ratios of H, C, N and S with $^{87}Sr/^{86}Sr$ for geographical origin assignment of orange juices. *Food Chemistry* 118: 890–900.

Santato, A., D. Bertoldi, M. Perini, F. Camin and R. Larcher. 2012. Using elemental profiles and stable isotopes to trace the origin of green coffee beans on the global market. *Journal of Mass Spectrometry* 24: 1132–1140.

Serra, F., C. G. Guillou, F. Reniero, L. Ballarin, M. I. Cantagallo, M. Wieser, S. S. Iyer, K. Héberger and F. Vanhaecke. 2005. Determination of the geographical origin of green coffee by principal component analysis of carbon, nitrogen and boron stable isotope ratios. *Rapid Communications in Mass Spectrometry* 19: 2111–2115.

Shen, S., L. Xia, N. Xiong, Z. Liu and H. Sun. 2013. Determination of the geographic origin of rice by element fingerprints and correlation analyses with the soil of origin. *Analytical Methods* 5: 6177–6185.

Smith, B. N. 1975. Carbon and hydrogen isotopes of sugars from various sources. *Naturwissenschaften* 62: 390.

Song, B.-Y., J.-S. Ryu, H. S. Shin and K.-S. Lee. 2014. Determination of the source of bioavailable Sr using $^{87}Sr/^{86}Sr$ tracers: A case study of hot pepper and rice. *Journal of Agricultural and Food Chemistry* 62: 9232–9238.

Swoboda, S., M. Brunner, S. F. Boulyga, P. Galler, M. Horacek and T. Prohaska. 2008. Identification of Marchfeld asparagus using Sr isotope ratio measurements by MC-ICP-MS. *Analytical and Bioanalytical Chemistry* 390: 487–494.

Techer, I., J. Lancelon, F. Descroix and B. Guyot. 2011. About Sr isotopes in coffee 'Bourbon Pointu' of the Réunion Islands. *Food Chemistry* 126: 718–724.

Trincherini, P. R., C. Baffi, P. Barbero, E. Pizzoglio and S. Spalla. 2014. Precise determination of strontium isotope ratios by TIMS to authenticate tomato geographical origin. *Food Chemistry* 145: 349–355.

Valentin, J. L. and R. J. Watling. 2013. Provenance establishment of coffee using solution ICP-MS and ICP-AES. *Food Chemistry* 141: 98–104.

Versini, G., F. Camin, M. Ramponi and E. Dellacassa. 2006. Stable isotope analysis in grape products: ^{13}C-based internal standardisation methods to improve the detection of some types of adulterations. *Analytica Chimica Acta* 563: 325–330.

Weckerle, B., E. Richling, S. Heinrich and P. Schreier. 2002. Origin assessment of green coffee (*Coffea arabica*) by multi-element stable isotope analysis of caffeine. *Analytical and Bioanalytical Chemistry* 374: 886–890.

Wei, F., K. Furihata, M. Koda, F. Hu, R. Kato, T. Miyakawa and M. Tanokura. 2012. ^{13}C NMR-based metabolomics for the classification of green coffee beans according to variety and origin. *Journal of Agricultural and Food Chemistry* 60: 10118–10125.

CHAPTER 8

Alcoholic Beverages I—Wine

James F. Carter[1,*] and *Federica Camin*[2]

8.1 Introduction

8.1.1 Everyone likes a drink!

In 2014, the world's population spent approximately 17% of its total food budget on alcoholic beverages, which in the USA alone was valued at US$17.6 billion (USDA 2014). In some countries (especially Islamic countries) spending on alcoholic beverages was as low as 2% of total food budgets, but in other countries, such as Ireland and Poland, it was as high as 50%. Alcoholic beverages are not only popular, but drinks such as vintage wines and super-premium spirits comprise some of the most expensive retail products available, easily exceeding the price of caviar, truffles, saffron, etc.

The value of these premium drinks is determined, almost exclusively, by the brand/origin and age/vintage and for this reason the provenance of alcoholic drinks has generated more published research than any other area of food forensics. The following sections briefly describe the commercial markets and the potential for fraud of the most popular alcoholic beverages: wines, spirits, beers, and ciders. Each section outlines how the isotopic and elemental signatures of the beverages are derived and discusses techniques developed to provenance specific alcoholic beverages based on a review of literature published, primarily, between 2000 and 2016. Evolving techniques such as DNA, proteomic, metabolomics, and *foodomics* are not discussed in any detail as these techniques are not well established at this time (and may provide an excuse for a 2nd edition of this volume!).

[1] Forensic and Scientific Services, Health Support Queensland, 39 Kessels Road, Coopers Plains QLD 4108, AUSTRALIA.
[2] Department of Food Quality and Nutrition, Research and Innovation Centre, Fondazione Edmund Mach (FEM), Via E. Mach 1, 38010 San Michele all'Adige, ITALY.
Email: federica.camin@fmach.it
* Corresponding author: Jim.Carter@health.qld.gov.au

8.1.2 *The sources of metallic elements in alcoholic beverages*

Detailed reviews exist concerning the various sources, sinks, and roles of metallic elements in alcoholic beverages (Ibanez et al. 2008) and specifically wines (Pohl 2007); the following text provides a summary.

The presence of metallic elements in alcoholic beverages can have both negative and positive effects. The negative effects include: spoilage, hazing, unpleasant sensory perception, and possible health problems. The primary positive effect is the role that metallic elements play in alcoholic fermentation (especially Ca, K, Mg, and Na) through the metabolism of yeasts, which is controlled by ionic strength and pH.

Metals in alcoholic beverages are sometimes described as having two sources: *primary* and *secondary*. *Primary* metals are natural in origin and derived from the fruits, cereals, and other plant materials from which the beverages are manufactured. For drinks manufactured from fruit juices (wines, ciders, etc.), the metals of primary origin comprise the largest part of the total metal concentration and are potentially the most characteristic. The relative concentrations of these metals reflect the geological characteristics of the soil in which the crops were grown and also the plant variety, the maturity of the crop at harvest, and the climatic conditions during growth. In many ways, it is the *primary metal signature* that is of most interest in food forensics as it is derived from geographic location and records other potentially useful information such as vintage.

Metals of *secondary* origin are derived from any number of external sources that contact the crop or the beverage during any stage of manufacture, from harvest to bottling and subsequent storage. Some secondary metals are classed as *geogenic*, also derived from soil, but as a result of the application of fertilizers (notably K, Ca, and Cu), fungicides, and pesticides (notably Cd, Cu, Mn, Pb, and Zn). Other examples include high sodium concentrations observed in crops from coastal regions, due to marine spray (e.g., Diaz et al. 2003) and relatively high concentrations of lead and cadmium from crops grown close to road traffic or industrial areas (e.g., Medina et al. 2000). Metallic elements may also be intentionally added to beverages, for example to control water hardness or to regulate acidity or sweetness. This category would also include clarifying and fining agents (flocculants, such as bentonite clay) used to precipitate suspended particles and reduce turbidity.

During manufacture, a beverage will come into contact with many different materials, such as aluminium, brass, stainless steel, glass, and wood, which form parts of the manufacturing machinery, pipes, barrels, etc. The extended contact times of beverages and raw materials with the surfaces of storage vessels have the potential to introduce significant metal content—for example, the storage of grape pulp and must in concrete tanks and storage of vodka in low quality steel containers are reported to result in corrosion with a concomitant increase in Fe content. Finally, the finished beverages are sold in a variety of retail packaging including: metal cans, glass bottles, plastic containers, and paperboard cartons, all of which may be a source of metals in the beverage. To further complicate matters, secondary metals may come from different sources in different beverages—for example, the majority of copper in beer is derived from the raw materials, whereas copper present in wine may be due to the addition of copper sulfate (to remove sulfurous odors) and the trace amounts of copper present in a distilled spirit will be largely derived from metallic surfaces in the still (often made of copper or bronze).

From one perspective, the *secondary metal signature* simply masks or obscures the *primary metal signature*, which contains all the useful information for geo-location. From another perspective, the *secondary metal signature* contains additional information that further serves to characterize an alcoholic product.

8.1.3 The sources of isotope ratio variations in alcoholic beverages

Very much akin to the metallic elemental composition, variations in the isotopic composition of alcoholic beverages can be regarded as having two components, a *primary signature* and a *secondary signature*. Also akin to elemental composition, the primary isotopic signature of a beverage is derived from the ingredients, fruits, and/or cereals used in the manufacture.

To summarize some of the key points from earlier chapters:

- the carbon isotopic composition is chiefly a function of plant metabolism, which is affected by parameters such as climate.
- both nitrogen and sulfur isotopic compositions are chiefly functions of nutrient availability, principally fertilization practices, although, to date, neither have been studied for alcoholic beverage authentication.

Intuitively, the isotopic composition of (baked) food is simply the arithmetic sum of the isotopic composition of the ingredients and little change occurs as a result of cooking (Bostic et al. 2015). The isotopic composition of a beverage may, however, change during manufacture due to loss of volatile components, notably water and CO_2, or to the formation of precipitates, which are then removed. In many countries, the incorporation of a limited range of food additives to alcoholic beverages is permitted and/or regulated, including, but not limited to: fruit acids, natural colors, preservatives, and thickening agents. Most of these substances will only be used in amounts equivalent to parts-per-million (ppm) concentrations of the final beverage and are, therefore, unlikely to make significant contributions to the overall isotopic signature of the finished product, with two exceptions: sugars and/or sulfite.

Legislation for some types of alcoholic beverages allows for the addition of sugar, post-fermentation, as a sweetener or to promote secondary fermentation (not for the primary production of alcohol). Much commercial sugar is sugar cane (sucrose), or high fructose corn syrup (HFCS), both of which are derived from C_4 plants with carbon isotopic signature distinct from typical C_3 beverage ingredients: barley, apples, grapes, etc. or beet sugar (a C_3 plant with lower $^2H/^1H$ in the methyl group of ethanol compared to cereal and fruit sugar). Once sugar is present in a beverage it becomes susceptible to microbial activity and to prevent spoilage, preservatives (typically sulfur dioxide/sodium metabisulfite) are added. Even at ppm concentrations, added sulfite is likely to be the major source of sulfur in an alcoholic beverage, negating the potentially useful information from sulfur isotopic analysis.

8.1.4 Fractions and lines—isotopic variations in water

Most foodstuffs contain a high proportion of water and the source, addition, and loss of water all play important roles in the production and subsequent authentication of alcoholic beverages. This short section provides some explanation of why and how the isotopic composition of water changes.

Rayleigh fractionation (Equation 8.1) (first described by Lord Rayleigh, 1842–1919) describes the exponential partitioning of isotopes as a material is continuously removed from a system containing two or more isotopes—e.g., the evaporation of water containing ^{18}O and ^{16}O. In such a system R is the isotopic ratio in the residue, R_0 is the initial isotopic ratio, f is the fraction of the material remaining, and α is the fractionation factor. This equation predicts that, as the reservoir decreases in size, it becomes increasingly enriched in the heavier isotope, which has some important consequences for food forensics.

$$R = R_0 \cdot f^{(\alpha-1)}$$ Equation 8.1

First, the partitioning of the stable isotopes of hydrogen and oxygen ($^2H/^1H$ and $^{18}O/^{16}O$) in the global water cycle with increasing latitude, altitude, and continental effects (the ratios decrease inland from the coast) creates consistent patterns of global hydrogen and oxygen isotopic distributions in water. At a given location the average and seasonal variations in δ^2H and $\delta^{18}O$ signatures of water remain fairly constant because climatic conditions remain fairly uniform and these spatial patterns or *isoscapes* have found application across a wide range of disciplines, including food forensics. Precipitation is the ultimate source of groundwater in virtually all systems and will interact with geological, biological, and man-made materials and become incorporated into these materials, together with some imprint of its isotopic composition. Although the overall process can be complex (Bowen and Revenaugh 2003), the acquired isotopic signature can be attributed to a geographical source through comparison with models based on precipitation.

In 1961, Harmon Craig (Craig 1961) observed that the δ^2H and $\delta^{18}O$ compositions of precipitation water (that had not undergone significant evaporation) where related by Equation 8.2.

$$\delta^2H = 8 \times \delta^{18}O + 10 \qquad \text{Equation 8.2}$$

Known as the Global Meteoric Water Line (GMWL), this equation was based on precipitation data from locations around the globe and has a high correlation coefficient, $R^2 > 0.95$. The slope and intercept of a Local Meteoric Water Line (LMWL), derived from the precipitation at a specific site, can be significantly different from the GMWL and slopes in the range of 5 to 9 are not uncommon. Evaporated water typically plots below the GMWL and slopes of 2 to 5 are common. Vegetal water is subjected to an enrichment effect in heavier isotopes due to evapotranspiration in the leaves.

Rayleigh fractionation makes a further contribution to the isotopic signatures of foodstuffs through evaporation as a result of drying or cooking processes. Experimental data has shown that boiling water to make a hot drink raised the $\delta^{18}O$ value by 0.4 to 0.5‰ (Brettell et al. 2012; Chesson et al. 2014) whereas slow-cooking using large stew pots increased the oxygen isotope composition of the "pottage" by greater than 20‰ after 3 hours (Brettell et al. 2012). Brewing may increase the $\delta^{18}O$ value of ale by 1.3‰ over that of the initial water and other research has suggested that isotopic enrichment also occurs during long, slow fermentation in open vats, such as occurs during the production of ciders (see Chapter 9). Although Lord Rayleigh's original experiments concerned the *distillation of mixed liquids,* changes in isotopic composition during the distillation of potable spirits have received little attention other than a comprehensive study of cherry brandy (Baudler et al. 2006) (see Chapter 9). The natural *breathability* of the wooden cask used to mature many wines, beers, and distilled spirits must also result in the loss of a significant volume of alcohol and/or water during maturation, but no research appears to have addressed changes in isotope ratio through this process.

8.2 Sample preparation and analysis

Alcoholic beverages are typically low-viscosity, homogeneous liquids and require little sample preparation prior to analysis, as compared to other foodstuffs. It is, however, important to bear in mind that the metal content of a wine, beer, or cider is unlikely to be evenly distributed between the liquid and any sediment present. Samples with sediments can simply be shaken to "homogenize" them but are better filtered, to present a more consistent analyte.

Samples of alcoholic beverages can be stored at room temperature in the original, unopened containers but once opened or if sub-samples are presented for analysis, these should be stored refrigerated at 3–4°C to prevent evaporation or microbial activity. Samples intended for elemental concentration analysis can be further preserved by the addition of toluene but this will obviously affect the hydrogen and carbon isotopic compositions.

Although samples will be liquids, they are not suitable for direct analysis by ICP-OES or ICP-MS because of the high alcohol and sugar content and the preferred method of sample preparation for elemental analysis is microwave digestion (Grindlay et al. 2008).

Isotopic analysis can be (and has been) performed on whole, untreated samples but, as discussed in Chapter 2, potentially valuable information may be "mixed" when components are not analyzed separately. Most commonly, alcoholic beverages are divided into alcohol and water components, which are analyzed separately. Any method for ethanol extraction can be used as long as the alcohol is recovered without isotopic fractionation; this preparation step is mostly performed by distillation, collecting the fraction with a vapor temperature of 78°C (the ethanol/water azeotrope). The $\delta^{13}C$ composition of ethanol, determined by distillation, has been compared with direct measurements of beverages using GC-IRMS and LC-IRMS (Cabañero et al. 2008) with coupled techniques found to be faster and equally precise.

The δ^2H, $\delta^{13}C$, and $\delta^{18}O$ compositions of water and ethanol are (almost exclusively) measured using conventional IRMS methodology but additional information is often obtained using Stable Natural Isotope Fractionation (SNIF)-NMR. During fermentation, the deuterium (^2H or D) originally present in sugars and in water will be redistributed between ethanol and water (Table 8.1). SNIF-NMR is used to measure the deuterium isotope ratios (D/H) of the methyl and methylene groups of ethanol (note these are true ratios and not δ-values). The technique was developed by Gerard

Table 8.1 The nomenclature used when reporting the relative distribution of deuterium (D) in ethanol.

Designation	Molecule	Definition	Typical standard deviation
I	$CH_2D.CH_2.OH$	$(D/H)_I$	0.26
II	$CH_3.CH\ D.OH$	$(D/H)_{II}$	0.30
III[A]	$CH_3.CH_2.OD$		
IV[A]	HOD	$(D/H)_W$	
R[B]		$2(D/H)_{II}/(D/H)_I$ [A]	0.005

[A] Molecules III and IV are in equilibrium and only $(D/H)_W$ needs to be measured.
[B] R is measured directly from the peak heights of the NMR signals: $R = 3h_{II}/h_I$.

Martin at the University of Nantes, specifically to detect the enrichment of grape musts (Martin et al. 1982) and has been adopted as an official method for the authentication of wines by the Organisation Internationale de la Vigne et du vin (OIV) and by the European Union (OVI-MA-311-05: R2011). Table 8.1 details the nomenclature used to report analyses by SNIF-NMR. Recent research has examined the use of position specific ^{13}C-NMR but concluded that it did not provide additional information compared to the direct measurement of the δ^{13}C signature performed by IRMS (Guyon et al. 2015). As an aside, the SNIF-NMR technique is also applicable to fruit juice, vinegar, vanillin, and other flavor compounds.

Despite the many potential advantages of stable isotope measurements by isotope ratio infrared spectroscopy (IRIS), such as portability, few publications have used this technique to study the δ^2H/δ^{18}O composition of beverage-water. The most comprehensive study, to date, measured the isotopic composition of beer, soda, fruit juices, and milk by IRIS and IRMS (Chesson et al. 2010) and found excellent correlations between δ^2H and δ^{18}O values measured for extracted waters and untreated beverages. This study did, however, urge caution as measurements by IRIS can be affected by the presence of both sugar and alcohol. This warning was (possibly) confirmed by a more recent study of the δ^{18}O composition of wine-water (Gupta et al. 2013) which found that results obtained by IRIS were not in close agreement with those obtained by IRMS: standard deviation $(1\sigma) = \pm 0.63\text{‰}$.

8.3 Wine

8.3.1 The global market for wine

There are more than 1 million wine producers in the world, cultivating over 7.4 million hectares of vineyards (over 28,000 square miles), producing approximately 38 billion 750-ml bottles of wine each year with a wholesale, pre-tax production value of US$98 billion (OVI 2014). It is estimated that the equivalent of a further 1.8 billion bottles are used for non-wine products such as vermouth, brandy, etc. (Wittwer and Anderson 2009). When assessing the commercial value, wines are often classified by quality categories shown in Table 8.2.

The wine export trade has grown dramatically in the past 30 years and is now worth an estimated US$30 billion (excluding internal EU sales) and accounts for about 35% of global production, compared to 15% of global production 30 years ago. New World wines now supply approximately 30% of global exports, compared to less than 3% in the early 1980s.

The growth in exports has been dominated by demand, primarily from Asian and South American countries as well as South Africa, for commercial-premium and super-premium wines. France is the leading supplier of super-premium still (as opposed to sparkling) wine, supplying 40% of exports by value, followed by Italy (17%),

Table 8.2 The quality (and price) categories of wines: based on Wittwer and Anderson 2009.

Category	Price per liter (US$)*	Sales percentage	
		by volume	by value
non-premium	<2.50	47	14
commercial-premium	2.50 to 7.50	37	40
super-premium	>7.50	9	33
sparkling		7	13

* Pre-tax value before any wholesale or retail mark-up.

New Zealand (9%), Portugal (6%), Australia (3%), and Spain (3%) (Wittwer and Anderson 2009). Since 2007, declining production and rising consumption have resulted in an overall under-supply of almost 3.6 billion bottles of wine per year. This shortage inevitably leads unscrupulous vendors to fill the gap in the market with bogus products, especially at the commercial-premium and super-premium end of the market.

Wine fraud comes in many forms but the most common are adulteration with water or cheaper products, and the re-labelling of cheaper, inferior wines as a more expensive brand. Wine is defined (by EC Regulations 479/2008) as the product obtained exclusively from the alcoholic fermentation of fresh grapes, whether crushed or not, or from *grape must* (freshly pressed juice containing skin, seeds, and stems). According to this definition, wine cannot be diluted with water or contain ethanol derived from petroleum, the pyrolysis of wood (synthetic ethanol), or from the fermentation of non-grape sugars, except in some regions of northern Europe. Water can be (illegally) added at the start of the winemaking process to compensate for the loss of natural water through dehydration or used to dilute or "stretch out" finished wine to make it more palatable or to reduce the alcoholic strength. Sometimes coined "adding Jesus units" (because water is turned into wine), *some of these processes are legal and others occupy a grey area between accepted practice and fraud.*

Counterfeit wine was a problem in classical Rome and continues today across all sectors of the market. At the high end of the market, some estimates claim that over 5% of wine sold at auction is fake, while at the corner shop end of the market, forged alcohol now accounts for 73% of all UK Trading Standards investigations (www.thedrinksbusiness.com 2014). Possibly the most infamous wine fraudster of recent times was Rudy Kurniawan, who would buy good (but not great) vintage French Burgundy and create a "better" vintage by mixing in some newer, vibrant wine. He was arrested in 2012 by the FBI after attempting to sell more than US$2 million of wine at auction and, in 2013, was sentenced to 10 years in jail for committing fraud (BBC 2014).

8.3.2 Wine making (vinification)

Figure 8.1 outlines the wine making process (vinification). For a detailed discussion of this process the reader is directed to Hornsey 2007.

Most regulatory bodies have very strict rules, which preclude the addition (or abstraction) of any substances to (or from) unfermented grape *must*. For this reason, the isotopic composition of a wine should be very similar to that of the grape juice, with some redistribution of isotopes between water, alcohol, sugar, and other minor components. Practices that increase naturally the sugar content of grape juice—withering and must concentration—are discussed below.

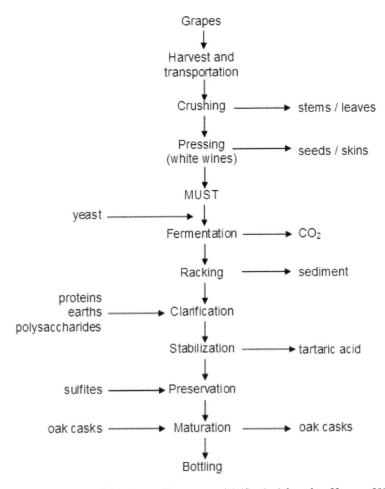

Figure 8.1 Schematic of the wine making process (vinification): based on Hornsey 2007.

Following fermentation a range of materials can be added, such as proteins (gelatine, isinglass, casein, etc.), earths (bentonite, kaolin), and polysaccharides (gum Arabic), while sediments are removed. All of these processes are likely to affect elemental composition (especially trace and ultra-trace elements), as will long-term interaction with storage casks and glass bottles, but not the isotopic ratios of H, C, and O.

The concentration of metallic elements affects the flavor, taste, aroma, and color of wine, mainly through the formation of precipitates. The major metallic elements in wine are Ca, K, Na, and Mg, which are typically present at concentrations in the range 10 to 1,000 mg/L. The concentrations of Ca, Mg, and Na are typically comparable and lower than the concentration of K. Minor elements present include Al, Cu, Fe, Mn, Rb, Sr, and Zn at concentrations in the range 0.1 to 10 mg/L. Trace elements, including Ba, Cd, Co, Cr, Li, Ni, Pb, and V are present at concentration below 0.1 mg/L. Typically, the major elements exhibit the smallest range of concentrations and trace elements the widest (especially Ni and Pb), with a concomitant potential for discrimination between samples.

The "fingerprinting" provided by elemental concentrations in finished wines is useful mainly because they are controlled by a number of factors during wine manufacture but remain largely unchanged thereafter. The concentrations of Ba, Ca, Cr, Li, Mg, Mn, Na, Rb, and Sr are typically characteristic of a geographical region of origin, whereas Na content reflects naturally occurring ocean sprays and Fe concentration is affected by both soil composition and technological processing. K and Mg contents are important variables related to grape maturity and can differentiate sweet and dry wines. Copper concentration is primarily affected by oenological treatments such as the addition of copper sulfate to remove odors.

Early research (Dunbar and Wilson 1983) showed that water in the stems of grape vines (New Zealand) was similar in isotopic composition to that of the groundwater available to the plant during growth, whereas water in the fruit was enriched in ^{18}O. The same authors also examined different varieties of grape, grown at the same location, and did not find significant differences in the extent of ^{18}O enrichment. The enrichment was attributed to the combined effects of evapotranspiration and equilibration with atmospheric moisture, which gave a relationship between δ^2H and $\delta^{18}O$ significantly different to meteoric waters (Equation 8.3):

$$\delta^2H = 3.9 \times \delta^{18}O - 6.1 \qquad \text{Equation 8.3}$$

Subsequent work has shown that the $\delta^2H/\delta^{18}O$ composition of wine-water from the Napa Valley (northern California, USA) was described by a slope of 3.4 (Ingraham and Caldwell 1999) and that the stable isotopic composition of the grape water was controlled by transpiration in the few weeks prior to harvest. Due to the later time of grape harvest, red wines were found to be 4 to 5‰ enriched in ^{18}O compared to white wines. Surprisingly, this research proposed the isotopic composition of wine as a tool to reconstruct past weather, rather than vice versa. A study of 775 wines from six vintages in Germany showed that the relative humidity and δ^{18}O composition of the atmospheric humidity, for a period of 30 days prior to harvest, played the predominant role in the isotopic composition of wine (Hermann and Voerkelius 2008). The predictable relationship between the isotopic composition of precipitation and wine-water provides a powerful technique for wine authentication.

Factors, including oenological practices, which affect the variability of δ^{18}O composition of wine-water and the δ^{13}C and D/H values of ethanol have been studied since the 1990s; these parameters are now the basis of the official methods of analysis for identifying the authenticity of wine in terms of watering (down), sugar addition, and mislabelling (OIV methods MA-AS311-05, MA-AS312-07, and MA-AS2-12, EU reg. 555/2000). These parameters are derived primarily from the water and carbon dioxide adsorbed by the grape vines, which are, in turn, influenced by the geographical and climatic characteristics of the production area (Camin et al. 2015). The carbon isotopic composition of ethanol originates from that of sugar, $(D/H)_I$ is derived from glucose, and $(D/H)_{II}$ is derived from grape-water (Martin et al. 1986). Researchers have studied a number of factors, including: the strain of yeast, temperature of fermentation, duration of fermentation, the concentration of sugar and nutrients, and variation in the reaction rates induced by the medium. If fermentation proceeds to 70% conversion of the sugar to alcohol, none of these factors have been found to affect either the δ^{13}C or $(D/H)_I$ composition of wine-ethanol, whereas $(D/H)_{II}$ may exhibit significant variations (Vallet et al. 1996; Zhang et al. 1997; Fauhl and Wittkowski 2000; Perini et al. 2014). Recently, a study of the effect of adding gum Arabic to wine (Sprenger et al. 2015) concluded that only the δ^{18}O composition of the colloid was altered.

Only during the production of some particular types of wine or when particular oenological practices are employed are the isotopic ratios of wine-water and/or wine-ethanol altered. The first technique to consider is the practice of *grape withering* in which the fruit is dried, either on the vine (*en plein air*) or in ventilated or unventilated fruit drying rooms (*fruttaio*). This process has been studies in the production of Italian *passito* wines using a combination of SNIF-NMR, δ^{13}C and δ^{18}O signatures of wine ethanol, and the δ^{18}O signature of wine-water (Perini et al. 2015). The conclusion of this study was that the δ^{18}O signature of both wine-water and ethanol decreased in the case of passito wines produced in northern and central Italy (using post-harvest drying of the grapes) whereas for wines produced in southern Italy (where the main technique involves withering on the plant), δ^{18}O signatures tended to increase. As a result of these findings the authors recommend that particular attention be paid in the evaluation of the $\delta^{18}O$ data of passito wines for fraud detection, a recommendation that may have wider relevance.

Another process that can affect the isotopic composition of a wine is the more controversial practice of *must concentration*, to control the concentration of grape juice sugars prior to fermentation. Both reverse osmosis (RO) and high-vacuum evaporation (HVE) must concentration techniques have been studied (Guyon et al. 2006) and the RO technique was found to have very little effect on isotopic composition. In contract, HVE concentration (which involves water evaporation) caused a change in the $\delta^{18}O$ composition of the must and in the ethanol deuterium concentration.

Some producers will seek to reduce the alcoholic strength, prior to bottling, in order to maintain a character demanded by consumers or to supply the market for low alcohol or alcohol-free wines. As noted above, this can be simply achieved by the addition of water, although most regulatory bodies forbid this practice. De-alcoholization has, however, become a common practice in wine production and can be applied on an industrial scale using membrane contactors with water as the stripping agent. Researchers agree that this method of de-alcoholization affects the composition of wine (aroma, flavor, etc.) and has been found to increase the $\delta^{13}C$ composition of ethanol (Fedrizzi et al. 2014), but have no measurable effect on the $\delta^{18}O$ signature of the wine water (Ferrarini et al. 2016).

8.3.3 Wine authentication—In vino veritas

A review of published research related to wine authentication over the past 15 or so years (summarized in Tables 8.3 to 8.5) would suggest that the topic is divided roughly equally between methods based on stable isotopes, methods based on elemental composition, and methods based on chemical profiling: spectroscopic and/or chromatographic techniques. Although this division may reflect the emphasis of research within the subject, following the development of SNIF-NMR in 1982 the majority of samples are undoubtedly analyzed using stable isotope techniques. Since 1991, the Joint Research Centre of the European Community (EC-JRC) has compiled a database of the isotopic compositions of many thousands of wines from European and non-European countries together with information about the geographical origin, botany, and method of production; the current European Regulation is 555/2008. This database is available to official public laboratories for comparison with potentially counterfeit or substitute wines. The EC-JRC has also developed a number of European Reference Materials (ERMs) [certified for $\delta^{13}C$, $(D/H)_I$, $(D/H)_{II}$ and R composition] specifically to calibrate and validate instrumentation for the authentication of wine.

Measurements of $\delta^{18}O$ of wine-water, $\delta^{13}C$ of alcohol, and the relative deuterium substitution of methyl and methylene groups (R) are considered mature techniques, with official methods of analysis (EU, OIV, AOAC), and not in need of further research or improvement.

It may, therefore, appear that the matter of wine authentication is finalized:

- if the isotopic profile matches the EC-JRC database, it is likely to be genuine,
- if the isotopic profile does not match the EC-JRC database, it is likely to be bogus.

Any assessment against the EC-JRC database must take into account all of the possible oenological caveats discussed above.

Table 8.3 summarizes recent research publications of wine authentication using stable isotope techniques. While earlier papers studied European wines, following the establishment of the EC-JRC database the majority of research has studied wine from non-EU countries (Hungary and Slovenia joined the EU in 2004, Romania in 2007, and Croatia in 2013). It is also interesting to note that few of these authors used techniques in addition to stable isotopes, possibly confirming the pre-eminence of IRMS and SNIF-NMR.

Rather than improve the analytical technology, many of the publications listed in Table 8.3 have attempted to improve the functionality of the EC-JRC database through comparisons based on multivariate statistics such as Principal Component Analysis (PCA), Linear Discriminant Analysis (LDA) (Ogrinc et al. 2001; Dordevic et al. 2012), and Monte-Carlo substitution (Godelmann et al. 2013). Other authors have attempted to correlate the measurements in the database with parameters such as grape cultivar and vintage (Charlton et al. 2010; Costinel et al. 2011) or geographical location, precipitation, and temperature (Camin et al. 2015). These studies have the potential to extend the application of the database to a questioned sample for which no authentic comparison sample exists by answering the question: "Is the isotopic composition of this sample consistent with the claimed vintage, location, and grape variety?"

Some researchers have taken an alternative approach to compiling large databases of analytical measurements by making use of the largest (and global) collection of isotopic composition, the Global Network of Isotopes in Precipitation (GNIP) database of precipitation isotope ratios. The relationship between wine-water $\delta^{18}O$ and spatial climate and precipitation $\delta^{18}O$ signatures was evaluated across the wine growing regions of Washington, Oregon, and California (USA) (West et al. 2007). Figure 8.2a shows the geographical distribution of samples from that study superimposed on a digital elevation map. Significant correlations were observed between the measured wine-water $\delta^{18}O$ values from 2002 and the long-term average precipitation $\delta^{18}O$ values and late season climate.

Table 8.3 Summary of research on the authentication of wines using stable isotope techniques.

Author	Year	$\delta^{13}C$ alcohol	$\delta^{18}O$ water	SNIF-NMR	Other techniques	Details—regions, etc.
Martin et al.	1982			✓		France
Breas et al.	1994	✓	✓			EU countries
Roßmann et al.	1995	✓		✓		Italy, France, and Germany 1991–1993
Gimenez-Miralles et al.	1999	✓		✓		Spain
Ogrinc et al.	2001	✓	✓	✓		Slovenia 1996–1998
Kosir et al.	2001	✓		✓		Slovenia
Gremaud et al.	2002		✓	✓		Switzerland
Christoph et al.	2003	✓	✓	✓		Hungary and Croatia
Gremaud et al.	2004		✓	✓	elemental, chemical, IR	Switzerland
Adami et al.	2010	✓	✓			Brazil
Aghemo et al.	2011	✓		✓		Italy (Piedmont)
Costinel et al.	2011	✓	✓	✓		Romania
Dutra et al.	2011	✓	✓			Brazil
Dordevic et al.	2012					Italy 2010–2011
Magdas et al.	2012	✓	✓			Romania 2003–2004
Dutra et al.	2013	✓	✓		elemental	Brazil 2007–2008
Zyakun et al.	2013	✓			$\delta^{13}C$ dry residue	Russia
Godelmann et al.	2013			✓		Germany
Papotti et al.	2013			✓		Italy (Modena)
Pirnau et al.	2013	✓		✓		Romania
Hermann	2014		✓			German ice wines
Monakhova et al.	2014	✓	✓	✓		Germany
Camin et al.	2015	✓	✓	✓		Italy
Geana et al.	2016	✓	✓	✓	chemical	Romania

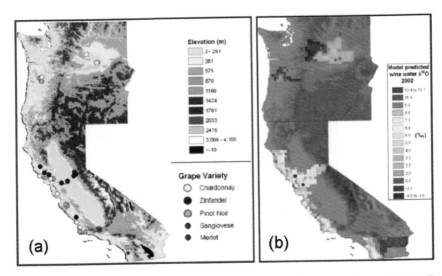

Figure 8.2 (a) the geographical distribution of samples from the study of West et al. (2007) superimposed on a digital elevation map and (b) the output of the regression model of wine water oxygen isotopic composition that was implemented spatially in a geographic information system (GIS) for 2002.

Reprinted with permission from West et al. 2007. Copyright 2007 American Chemical Society.

From this the authors developed a regression model that was implemented spatially in a geographic information system (GIS). Figure 8.2b shows the output of the model for 2002. The model was able to predict both the location and vintage of wines from the Napa and Livermore Valleys (California, USA).

From a purely scientific viewpoint, it may be disappointing that an elegant and flexible tool such as isoscapes has been over-shadowed by the monolith of the EC-JRC database. From a pragmatic (and forensic) viewpoint it is easy to understand how comparisons with this vast database are easily defendable in legal proceedings.

Some authors have used isotopic techniques to address directly issues of wine adulteration such as the addition of sugar and/or water (Dordevic et al. 2013; Geana et al. 2016) or added ethanol (Perini et al. 2013). A few researchers have studied the isotopic composition of components, other than ethanol or water, using more technically challenging and less proven techniques. Researchers have studied possible adulteration with glycerol using GC-IRMS (Calderone et al. 2004) and LC-IRMS (Cabañero et al. 2009), sugars (Guyon et al. 2011), and organic acids such as tartaric acid (Rojas et al. 2007; Guyon et al. 2013).

Tables 8.4 and 8.5 summarize recent publications of wine authentication using elemental and chemical profiling, respectively. All of these techniques can be regarded as *non-targeted*—i.e., the data are not acquired to address a specific question (in this case the authenticity of wine). Instead, the data are acquired and then explored using *chemometric* techniques to determine whether the data can answer the question: "Is this wine authentic?"

Although much of this research might be viewed as the work of laboratories without access to IRMS or SNIF-NMR equipment, some more positive features are apparent in these tables. First, approximately half of these studies concern non-EU wines, which may suggest that researchers in non-EU countries are willing to explore techniques beyond the standard of IRMS/SNIF-NMR. Second, approximately half of these studies concern EU-wines, but seek to resolve authentication at small regional, or even vineyard levels. It is very likely that variations in the style of wine, grape cultivar, etc. will be be better addressed by techniques that focus on color (IR/UV-vis) or aroma [electronic sensors or volatile organic compounds (VOCs)].

Although many of these studies showed great promise, the number of samples in any given project is tiny when compared to the EC-JRC database. Any up-and-coming technique will need to make a very a good case to be included in the EC-JRC or another global database. One interesting trend in Table 8.4 is the increasing, albeit still limited, use of strontium isotope ratios as a guide to wine authenticity (e.g., Durante et al. 2013; Durante et al. 2015; Petrini et al. 2015; Victor et al. 2015). This trend may suggest that the interpretation of database information through isoscapes is still the future of food forensics.

8.3.4 Sparkling wines—bubbles or fizz?

Wines with bubbles are associated with festivities, and consumers will happily pay a premium price to celebrate special birthdays, wedding anniversaries, etc. The appearance of a bottle, presented in a splendid wine cooler, and the pop of a cork makes any event special, regardless of the quality of the wine. Sparkling wines are a prime example of a premium product of which consumers have little, regular brand experience—a circumstance that invites deception.

Sparkling wines can be categorized according to production method, which also has a significant effect on consumer aspiration and, therefore, on the retail price.

- Champagne—traditional fermentation and aging in the same bottle
- sparkling wine—natural fermentation in large closed tanks (Charmat method)
- carbonated wine—CO_2 added from a source other than wine fermentation.

Table 8.4 Summary of research on the authentication of wines using elemental composition.

Authors	Year	Elements					Isotope ratio	Details—regions, etc.
		Major	Minor	Trace	Ultra-trace	#		
Rebolo et al.	2000		✓	✓		3		Spain: Galicia
Frias et al.	2001	✓	✓			6		Canary Islands: Lanzarote and La Palma islands
Barbaste et al.	2001						Pb	Unknown
Galani-Nikolakaki et al.	2002		✓	✓		10		Crete: Chania
Diaz et al.	2003	✓	✓	✓		11		Canary Islands
Larcher et al.	2003						Pb	Italy
Castineira et al.	2004		✓	✓		13		Germany: Baden, Rheingau, Rheinhessen, Pfalz
Jos et al.	2004	✓	✓			12		Cava/Champagne
Kment et al.	2005	✓	✓	✓		27		Czech Republic
Korenovska and Suhaj	2005	✓	✓	✓		9		Slovenia
Coetzee and Vanhaeke	2005						B	South Africa, France, Italy
Marini et al.	2006	✓	✓	✓	✓	56		Australia
Álvarez et al.	2007	✓	✓			12		Andalusia
Jaganathan et al.	2007			✓	✓	9		USA
Moreno et al.	2007	✓	✓			11		Czech Republic: Tacoronte-Acentejo, Valle de la Orotava
Razic et al.	2007		✓	✓		5		Serbia
Garcia-Ruiz et al.	2007	✓	✓	✓	✓	34	Sr	Spain, France, Switzerland, England
Galgano et al.	2008	✓	✓	✓	✓	29		Italy: Basilicata, Calabria, Campania
Gonzalvez et al.	2009	✓	✓	✓	✓	38		Spain: Utiel-Requena, Jumilla, Yecla, Valencia

	Year					Sr	Region
Fabani et al.	2010	✓	✓		7		Argentina
Catarino et al.	2011			✓	14		Portugal: Dao, Obidos, Palmela
Rodrigues et al.	2011	✓	✓	✓	17		Portugal: Alentejo, Bairrada, Dão, Vinho Verde
Di Paola-Naranjo et al.	2011					Sr	Argentina
Durante et al.	2013	✓	✓	✓		Sr	Italy: Modena (Lambrusco wines)
Geana et al.	2013		✓	✓	11		Romania: Valea Calugareasca, Murfatlar, Moldova
Shen et al.	2013	✓	✓	✓	19		Chinese rice wines
Marchionni et al.	2013					Sr	Italy
Azcarate et al.	2015		✓		5		Argentina: Mendoza, Rio Negro, San Juan, Salta
Durante et al.	2015					Sr	Italy: Modena (Lambrusco)
Petrini et al.	2015					Sr	Italy: northern, Glera vineyards (Prosecco)
Victor et al.	2015					Sr	Canada

Table 8.5 Summary of research on the authentication of wines using chemical composition.

Authors	Year	Sensory		IR	UV-Vis	Chemical		Other	Details—country, variety, etc.
		Human	Electronic			Phenolic	VCO		
Kosir et al.	1998							2D ^{1}H and ^{13}C NMR	Slovenia
Fischer et al.	1999	✓							German Riesling
Kotseridis et al.	1998						✓		Merlot noir cultivar clones
Edelmann et al.	2001			✓	✓				Austria: Lemberger, St. Laurent, Zweigelt
Kallithraka et al.	2001	✓				✓			Greece
Marengo et al.	2002						✓		Italy: Piedmont
Kiss and Sass-Kiss	2005							amines/acids	Hungary: Tokaj region
Bevin et al.	2006			✓					Australia: Cabernet Sauvignon, Shiraz and Merlot, Chardonnay, Riesling, Sauvignon Blanc, Viognier
Pennington et al.	2007			✓			✓		USA
Bevin et al.	2008			✓					Australia: Shiraz, Cabernet Sauvignon, Merlot
Liu et al.	2008			✓	✓				Riesling: Australia, New Zealand, France, Germany
Viggiani et al.	2008							2D ^{1}H and ^{13}C NMR	Italy: Aglianico wines from Basilicata and Campania
Chambery et al.	2009							peptides	Italy: Campania
Louw et al.	2009			✓					South Africa
Casale et al.	2010		✓	✓					Italy: Barbera d'Alba, Dolcetto d'Alba

Reference	Year							Method	Region / Variety
Cynkar et al.	2010					✓			Australia, Spain: Tempranillo
Fischer et al.	2010	✓							Germany: Rheingau
Jaitz et al.	2010				✓				Austria
Mazzei et al.	2010							¹H NMR	Italy: Aglianico red wines from the Campania region
Bauer et al.	2011	✓							German Riesling
Cozzolino et al.	2011			✓		✓			Australia, New Zealand: Sauvignon Blanc
Gutierrez et al.	2011				✓				Catalan
Ioannou-Papayianni et al.	2011			✓					Cyprus
Anga et al.	2012					✓			India
Kumsta et al.	2012				✓				Czech Republic
Bednarova et al.	2013							classical	Slovenia
Fabani et al.	2013		✓						Argentina: Tulum
Lampir	2013				✓				Czech Republic
Lampir and Pavlousek	2013				✓				Czech Republic and Perna
Pavlousek and Kumsta	2013				✓				Czech Republic
Salvatore et al.	2013				✓				Italy: Lambrusco (Grasparossa, Salamino, Sorbara)
Ceto et al.	2014						✓		Cava
Gomez-Meire et al.	2014		✓						Spain: Galicia

Table 8.5 contd. ...

...Table 8.5 contd.

Authors	Year	Sensory		IR	UV-Vis	Chemical		Other	Details—country, variety, etc.
		Human	Electronic			Phenolic	VCO		
Martelo-Vidal and Vazquez	2014			✓	✓				Galicia: godello, treixadura, albarino and palomino
Sen and Tokatli	2014					✓			Turkey
Springer et al.	2014						✓		Germany: Riesling, Silvaner, Pinot Gris, Pinot Blanc
Vergara et al.	2011					✓			Chile: Carménère, Merlot
Sen and Tokatli	2016				✓				Syrah, Cabernet Sauvignon, Chardonnay, Muscat, Emir

At the super-premium end of the sparkling wine market is Champagne. Although many wines are produced using the *Champagne method* (such as cava), only wines produced in the designated region of France can make this claim of origin. At the other end of the market, carbonated wine (often termed *spumante*) are inexpensive and of variable quality. Regardless of the production method, all sparkling wines begin as a *base wine*, a dry wine from traditional fermentation. The bubbles in both Champagnes and sparkling wines are produced by secondary fermentation and a source of fermentable sugar must be added. In the case of Champagne this is added as a mixture of the base wine, sugar, and yeast (*liqueur de tirage*) and in the case of sparkling wine as a mixture of sugar and yeast. During secondary fermentation, the ethanol content of the wine typically increases by 1.3 to 1.4%. At the end of secondary fermentation the wine is typically dry with high acidity, and benefits from sweetening. Champagne is typically sweetened with sucrose dissolved in the base wine (*dosage*) and sparkling wine with sugar in combination with ascorbic acid and/or SO_2, to prevent further fermentation.

Early research demonstrated that, based on $\delta^{13}C$ measurements, it was straightforward to determine whether the CO_2 from New Zealand (Dunbar 1982) or Spanish (González-Martin et al. 1997) sparkling wines was derived from a C_3 or C_4 plant or industrial sources. By comparing the $\delta^{13}C$ signature of the wine alcohol with the CO_2 it was also possible to make inferences about the manufacturing process, as illustrated in Table 8.6 (Martinelli et al. 2003). The data in this table show that the CO_2 present in Brazilian sparkling wines was highly enriched in ^{13}C, indicating that C_4 sugar (most likely cane sugar) had been added to promote secondary fermentation. The ^{13}C content of the sweeter Brazilian sparkling wines (predominantly sugar) was also enriched, but not to the same extent as the CO_2. This would suggest that the sweetness of the wines was due to a combination of natural (C_3) grape sugars and added C_4 sugar.

The $\delta^{13}C$ signature of CO_2 can also indicate if a sparkling wine has been *gasified*—the illegal addition of industrial CO_2 derived from the combustion of fossil fuels. Industrial sources of CO_2 have $\delta^{13}C$ values lower than $-29‰$ and, therefore very different from bubbles deriving from either grape or cane and beet sugar (OIV MA AS314-03).

Table 8.6 The carbon isotopic composition of wines and CO_2: adapted from Martinelli et al. 2003.

	$\delta^{13}C$ – wine (‰)	$\delta^{13}C$ – CO_2 (‰)
Brazilian brut	20.5 ± 1.2	−10.8 ± 1.2
Brazilian demi-sec	−18.1 ± 1.3	
Brazilian doux	−15.8	
Argentina and Chile	−26.1 ± 1.6	
Europe	−25.5 ± 1.2	−22.0 ± 1.2

References

Baudler, R., L. Adam, A. Rossmann, G. Versini and K. -H. Engel. 2006. Influence of the distillation step on the ratios of stable isotopes of ethanol in cherry brandies. *Journal of Agricultural and Food Chemistry* 54: 864–869.

BBC News. 2014. Wine fraud: How easy is it to fake a 50-year-old bottle? http://www.bbc.com/news/business-28697721 (accessed Feb. 2016).

Bostic, J. N., S. J. Palafox, M. E. Rottmueller and A. H. Jahren. 2015. Effect of baking and fermentation on the stable carbon and nitrogen isotope ratios of grain-based food. *Rapid Communications in Mass Spectrometry* 29: 937–947.

Bowen, G. J. and J. Revenaugh. 2003. Interpolating the isotopic composition of modern meteoric precipitation. *Water Resource Research* 39: 1299–1311.

Brettell, R., J. Montgomery and J. Evans. 2012. Brewing and stewing: The effect of culturally mediated behaviour on the oxygen isotope composition of ingested fluids and the implications for human provenance studies. *Journal of Analytical Atomic Spectrometry* 27: 778–785.

Cabañero, A. I., J. L. Recio and M. Ruperez. 2008. Isotope ratio mass spectrometry coupled to liquid and gas chromatography for wine ethanol characterization. *Rapid Communications in Mass Spectrometry* 22: 3111–3118.

Cabañero, A. I., J. L. Recio and M. Rupérez. 2009. Simultaneous stable carbon isotopic analysis of wine glycerol and ethanol by liquid chromatography coupled to isotope ratio mass spectrometry. *Journal of Agricultural and Food Chemistry* 58: 722–728.

Calderone, G., N. Naulet, C. Guillou and F. Reniero. 2004. Characterization of European wine glycerol: Stable carbon isotope approach. *Journal of Agricultural and Food Chemistry* 52: 5902–5906.

Camin, F., N. Dordevic, R. Wehrens, M. Neteler, L. Delucchi, G. Postma and L. Buydens. 2015. Climatic and geographical dependence of the H, C and O stable isotope ratios of Italian wine. *Analytica Chimica Acta* 853: 384–390.

Charlton, A. J., M. S. Wrobel, I. Stanimirova, M. Daszykowski, H. H. Grundy and B. Walczak. 2010. Multivariate discrimination of wines with respect to their grape varieties and vintages. *European Food Research and Technology* 231: 733–743.

Chesson, L. A., L. O. Valenzuela, S. P. O'grady, T. E. Cerling and J. R. Ehleringer. 2010. Links between purchase location and stable isotope ratios of bottled water, soda, and beer in the United States. *Journal of Agricultural and Food Chemistry* 58: 7311–7316.

Chesson, L. A., B. J. Tipple, J. Howa, G. J. Bowen, J. E. Barnette, T. E. Cerling and J. R. Ehleringer. 2014. Stable isotopes in forensic applications. pp. 285–317. *In*: H. Holland and K. Turekian (eds.). *Treatise on Geochemistry 2nd Ed.* Oxford, UK: Elsevier Ltd.

Compendium of International Analysis of Methods—OIV-MA-311-05: R2011. *Determination of the deuterium in ethanol by SNIF-NMR.*

Costinel, D., A. Tudorache, R. E. Ionete and R. Vremera. 2011. The impact of grape varieties to wine isotopic characterization. *Analytical Letters* 44: 2856–2864.

Craig, H. 1961. Isotopic variation in meteoric waters. *Science* 133: 1702–1703.

Diaz, C., J. E. Conde, D. Estevez, S. J. Perez Olivero and J. P. Perez Trujillo. 2003. Application of multivariate analysis and artificial neural networks for the differentiation of red wines from the Canary Islands according to the island of origin. *Journal of Agricultural and Food Chemistry* 51: 4303–4307.

Dordevic, N., F. Camin, R. M. Marianella, G. J. Postma, L. M. C. Buydens and R. Wehrens. 2013. Detecting the addition of sugar and water to wine. *Australian Journal of Grape and Wine Research* 19: 324–330.

Dordevic, N., R. Wehrens, G. J. Postma, L. M. C. Buydens and F. Camin. 2012. Statistical methods for improving verification of claims of origin for Italian wines based on stable isotope ratios. *Analytica Chimica Acta* 757: 19–25.

Dunbar, J. 1982. Use of $^{13}C/^{12}C$ ratios for studying the origin of CO_2 in sparkling wines. *Fresenius' Z. Analytical Chemistry* 311: 578–580.

Dunbar, J. and A. T. Wilson. 1983. Oxygen and hydrogen isotopes in fruit and vegetable juices. *Plant Physiology* 72: 725–727.

Fedrizzi, B., E. Nicolis, F. Camin, E. Bocca, C. Carbognin, M. Scholz, P. Barbieri, F. Finato and R. Ferrarini. 2014. Stable isotope ratios and aroma profile changes induced due to innovative wine dealcoholisation approaches. *Food and Bioprocess Technology* 7: 62–70.

Ferrarini, R., G. M. Ciman, F. Camin, S. Bandini and C. Gostoli. 2016. Variation of oxygen isotopic ratio during wine dealcoholization by membrane contactors: Experiments and modelling. *Journal of Membrane Science* 498: 385–394.

Fauhl, C. and R. Wittkowski. 2000. Oenological influence on the D/H ratios of wine ethanol. *Journal of Agriculture and Food Chemistry* 48: 3979–3984.

Geana, E. I., R. Popescu, D. Costinel, O. R. Dinca, I. Stefanescu, R. E. Ionete and C. Bala. 2016. Verifying the red wines adulteration through isotopic and chromatographic investigations coupled with multivariate statistic interpretation of the data. *Food Control* 62: 1–9.

Godelmann, R., F. Fang, E. Humpfer, B. Schuetz, M. Bansbach, H. Schaefer and M. Spraul. 2013. Targeted and nontargeted wine analysis by ^{1}H NMR spectroscopy combined with multivariate statistical analysis differentiation of important parameters: grape variety, geographical origin, year of vintage. *Journal of Agricultural and Food Chemistry* 61: 5610–5619.

González-Martin, I., C. González-Pérez and E. Marqués-Macias. 1997. Contribution to the study of the origin of CO_2 in Spanish sparkling wines by determination of the $^{13}C/^{12}C$ isotope ratio. *Journal of Agricultural and Food Chemistry* 45: 1149–1151.

Grindlay, G., J. Mora, S. Maestre and L. Gras. 2008. Application of a microwave-based desolvation system for multi-elemental analysis of wine by inductively coupled plasma based techniques. *Analytica Chimica Acta* 629: 24–37.

Gupta, M., J. B. Leen, E. S. F. Berman and A. Ciambotti. 2013. Laser-based measurements of $^{18}O/^{16}O$ stable isotope ratios ($\delta^{18}O$) in wine samples. *International Journal of Wine Research* 5: 47–54.

Guyon, F., C. Douet, S. Colas, M. -H. Salagoïty and B. Medina. 2006. Effects of must concentration techniques on wine isotopic parameters. *Journal of Agricultural and Food Chemistry* 54: 9918–9923.

Guyon, F., L. Gaillard, A. Brault, N. Gaultier, M. -H. Salagoity and B. Medina. 2013. Potential of ion chromatography coupled to isotope ratio mass spectrometry via a liquid interface for beverages authentication. *Journal of Chromatography A* 1322: 62–68.

Guyon, F., L. Gaillard, M. -H. Salagoity and B. Medina. 2011. Intrinsic ratios of glucose, fructose, glycerol and ethanol $^{13}C/^{12}C$ isotopic ratio determined by HPLC-co-IRMS: Toward determining constants for wine authentication. *Analytical and Bioanalytical Chemistry* 401: 1551–1558.

Guyon, F., C. Van Leeuwen, L. Gaillard, M. Grand, S. Akoka, G. S. Remaud, N. Sabathie and M. -H. Salagoity. 2015. Comparative study of ^{13}C composition in ethanol and bulk dry wine using isotope ratio monitoring by mass spectrometry and by nuclear magnetic resonance as an indicator of vine water status. *Analytical and Bioanalytical Chemistry* 407: 9053–9060.

Hermann, A. and S. Voerkelius. 2008. Meteorological impact on oxygen isotope ratios of German wines. *American Journal of Enology and Viticulture* 59: 194–199.

Hornsey, I. S. 2007. *The Chemistry and Biology of Winemaking.* Cambridge, UK: Royal Society of Chemistry.

Ibanez, J. G., A. Carreon-Alvarez, M. Barcena-Soto and N. Norberto Casillas. 2008. Metals in alcoholic beverages: A review of sources, effects, concentrations, removal, speciation, and analysis. *Journal of Food Composition and Analysis* 21: 672–683.

Ingraham, N. L. and E. A. Caldwell. 1999. Influence of weather on the stable isotopic ratios of wines: Tools for weather/climate reconstruction? *Journal of Geophysical Research: Atmospheres* 104: 2185–2194.

Martin, G. J., M. L. Martin, F. Mabon and M. -J. Michon. 1982. Identification of the origin of natural alcohols by natural abundance hydrogen-2 Nuclear Magnetic Resonance. *Analytical Chemisty* 54: 2382–2384.

Martin, G. J., B. L. Zhang, N. Maulet and M. L. Martin. 1986. Deuterium transfer in the bioconversion of glucose to ethanol studied by specific isotope labelling at the natural abundance level. *Journal of the American Chemical Society* 108: 5116–5122.

Martinelli, L. A., M. Z. Moreira, J. P. H. B. Ometto, A. R. Alcarde, L. A. Rizzon, E. Stange and J. R. Ehleringer. 2003. Stable carbon isotopic composition of the wine and CO_2 bubbles of sparkling wines: Detecting C_4 sugar additions. *Journal of Agricultural and Food Chemistry* 51: 2625–2631.

Medina, B., S. Augagneur, M. Barbaste, F. E. Grouset and P. Buat-Meard. 2000. Influence of atmospheric pollution on the lead content of wines. *Food Additives and Contaminants* 17: 435–445.

Organisation internationale de la Vigne et du Vin. State of the Vitiviniculture world Market May 2014. http://www.oiv.int/en/ (accessed Feb. 2016).

Ogrinc, N., I. J. Košir, M. Kocjančič and J. Kidrič. 2001. Determination of authenticy, regional origin, and vintage of Slovenian wines using a combination of IRMS and SNIF-NMR analyses. *Journal of Agricultural and Food Chemistry* 49: 1432–1440.

Perini, M. and F. Y. Camin. 2013. $\delta^{18}O$ of ethanol in wine and spirits for authentication purposes. *Journal of Food Science* 78: C839–C844.

Perini, M., R. Guzzon, M. Simoni, M. Malacarne, R. Larcher and F. Camin. 2014. The effect of stopping alcoholic fermentation on the variability of H, C and O stable isotope ratios of ethanol. *Food Control* 40: 368–373.

Perini, M., L. Rolle, P. Franceschi, M. Simoni, F. Torchio, V. Di Martino, R. M. Marianella, V. Gerbi and F. Camin. 2015. H, C, and O stable isotope ratios of passito wine. *Journal of Agricultural and Food Chemistry* 63: 5851–5857.

Pohl, P. 2007. What do metals tell us about wine? *Trends in Analytical Chemistry* 26(9): 941–949.

Rojas, J. M. M., S. Cosofret, F. Reniero, C. Guillou and F. Serra. 2007. Control of oenological products: Discrimination between different botanical sources of L-tartaric acid by isotope ratio mass spectrometry. *Rapid Communications in Mass Spectrometry* 21: 2447–2450.

Sprenger, S., K. Meylahn, A. Zaar, H. Dietrich and F. Will. 2015. Identification of gum Arabic in white wine based on colloid content, colloid composition and multi-element stable isotope analysis. *European Food Research Technology* 240: 909–921.

The Drinks Business 2014. Fake big-brand wine on the rise. http://www.thedrinksbusiness.com/2014/07/fake-wine-crime-on-the-rise/(accessed Feb. 2016).

United States Department of Agriculture (USDA) Economic Research Service http://www.ers.usda.gov/data-products/food-expenditures.aspx (accessed Feb. 2016).

Vallet C, R. Said, C. Rabiller and M. L. Martin. 1996. Natural abundance isotopic fractionation in the fermentation reaction: Influence of the nature of the yeast. *Bioorganic Chemistry* 24: 319–330.

West, J. B., J. R. Ehleringer and T. E. Cerling. 2007. Geography and vintage predicted by a novel GIS model of wine $\delta^{18}O$. *Journal of Agricultural and Food Chemistry* 55: 7075–7083.

Wittwer, G. and K. Anderson. 2009. *Global Wine Markets, 1961 to 2003: A Statistical Compendium.* Adelaide: University of Adelaide Press.

Zhang, B. -L. Y., C. Vallet, Y. L. Martin and M. L. Martin. 1997. Natural abundance isotopic fractionation in the fermentation reaction: Influence of the fermentation medium. *Bioorganic Chemistry* 25: 117–129.

References in Tables

Table 8.3

Adami, L., S. V. Dutra, A. R. Marcon, G. J. Carnieli, C. A. Roani and R. Vanderlinde. 2010. Geographic origin of southern Brazilian wines by carbon and oxygen isotope analyses. *Rapid Communications in Mass Spectrometry* 24: 2943–2948.

Aghemo, C., A. Albertino, R. Gobetto and F. Spanna. 2011. Correlation between isotopic and meteorological parameters in Italian wines: A local-scale approach. *Journal of the Science of Food and Agriculture* 91: 2088–2094.

Breas, O., F. Reniero and G. Serrini. 1994. Isotope ratio mass spectrometry: Analysis of wines from different European countries. *Rapid Communications in Mass Spectrometry* 8: 967–970.

Camin, F., N. Dordevic, R. Wehrens, M. Neteler, L. Delucchi, G. Postma and L. Buydens. 2015. Climatic and geographical dependence of the H, C and O stable isotope ratios of Italian wine. *Analytica Chimica Acta* 853: 384–390.

Christoph, N., A. Rossmann and S. Voerkelius. 2003. Possibilities and limitations of wine authentication using stable isotope and meteorological data, data banks and statistical tests. Part 1: Wines from Franconia and Lake Constance 1992 to 2001. *Mitt Klosterneuburg* 53: 23–40.

Costinel, D., A. Tudorache, R. E. Ionete and R. Vremera. 2011. The impact of grape varieties to wine isotopic characterization. *Analytical Letters* 44: 2856–2864.

Dordevic, N., R. Wehrens, G. J. Postma, L. M. C. Buydens and F. Camin. 2012. Statistical methods for improving verification of claims of origin for Italian wines based on stable isotope ratios. *Analytica Chimica Acta* 757: 19–25.

Dutra, S. V., L. Adami, A. R. Marcon, G. J. Carnieli, C. A. Roani, F. R. Spinelli, S. Leonardelli, C. Ducatti, M. Z. Moreira and R. Vanderlinde. 2011. Determination of the geographical origin of Brazilian wines by isotope and mineral analysis. *Analytical and Bioanalytical Chemistry* 401: 1571–1576.

Dutra, S. V., L. Adami, A. R. Marcon, G. J. Carnieli, C. A. Roani, F. R. Spinelli, S. Leonardelli and R. Vanderlinde. 2013. Characterization of wines according the geographical origin by analysis of isotopes and minerals and the influence of harvest on the isotope values. *Food Chemistry* 141: 2148–2153.

Geana, E. I., R. Popescu, D. Costinel, O. R. Dinca, I. Stefanescu, R. E. Ionete and C. Bala. 2016. Verifying the red wines adulteration through isotopic and chromatographic investigations coupled with multivariate statistic interpretation of the data. *Food Control* 62: 1–9.

Gimenez-Miralles, J. E., D.M. Salazar and I. Solana. 1999. Regional origin assignment of red wines from Valencia (Spain) by ^{2}H NMR and ^{13}C IRMS stable isotope analysis of fermentative ethanol. *Journal of Agricultural and Food Chemistry* 47: 2645–2652.

Godelmann, R., F. Fang, E. Humpfer, B. Schuetz, M. Bansbach, H. Schaefer and M. Spraul. 2013. Targeted and nontargeted wine analysis by ^{1}H NMR spectroscopy combined with multivariate statistical analysis; differentiation of important parameters: Grape variety, geographical origin, year of vintage. *Journal of Agricultural and Food Chemistry* 61: 5610–5619.

González-Martin, I., C. González-Pérez and E. Marqués-Macias. 1997. Contribution to the study of the origin of CO_2 in Spanish sparkling wines by determination of the $^{13}C/^{12}C$ isotope ratio. *Journal of Agricultural and Food Chemistry* 45: 1149–1151.

Gremaud, G., E. Pfammatter, U. Piantini and S. Quaile. 2002. Classification of Swiss wines on a regional scale by means of multi-isotopic analysis combined with chemometric methods. *Mitt. Lebensmittelunters Hyg.* 93: 44–56.

Gremaud, G., S. Quaile, U. Piantini, E. Pfammatter and C. Corvi. 2004. Characterization of Swiss vineyards using isotopic data in combination with trace elements and classical parameters. *European Food Research and Technology* 219: 97–104.

Hermann, A. 2014. Icewine authentication by ^{18}O measurements. *American Journal of Enology and Viticulture* 65: 499–503.

Kosir, I. J., M. Kocjancic, N. Ogrinc and J. Kidric. 2001. Use of SNIF-NMR and IRMS in combination with chemometric methods for the determination of chaptalisation and geographical origin of wines (the example of Slovenian wines). *Analytica Chimica Acta* 429: 195–206.

Magdas, D. A., S. Cuna, G. Cristea, R. E. Ionete and D. Costinel. 2012. Stable isotopes determination in some Romanian wines. *Isotopes in Environmental and Health Studies* 48: 345–353.

Martin, G. J., M. L. Martin, F. Mabon and M. -J. Michon. 1982. Identification of the origin of natural alcohols by natural abundance hydrogen-2 Nuclear Magnetic Resonance. *Analytical Chemistry* 54: 2382–2384.

Monakhova, Y. B., R. Godelmann, A. Hermann, T. Kuballa, C. Cannet, H. Schaefer, M. Spraul and D. N. Rutledge. 2014. Synergistic effect of the simultaneous chemometric analysis of ^{1}H NMR spectroscopic and stable isotope (SNIF-NMR, ^{18}O, ^{13}C) data: Application to wine analysis. *Analytica Chimica Acta* 833: 29–39.

Ogrinc, N., I. J. Kosir, M. Kocjancic and J. Kidric. 2001. Determination of authenticity, regional origin and vintage of Slovenian wines using a combination of IRMS and SNIF-NMR analyses. *Journal of Agricultural and Food Chemistry* 49: 1432–1440.

Papotti, G., D. Bertelli, R. Graziosi, M. Sivestri, L. Bertacchini, C. Durante and M. Plessi. 2013. Application of one- and two-dimensional NMR spectroscopy for the characterization of protected designation of origin Lambrusco wines of Modena. *Journal of Agricultural and Food Chemistry* 61: 1741–1746.

Pirnau, A., M. Bogdan, D. A. Magdas and D. Statescu. 2013. Isotopic analysis of some Romanian wines by ^{2}H NMR and IRMS. *Food Biophysics* 8: 24–28.

Roßmann, A., H. L. Schmidt, F. Reniero, G. Versini, I. Moussa and M. H. Merle. 1995. Stable carbon isotope content in ethanol of EC data bank wines from Italy, France and Germany. *Z Lebensm Unters Forsch A* 203: 293–301.

Zyakun, A. M., L. A. Oganesyants, A. L. Panasyuk, E. I. Kuz'mina, A. A. Shilkin, B. P. Baskunov, V. N. Zakharchenko and V. P. Peshenko. 2013. Mass spectrometric analysis of the ^{13}C/^{12}C abundance ratios in vine plants and wines depending on regional climate factors (Krasnodar krai and Rostov oblast, Russia). *Journal of Analytical Chemistry* 68: 1136–1141.

Table 8.4

Álvarez, M., I. M. Moreno, Á. Jos, A. M. Cameán and A. Gustavo González. 2007. Differentiation of two Andalusian DO fino wines according to their metal content from ICP-OES by using supervised pattern recognition methods. *Microchemical Journal* 87: 72–76.

Azcarate, S. M., L. D. Martinez, M. Savio, J. M. Camina and R. A. Gil. 2015. Classification of monovarietal Argentinean white wines by their elemental profile. *Food Control* 57: 268–274.

Barbaste, M., L. Halicz, B. Medina, H. Emteborg, F. C. Adams and R. Lobinski. 2001. Evaluation of the accuracy of the determination of lead isotope ratios in wine by ICP MS using quadrupole, multicollector magnetic sector and time-of-flight analyzers. *Talanta* 54: 307–317.

Castineira Gomez, M. D. M., I. Feldmann, N. Jakubowski and J. T. Andersson. 2004. Classification of German white wines with certified brand of origin by multielement quantitation and pattern recognition techniques. *Journal of Agricultural and Food Chemistry* 52: 2962–74.

Catarino, S., I. M. Trancoso, M. Madeira, F. Monteiro, R. Bruno De Sousa and A. S. Curvelo-Garcia. 2011. Rare earths data for geographical origin assignment of wine: A Portuguese case study. *Bulletin O.I.V.* 84: 333–346.

Coetzee, P. P. and F. Vanhaecke. 2005. Classifying wine according to geographical origin via quadrupole-based ICP-mass spectrometry measurements of boron isotope ratios. *Analytical and Bioanalytical Chemistry* 383: 977–984.

Di Paola-Naranjo, R. D., M. a. V. Baroni, N. S. Podio, H. C. R. Rubinstein, M. a. P. Fabani, R. L. G. Badini, M. Inga, H. C. A. Ostera, M. Cagnoni, E. Gallegos, E. Gautier, P. Peral-GarcíA, J. Hoogewerff and D. A. Wunderlin. 2011. Fingerprints for main varieties of Argentinean wines: Terroir differentiation by inorganic, organic, and stable isotopic analyses coupled to chemometrics. *Journal of Agricultural and Food Chemistry* 59: 7854–7865.

Diaz, C., J. E. Conde, D. Estevez, S. J. Perez Olivero and J. P. Perez Trujillo. 2003. Application of multivariate analysis and artificial neural networks for the differentiation of red wines from the Canary Islands according to the island of origin. *Journal of Agricultural and Food Chemistry* 51: 4303–4307.

Durante, C., C. Baschieri, L. Bertacchini, M. Cocchi, S. Sighinolfi, M. Silvestri and A. Marchetti. 2013. Geographical traceability based on ^{87}Sr/^{86}Sr indicator: A first approach for PDO Lambrusco wines from Modena. *Food Chemistry* 141: 2779–2787.

Durante, C., C. Baschieri, L. Bertacchini, D. Bertelli, M. Cocchi, A. Marchetti, D. Manzini, G. Papotti and S. Sighinolfi. 2015. An analytical approach to Sr isotope ratio determination in Lambrusco wines for geographical traceability purposes. *Food Chemistry* 173: 557–563.

Fabani, M. P., R. C. Arrúa, F. Vázquez, M. P. Diaz, M. V. Baroni and D. A. Wunderlin. 2010. Evaluation of elemental profile coupled to chemometrics to assess the geographical origin of Argentinean wines. *Food Chemistry* 119: 372–379.

Frias, S., J. P. P. Trujillo, E. M. Pena and J. E. Conde. 2001. Classification and differentiation of bottled sweet wines of Canary Islands (Spain) by their metallic content. *European Food Research and Technology* 213: 145–149.

Galani-Nikolakaki, S., N. Kallithrakas-Kontos and A. A. Katsanos. 2002. Trace element analysis of Cretan wines and wine products. *Science of the Total Environment* 285: 155–163.

Galgano, F., F. Favati, M. Caruso, T. Scarpa and A. Palma. 2008. Analysis of trace elements in southern Italian wines and their classification according to provenance. *LWT—Food Science and Technology* 41: 1808–1815.

Garcia-Ruiz, S., M. Moldovan, G. Fortunato, S. Wunderli and J. I. G. Alonso. 2007. Evaluation of strontium isotope abundance ratios in combination with multi-elemental analysis as a possible tool to study the geographical origin of ciders. *Analytica Chimica Acta* 590: 55–66.

Geana, I., A. Iordache, R. Ionete, A. Marinescu, A. Ranca and M. Culea. 2013. Geographical origin identification of Romanian wines by ICP-MS elemental analysis. *Food Chemistry* 138: 1125–1134.

Gonzalvez, A., A. Llorens, M. L. Cervera, S. Armenta and M. De La Guardia. 2009. Elemental fingerprint of wines from the protected designation of origin Valencia. *Food Chemistry* 112: 26–34.

Jaganathan, J., A. Mabud and S. Dugar. 2007. Geographic origin of wine via trace and ultra-trace elemental analysis using inductively coupled plasma mass spectrometry and chemometrics. *ACS Symposium Series* 952: 200–206.

Jos, A., I. Moreno, A. G. Gonzalez, G. Repetto and A. M. Camean. 2004. Differentiation of sparkling wines (cava and champagne) according to their mineral content. *Talanta* 63: 377–382.

Kment, P., M. Mihaljevič, V. Ettler, O. Šebek, L. Strnad and L. Rohlová. 2005. Differentiation of Czech wines using multielement composition—A comparison with vineyard soil. *Food Chemistry* 91: 157–165.

Korenovska, M. and M. Suhaj. 2005. Identification of some Slovakian and European wines origin by the use of factor analysis of elemental data. *European Food Research and Technology* 221: 550–558.

Larcher, R., G. Nicolini and P. Pangrazzi. 2003. Isotope ratios of lead in Italian wines by inductively coupled plasma mass spectrometry. *Journal of Agricultural and Food Chemistry* 51: 5956–5961.

Marchionni, S., E. Braschi, S. Tommasini, A. Bollati, F. Cifelli, N. Mulinacci, M. Mattei and S. Conticelli. 2013. High-precision [87]Sr/[86]Sr analyses in wines and their use as a geological fingerprint for tracing geographic provenance. *Journal of Agricultural and Food Chemistry* 61: 6822–6831.

Marini, F., R. Bucci, A. L. Magri and A. D. Magri. 2006. Authentication of Italian CDO wines by class-modeling techniques. *Chemometrics and Intelligent Laboratory Systems* 84: 164–171.

Moreno, I. M., D. González-Weller, V. Gutierrez, M. Marino, A. M. Cameán, A. G. González and A. Hardisson. 2007. Differentiation of two Canary DO red wines according to their metal content from inductively coupled plasma optical emission spectrometry and graphite furnace atomic absorption spectrometry by using probabilistic neural networks. *Talanta* 72: 263–268.

Petrini, R., L. Sansone, F. F. Slejko, A. Buccianti, P. Marcuzzo and D. Tomasi. 2015. The [87]Sr/[86]Sr strontium isotopic systematics applied to Glera vineyards: A tracer for the geographical origin of the Prosecco. *Food Chemistry* 170: 138–144.

Razic, S., D. Cokesa and S. Sremac. 2007. Multivariate data visualization methods based on elemental analysis of wines by atomic absorption spectrometry. *Journal of the Serbian Chemical Society* 72: 1487–1492.

Rebolo, S., R. M. Pena, M. J. Latorre, S. Garcia, A. M. Botana and C. Herrero. 2000. Characterisation of Galician (NW Spain) Ribeira Sacra wines using pattern recognition analysis. *Analytical Chimica Acta* 417: 211–220.

Rodrigues, S. M., M. Otero, A. A. Alves, J. Coimbra, M. A. Coimbra, E. Pereira and A. C. Duarte. 2011. Elemental analysis for categorization of wines and authentication of their certified brand of origin. *Journal of Food Composition and Analysis* 24: 548–562.

Shen, F., J. Wu, Y. Ying, B. Li and T. Jiang. 2013. Differentiation of Chinese rice wines from different wineries based on mineral elemental fingerprinting. *Food Chemistry* 141: 4026–4030.

Victor, V., S. Ross, P. Karine, P. Andre, H. Jean-Francois and W. David. 2015. Strontium isotope characterization of wines from the Quebec (Canada) terroir. *Procedia Earth and Planetary Science* 13: 252–255.

Table 8.5

Anga, H. K., P. C. Panchariya and A. L. Sharma. 2012. Authentication of Indian wines using voltammetric electronic tongue coupled with artificial neural networks. *Sensors and Transducers Journal* 145: 65–76.

Bauer, A., S. Wolz, A. Schormann and U. Fischer. 2011. Authentication of different terroirs of German Riesling applying sensory and flavor analysis. *ACS Symposium Series. 1081 (Progress in Authentication of Food and Wine)* 131–149.

Bednarova, A., R. Kranvogl, D. B. Voncina, T. Jug and E. Beinrohr. 2013. Characterization of Slovenian wines using multidimensional data analysis from simple enological descriptors. *Acta Chimica Slovenica* 60: 274–286.

Bevin, C. J., R. G. Dambergs, A. J. Fergusson and D. Cozzolino. 2008. Varietal discrimination of Australian wines by means of mid-infrared spectroscopy and multivariate analysis. *Analytica Chimica Acta* 621: 19–23.

Bevin, C. J., A. J. Fergusson, W. B. Perry, L. J. Janik and D. Cozzolino. 2006. Development of a rapid "fingerprinting" system for wine authenticity by mid-infrared spectroscopy. *Journal of Agricultural and Food Chemistry* 54: 9713–9718.

Casale, M., P. Oliveri, C. Armanino, S. Lanteri and M. Forina. 2010. NIR and UV-vis spectroscopy, artificial nose and tongue: Comparison of four fingerprinting techniques for the characterisation of Italian red wines. *Analytica Chimica Acta* 668: 143–148.

Ceto, X., J. Capdevila, A. Puig-Pujol and M. Del Valle. 2014. Cava wine authentication employing a voltammetric electronic tongue. *Electroanalysis* 26: 1504–1512.

Chambery, A., G. Del Monaco, A. Di Maro and A. Parente. 2009. Peptide fingerprint of high quality Campania white wines by MALDI-TOF mass spectrometry. *Food Chemistry* 113: 1283–1289.

Cozzolino, D., W. U. Cynkar, N. Shah and P. A. Smith. 2011. Can spectroscopy geographically classify Sauvignon Blanc wines from Australia and New Zealand? *Food Chemistry* 126: 673–678.

Cynkar, W., R. Dambergs, P. Smith and D. Cozzolino. 2010. Classification of Tempranillo wines according to geographic origin: Combination of mass spectrometry based electronic nose and chemometrics. *Analytica Chimica Acta* 660: 227–231.

Edelmann, A., J. Diewok, K. C. Schuster and B. Lendl. 2001. Rapid method for the discrimination of red wine cultivars based on mid-infrared spectroscopy of phenolic wine extracts. *Journal of Agricultural and Food Chemistry* 49: 1139–1145.

Fabani, M. P., M. J. A. Ravera and D. A. Wunderlin. 2013. Markers of typical red wine varieties from the Valley of Tulum (San Juan-Argentina) based on VOCs profile and chemometrics. *Food Chemistry* 141: 1055–1062.

Fischer, U., A. Bauer, S. Wolz, A. Schormann and H. -G. Schmarr. 2010. Authentication of different terroirs of German Riesling using sensory and flavor analysis. *Paper Read at 239th ACS National Meeting*, March 21–25, 2010, at San Francisco, CA.

Fischer, U., D. Roth and M. Christamann. 1999. The impact of geographic origin, vintage and wine estate on sensory properties of *Vitis vinifera* cv. Riesling wines. *Food Quality and Preference* 10: 281–288.

Gomez-Meire, S., C. Campos, E. Falque, F. Diaz and F. Fdez-Riverola. 2014. Assuring the authenticity of northwest Spain white wine varieties using machine learning techniques. *Food Research International* 60: 230–240.

Gutierrez, M., A. Llobera, A. Ipatov, J. Vila-Planas, S. Minguez, S. Demming, S. Buettgenbach, F. Capdevila, C. Domingo and C. Jimenez-Jorquera. 2011. Application of an e-tongue to the analysis of monovarietal and blends of white wines. *Sensors* 11: 4840–4857.

Ioannou-Papayianni, E., R. I. Kokkinofta and C. R. Theocharis. 2011. Authenticity of Cypriot sweet wine Commandaria using FT-IR and chemometrics. *Journal of Food Science* 76: C420–C427.

Jaitz, L., K. Siegl, R. Eder, G. Rak, L. Abranko, G. Koellensperger and S. Hann. 2010. LC-MS/MS analysis of phenols for classification of red wine according to geographic origin, grape variety and vintage. *Food Chem.* 122: 366–372.

Kallithraka, S., I. S. Arvanitoyannis, P. Kefalas, A. El-Zajouli, E. Soufleros and E. Psarra. 2001. Instrumental and sensory analysis of Greek wines; implementation of principal component analysis (PCA) for classification according to geographical origin. *Food Chemistry* 73: 501–514.

Kiss, J. and A. Sass-Kiss. 2005. Protection of originality of Tokaji Aszu: Amines and organic acids in botrytized wines by high-performance liquid chromatography. *Journal of Agricultural and Food Chemistry* 53: 10042–10050.

Kosir, I. J., M. Kocjancic and J. Kidric. 1998. Wine analysis by 1D and 2D NMR spectroscopy. *Analusis* 26: 97–101.

Kotseridis, Y., A. Anocibar Beloqui, A. Bertrand and J. P. Doazan. 1998. An analytical method for studying the volatile compounds of Merlot noir clone wines. *American Journal of Enology and Viticulture* 49: 44–48.

Kumsta, M., P. Pavlousek and J. Kupsa. 2012. Phenolic profile in Czech white wines from different terroirs. *Food Science and Biotechnology* 21: 1593–1601.

Lampir, L. 2013. Varietal differentiation of white wines on the basis of phenolic compounds profile. *Czech Journal of Food Sciences* 31: 172–179.

Lampir, L. and P. Pavlousek. 2013. Influence of locality on content of phenolic compounds in white wines. *Czech Journal of Food Sciences* 31: 619–626.

Liu, L., D. Cozzolino, W. U. Cynkar, R. G. Dambergs, L. Janik, B. K. O'neill, C. B. Colb and M. Gishen. 2008. Preliminary study on the application of visible-near infrared spectroscopy and chemometrics to classify Riesling wines from different countries. *Food Chemistry* 106: 781–786.

Louw, L., K. Roux, A. Tredoux, O. Tomic, T. Naes, H. H. Nieuwoudt and P. Van Rensburg. 2009. Characterization of selected South African young cultivar wines using FTMIR spectroscopy, gas chromatography, and multivariate data analysis. *Journal of Agricultural and Food Chemistry* 57: 2623–2632.

Marengo, E., M. Aceto and V. Maurino. 2002. Classification of Nebbiolo-based wines from Piedmont (Italy) by means of solid-phase microextraction-gas chromatography-mass spectrometry of volatile compounds. *Journal of Chromatography A* 943: 123–137.

Martelo-Vidal, M. J. and M. Vazquez. 2014. Rapid authentication of white wines Part 2: Classification by grape variety. *Agro Food Industry Hi-Tech* 25: 20–22.

Mazzei, P., N. Francesca, G. Moschetti and A. Piccolo. 2010. NMR spectroscopy evaluation of direct relationship between soils and molecular composition of red wines from Aglianico grapes. *Analytica Chimica Acta* 673: 167–172.

Pavlousek, P. and M. Kumsta. 2013. Authentication of Riesling wines from the Czech Republic on the basis of the non-flavonoid phenolic compounds. *Czech Journal of Food Science* 31: 474–482.

Pennington, N., F. Ni, A. Mabud and S. Dugar. 2007. A simplified approach to wine varietal authentication using complementary methods: Headspace mass spectrometry and FTIR spectroscopy. *ACS Symposium Series* 952: 180–199.

Salvatore, E., M. Cocchi, A. Marchetti, F. Marini and A. De Juan. 2013. Determination of phenolic compounds and authentication of PDO Lambrusco wines by HPLC-DAD and chemometric techniques. *Analytica Chimica Acta* 761: 34–45.

Sen, I. and F. Tokatli. 2014. Authenticity of wines made with economically important grape varieties grown in Anatolia by their phenolic profiles. *Food Control* 46: 446–454.

Sen, I. and F. Tokatli. 2016. Differentiation of wines with the use of combined data of UV-visible spectra and color characteristics. *Journal of Food Composition and Analysis* 45: 101–107.

Springer, A. E., J. Riedl, S. Esslinger, T. Roth, M. A. Glomb and C. Fauhl-Hassek. 2014. Validated modeling for German white wine varietal authentication based on headspace solid-phase microextraction online coupled with gas chromatography mass spectrometry Fingerprinting. *Journal of Agricultural and Food Chemistry* 62: 6844–6851.

Vergara, C., D. Von Baer, C. Mardones, L. Gutiérrez, I. Hermosín-Gutiérrez and N. Castillo-Muñoz. 2011. Flavanol profiles for varietal differentiation between Carménère and Merlot wines produced in Chile: HPLC and chemometric analysis. *Journal of the Chilean Chemical Society* 56: 827–832.

Viggiani, L. and M. a. C. Morelli. 2008. Characterization of wines by nuclear magnetic resonance: A work study on wines from the Basilicata region in Italy. *Journal of Agricultural and Food Chemistry* 56: 8273–8279.

CHAPTER 9

Alcoholic Beverages II— Spirits, Beer, Sake and Cider

James F. Carter

The original proposal for writing this book included a single chapter on alcoholic beverages. Very soon, however, I discovered what a large research effort has been invested in the authentication of wine—hence, the previous chapter is dedicated to the fruit of the vine and this chapter discusses the authentication of other alcoholic beverages: spirits, beers, sake, and cider.

9.1 Spirit drinks

"There is no bad whiskey. There are only some whiskeys that aren't as good as others."

—*Raymond Chandler (British-American novelist, 1888–1959)*

Spirit drinks—often referred to as *spirits, distilled spirit,* or *hard liquor*—encompass a wide range of products, manufactured from an equally wide range of ingredients by alcoholic fermentation followed by distillation to increase the alcohol (ethanol) content. Several of the most popular products are described in Table 9.1.

9.1.1 The market

Much akin to wines, spirits can be classified as *value, premium* (> US$20 per bottle), *super-premium* (> US$30 per bottle), and even *ultra-premium* (> US$40 per bottle) (Cunningham 2011).

---

[1] Forensic and Scientific Services, Health Support Queensland, 39 Kessels Road, Coopers Plains QLD 4108, AUSTRALIA.
Email: Jim.Carter@health.qld.gov.au

Spirits account for approximately one-third of the monetary value of the market for alcoholic beverage sales, a share that expanded by 7% in volume between 2010 and 2011 and, thanks to *premiumization*, increased by 10% in sales value over the same period. Brazil, Russia, India, and China (the BRIC countries) are the main markets for this growth, with sales of premium imported spirits [mostly whisk(e)y and brandy] growing by 16% and super-premium sales growing by 24% between 2010 and 2011. There is also a small but rising interest in *organic* spirits priced at the ultra-premium sector of the market (Wright 2011).

A comparison of Tables 9.1 and 9.2 reveals some interesting patterns about the national and international markets for spirits. Vodka is considered to be the largest global category of spirits and has its largest markets in Russia and the United States. In Russia, however, vodka is produced by a large number of small distilleries and the top three brands produce less

Table 9.1 The global production of spirits in 2010; based on Cunningham 2011.

Million liters	%	Style	Typical ingredients
4,325	22	vodka	neutral alcohol[a]
2,550	13	malt whisky	malted barley[b]
		grain whiskey	malted cereals[c]
1,440	7	rum	sugarcane juice, molasses
1,237	6	brandy/Weinbrand	wine spirits[d]
1,030	5	gin	neutral alcohol[a] flavored with juniper
874	4	liqueurs	neutral alcohol[a] containing 100 to 400 g/L of sugar
222	1	tequila	blue agave
8,333	42	others –	
		baijiu	sorghum
		sake	rice
		shōchū	barley, buckwheat, brown sugar, rice, or sweet potatoes
		soju	rice, wheat, or barley
		cachaça	sugarcane juice, molasses
		grappa	grape marc (pomace)[e]
		pastis/ouzo/ sambuca	neutral alcohol[a] flavored with aniseed
		bitters	neutral alcohol[a] flavored with cascarilla, cassia, gentian, etc.

[a] Derived from agricultural sources, typically grains or potatoes.
[b] See beer section for a description of the malting process.
[c] Indian whiskey brands are typically made from molasses and many consider these not to be whiskey.
[d] Wine-derived spirit that has been matured for at least one year.
[e] The solid remains of grapes after pressing.

than 8% of the spirits consumed. The largest selling branded spirits in the world are soju and shōchū, traditional colorless spirits of Korea and Japan, produced from a variety of ingredients including sorghum, rice, tapioca, and barley. China is currently the largest global spirits market and China and India are the two fastest growing markets. In China, the traditional white spirit baijiu dominates the alcoholic beverage industry in both volume and revenue. Like Russia, China is home to many thousand small distilleries and a vast number of local brands; however, the largest selling brands of spirits in China are not of baijiu, but imported brands of whisk(e)y and brandy. In a similar fashion, Brazil produces more than 1.3 billion liters of cachaça annually, from about 35,000 small distilleries, but less than 1% is currently exported. Despite the current pre-eminence of local distilleries, emerging markets are becoming increasingly important consumers of imported brands and are now key target markets for the global spirits industry's premium, super-premium, and ultra-premium brands, thanks to a focus on provenance and aspirational consumers. This trend in the value of sales for premium and super-premium brands is repeated globally, as illustrated in Table 9.3 (Cunningham 2011).

Table 9.2 The top ten spirit brands by sales volume; international brands are shown shaded.

Rank	Brand	Manufacturer	Style	Country of origin	Annual production (million liters)
1	Jinro	Jinro Ltd.	soju	Korea	576
2	Lotte Liquor	Lotte Liquor BG	soju	Korea	203
3	Smirnoff	Diageo	vodka		219
4	Pirassununga 51	Companhia Muller de Bebidas	cachaça	Brazil	175
5	Bacardi	Bacardi	rum		173
6	Tanduay	Tanduay Distillers	rum	Philippines	165
7	Bagpiper	United Spirits	whisky	India	147
8	Officer's Choice	Allied Blenders and Distillers	whisky	India	144
9	Johnnie Walker	Diageo	whisky		143
10	McDowell's No. 1	United Spirits	brandy	India	129

Table 9.3 Trends in general spirit sales *vs.* premium products.

	Period		Style	Overall sales	Premium sales
UK	2005	2009	whisky	−23%	+10%
USA	2003	2010	tequila	+47	+47%
global	2006	2011	gin	0%	+18%

As for any product, the rapid increase in demand for international premium and super-premium brands of spirits, combined with a local lack of knowledge or experience of genuine products, is an open invitation to counterfeiters.

9.1.2 Legislation

The permitted ingredients and final composition of most spirit drinks are controlled by legislation such as EEC 1576/89 (*laying down general rules on the definition, description and presentation of spirit drinks*). Note that, in Europe at least, the name *whisky* (as distinct from *whiskey*) is reserved for Scotch whisky, Irish whisky, and whisky Español—in this chapter the term whisk(e)y is used if no specific country of origin is implied.

Legislation typically defines spirits in terms of the primary ingredients, the time and means of maturation, the alcoholic strength at the point of distillation and at the point of sale, and the content of volatile substances (other than ethanol). For example, the definition of whisk(e)y requires the product to be:

- produced by the distillation of a mash of cereals,
- saccharified by the diastase of the malt contained therein, with or without other natural enzymes [*author's note*—i.e., *malted*],
- fermented by the action of yeast,
- distilled at less than 94.8% volume, so that the distillate has an aroma and taste derived from the raw materials used, and
- matured for at least three years in wooden casks not exceeding 700 liters capacity.

The definition of rum is similar but, in addition to requiring "organoleptic characteristics of rum," goes on to define the nature of the *organoleptic characteristics* as a content of volatile substances equal to or exceeding 225 grams per 100 liters (hectoliter) of the alcohol content of the finished product.

In contrast, vodka is defined as rectified ethyl alcohol of agricultural origin (also known as *neutral spirit*) purified by means of repeated distillation or filtered through activated charcoal, so that the organoleptic characteristics of the raw materials are selectively reduced.

Regardless of legislation, and in common with most other foodstuffs, the choice of ingredients is the first factor that determines the isotopic and chemical composition of sprits, the second factor being the addition of materials such as water, sugar, cream, etc. to the finished product. An important, additional factor that affects the isotopic and elemental concentrations of spirits is the distillation process.

9.1.3 Distillation and maturation

Stills are traditionally made of copper, or a copper alloy such as bronze, which reacts with and removes sulfur-containing compounds (that might otherwise give rise to an unpleasant odor or taste); stills inevitably introduce traces of metallic elements. The simplest still, commonly known as a *pot still*, consists of a single heated vessel and a collection vessel—capable of achieving an alcoholic strength of 40 to 50% alcohol by volume (ABV). Beverages may be distilled multiple times to achieve the desired alcoholic strength and character or distilled in a *column still*, which behaves like a series of vertical pot stills and can achieve a strength of 96% ABV (an azeotropic mixture of alcohol and water). As a further complication, a distillery will control the character of a beverage by discarding the *heads* and/or *tails* (the early and late distilling fractions, respectively), which contain flavor compounds such as ethyl acetate (heads) and amyl alcohol (tails). Distilleries producing vodka will discard both of these fractions to produce the desired clean taste, whereas beverages such as whisk(e)y and rum will retain some of these components to impart flavor.

Isotopic fractionation as a result of evaporation/condensation processes [vapor pressure isotope effect (VPIE)] have been studied for many decades (e.g., Jancso et al. 1994) but only one research group appears to have addressed isotopic fractionation during the distillation of alcoholic beverages (Baudler et al. 2006). In their unique experiments, the researchers distilled 130 liters of fermented cherry mash (7.2% ABV, $\delta^{13}C = -27.61‰$) in a traditional pot still, collecting 38 sequential fractions and measuring the $\delta^{13}C$, $(D/H)_I$ and $(D/H)_{II}$ composition of each fraction. The results of the $\delta^{13}C$ analysis are summarized in Figure 9.1, the first fraction being the most enriched ($-26.80‰$) and the final fraction being the most depleted ($-28.29‰$), which corresponded to an inverse VPIE.

Although this finding might suggest that distillation can cause significant isotopic differences, the authors also considered the practice of discarding the heads and tails of the distillate. When the fractions were combined into *cut1* and *cut2*, corresponding to 65 and 50% ABV, respectively, the difference in carbon isotopic composition was between 0.08 and 0.10‰. The authors, therefore, concluded that normal distillery practices can be expected to produce a product with a consistent $\delta^{13}C$ value. In contrast to the ^{13}C composition, the first distillate fraction exhibited the lowest $(D/H)_I$ and $(D/H)_{II}$ ratios, which corresponded to a normal VPIE. Comparing the $(D/H)_I$ and $(D/H)_{II}$ ratios of *cut1* and *cut2*, the authors found that the differences (0.27 to 0.28 ppm) where significantly smaller that the repeatability of the SNIF-NMR method (0.55 and 0.64 ppm, respectively).

Distillates will be clear, usually colorless liquids and *white spirits*, such as vodka and gin, can be bottled for immediate retail sales. Unlike wines

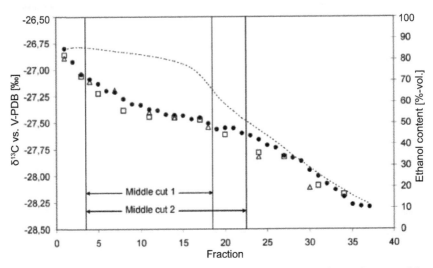

Figure 9.1 The carbon isotopic composition of ethanol in cherry brandy as a function of the distillation process. Dotted line represents the ethanol content (% volume) of the fraction. Symbols represent three replicate distillations. The decrease is carbon isotopic composition by fraction number/ethanol content corresponds to an inverse vapor pressure isotope effect (VPIE).

Reprinted with permission from Baudler et al. 2006. Copyright 2006 American Chemical Society.

spirits do not mature in the bottle and many styles of spirits are aged, prior to bottling, in wooden casks or stainless steel tanks for a period of time depending on the style and legal requirements. The most common vessels are white oak barrels and some manufacturers char the wood to impart a smoky flavor, or use barrels previously used with different kinds of alcohol [e.g., whisk(e)y or wine]. When aged in oak casks, spirits become brown or golden in color, whereas spirits aged in stainless steel tanks remains virtually colorless. In the barrels, spirits undergo numerous processes that contribute to the final flavor and to the isotopic and elemental compositions, principally through exchange for compounds with the wood and by evaporation.

MacNamara (2002) has presented a detailed study of the changes in the concentration of flavor components in un-peated Irish malt whisky stored in once-used American bourbon barrels for three and six years. The general conclusions were that compounds derived from the oxidation of ethanol (ethyl esters and acetates) increased with aging whereas sulfur compounds showed major decreases. The presence of phenolic compounds in the aged products was attributed to the cask wood, its pre-treatment, and fill history.

The natural breathability of the wooden casks used to mature many distilled spirits also results in the loss of a significant volume of alcohol during maturation. Known as the *Angels' share*, this loss of alcohol is estimated to be 2% per year, and significantly greater in hotter countries. This is very likely to result in a change in the overall isotopic composition of both alcohol and water components but the author is not aware of any studies of this effect. Finally, the mature spirit may be mixed with purified water to reduce the alcohol concentration prior to bottling with inevitable effects on both isotopic and elemental compositions.

9.1.4 Spirit authentication

Table 9.4 summarizes research into the authentication of spirits, mostly completed since 2000, but including some older references to the principle methods. It is interesting to note that the focus of research does not necessarily follow the trends in global production or sales (Tables 9.1 and 9.2), with most research focused on premium and super premium brands of whisk(e)y (especially Scotch), brandy/congac, or tequila/mezcal. A significant number of studies have also investigated popular regional products such as Brazilian cachaça, Asian soju, and other niche local market spirits such as Arrack (Sri Lanka) (Samarajeewa et al. 1981), Orujo (Spain) (Iglesias Rodriguez et al. 2010), Raki (Turkey) (Yucesoy and Ozen 2013), Tsipouro (Greece) (Fotakis and Zervou 2016), and Zivania (Cyprus) (Kokkinofta et al. 2003).

Although the total number of publications relating to the authentication of spirits is significantly fewer than those relating to the authentication of wine, both divide approximately equally between three broad categories of analytical techniques—both have been extensively analyzed using stable isotope and spectroscopic techniques. In contrast to wine, however, little research has addressed the elemental concentration of spirits, presumably because the primary elemental composition of spirits is removed by the distillation process, leaving only copper and related elements derived from the still. Research has, however, reported that vodkas manufactured at the same distillery exhibited relatively constant anion and cation compositions, which could serve as an identifying feature (Arbuzov and Savchuk 2002). The third, and largest, group of analytical techniques applied to spirits is the profiling of volatile organic compounds (VOCs), which are formed during fermentation and may be concentrated during distillation (see below).

Similar to wine, spectroscopic techniques applied to spirits authentication are typically in an un-targeted manner—i.e., the data are acquired without any knowledge of what parameters are likely to provide discrimination. Such techniques then rely on sophisticated statistical tools to delineate different classes.

Table 9.4 Recent (and not so recent) published research related to the authentication of spirit drinks.

Authors	Year	Technique	Samples
Filajdić and Djuković	1973	VOCs	Yugoslav plum brandy
Ng and Woo	1980	VOCs	Chinese alcoholic beverages
Samarajeewa et al.	1981	VOCs	Sri Lankan arrack
Simpkins and Rigby	1982	$\delta^{13}C$	whisky, brandy
Martin et al.	1983	SNIF-NMR	gin, rum, bourbon, malt whisky
Martin-Alvarez et al.	1988	VOCs, UV-vis, ABV	whisky
Headley and Hardy	1989	trace organic compounds	whisky
Aylott et al.	1994	VOCs	whisky
Bauer-Christoph et al.	1997	$\delta^{13}C$, SNIF-NMR	whisky, whiskey, fruit spirits
Parker et al.	1998	$\delta^{13}C$ of VOCs	whisky
González-Arjona et al.	1999	VOCs	whisky
Pissinatto et al.	1999	$\delta^{13}C$	Brazilian brandies
Panossian et al.	2001	VOCs	brandy
Savchuk et al.	2001	impurities	cognac
Aguilar-Cisneros et al.	2002	$\delta^{13}C$, $\delta^{18}O$	tequila
Arbuzov and Savchuk	2002	VOCs, anions, cations	vodka
Palma and Barroso	2002	IR	brandies
Bauer-Christoph et al.	2003	$\delta^{13}C$, SNIF-NMR	tequila
Kokkinofta et al.	2003	elemental	zivania
Lachenmeier et al.	2003	anions	vodka, rum
MacKenzie and Aylott	2004	UV-vis	whisky
Costa et al.	2004	refractive index	brandy, cachaça, rum, whiskey, vodka
Fernández et al.	2005	elemental	cachaças
Lachenmeier et al.	2005	congeners, FTIR	tequila
Lopez	2005	$\delta^{13}C$	tequila
Moller et al.	2005	saccharides and phenolics	whisky
Petrakis et al.	2005	^{1}H NMR	zivania
Pilar-Marti et al.	2005	VOCs, electronic nose	Cuban rum
Rau et al.	2005	$\delta^{13}C$	rice spirit
Villanueva et al.	2005	VOCs	tequila

Table 9.4 contd. ...

...Table 9.4 contd.

Authors	Year	Technique	Samples
De León-Rodríguez and González-Hernández	2006	VOCs	mezcal
González-Arjona et al.	2006	VOCs	whiskey
Lachenmeier et al.	2006	VOCs, anions	tequila, mezcal, sotol, bacanora
Picque et al.	2006	IR	cognac
Pontes et al.	2006	IR	whiskey, brandy, rum, vodka
Barbosa-García et al.	2007	UV-vis	tequila
de Souza et al.	2007	saccharides and phenolics	cachaça
Lachenmeier	2007	IR	fruit spirits
Savchuk and Kolesov	2007	phthalic acid esters	cognac
Vichi et al.	2008	diterpenoids	gin
Winterova et al.	2008	$\delta^{13}C$, SNIF-NMR, VOCs	fruit spirits
Ceballos-Magaña et al.	2009	elemental	tequila, mezcal
Jochmann et al.	2009	$\delta^{13}C$	whiskey, brandy, vodka, tequila
Lai et al.	2009	$\delta^{13}C$, SNIF-NMR	rice spirit
Sadecka et al.	2009	fluorescence	brandy
Contreras et al.	2010	UV-vis	tequila
Iglesias Rodriguez et al.	2010	elemental	orujo
Jelen et al.	2010	VOCs	raw spirit
Thomas et al.	2010	^{13}C-NMR	tequila
Heller et al.	2011	aldehydes	whiskey
McIntyre et al.	2011	IR	whisky
Meier-Augenstein et al.	2012	δ^2H, $\delta^{18}O$	whisky
Schipilliti et al.	2012	$\delta^{13}C$, VOCs	citrus liqueurs
Urickova et al.	2013	fluorescence	brandy
Yucesoy and Ozen	2013	IR	raki
Chen et al.	2014	IR NMR	rice spirit
de Souza et al.	2014	saccharides and phenolics	cachaça
Lai et al.	2014	SNIF-NMR	rice spirit
Nietner et al.	2014	δ^2H, $\delta^{13}C$, $\delta^{15}N$, $\delta^{18}O$, $\delta^{34}S$	distillers dry grains
Zhong et al.	2014	$\delta^{13}C$	rice spirit

Table 9.4 contd. ...

...Table 9.4 contd.

Authors	Year	Technique	Samples
Angel Cantarelli et al.	2015	UV-vis	whisky
Espinosa-Sanchez et al.	2015	surface plasmon resonance	tequila
Peng et al.	2015	electronic nose	rice spirit
Spanik et al.	2015	VOCs	brandy
Zhou et al.	2015	IR	distillers dry grains
Fotakis and Zervou	2016	NMR	tsipouro, tsikoudia

9.1.5 *Volatile Organic Compounds (VOCs)/congeners*

The study of VOCs in fermented and distilled beverages is as old as the technique that made such studies possible—gas chromatography (GC). A review published only a few years after GC became commercially available (Stevens 1960) reported over one hundred compounds that had been identified in *fusel oils*, the fraction containing higher alcohols that can be separated by distillation from a fermented beverage.

Many of these compounds are formed during fermentation and most of the higher alcohols are proposed to be formed by de-amination and de-carboxylation of amino acids: leucine yielding iso-amyl-alcohol, iso-leucine yielding amyl-alcohol, valine yielding iso-butyl alcohol, etc. Importantly, this means that the concentrations of higher alcohols in a beverage will reflect the protein content of the fermentation substrate and the organisms executing the fermentation. Thus, more than half a century ago it was recognised that the VOCs present in a distilled spirits could be used as a means of authentication—e.g., 3-methylbutan-2-ol was found to be present in alcohol fermented from cane molasses but was absent from the equivalent product fermented from beet molasses (Stevens 1960).

Typically, the higher alcohols comprise 80–90% of the VOCs present in a distilled spirit; other classes of compounds, such as terpene hydrocarbons, are derived directly from the fermentation substrate, carbonyl compounds are produced by the oxidation of ethanol, and phenolic compounds are leached from wooden casks during maturation. For various reasons it is experimentally difficult to study the minor components of fusel oil and the majority of studies have focused on ten higher alcohols, aldehydes, ethyl acetate, and methanol, universally described as *congeners* (Table 9.5) (somewhat ambiguously, this term might also apply to any component of fusel oil).

Early applications of *congener profiling* quickly demonstrated the strength of the technique. Considering only four congeners (2-methylpropan-1-ol,

Table 9.5 Congeners in spirits laid down in regulation (EC) No. 2870/2000.

Common name	IUPAC name
acetal	1,1-diethoxyethane
amyl alcohol	2-methylbutan-1-ol
iso-amyl alcohol	3-methylbutan-1-ol
methyl alcohol	methanol
ethyl acetate	ethyl ethanoate
n-butanol	butan-1-ol
sec-butanol	butan-2-ol
iso-butyl alcohol	2-methylpropan-1-ol
n-propanol	propan-1-ol
acetaldehyde	ethanal

3-methylbutan-1-ol, butan-1-ol, and butan-2-ol), Singer reported that "cheaper brandies...contained larger proportions of s-butanol" and "the expensive brandies contain more higher alcohol than the cheaper" (Singer 1966). This early study also recognized significant variability in the absolute concentration of congeners and proposed that relative, rather than absolute, concentrations were the key to product authentication—"re-expressed as the ratio...This consistency reflects the way in which the producers of spirits adhere to a routine of fermentation, distillation, ageing and blending result in a product of uniform composition."

Several decades later, the technique of congener profiling, as a means to authenticate spirit drinks, has been enshrined in legislation: EC 2870/2000 (*laying down community reference methods for the analysis of spirits drinks*) (19 December 2000), which states—

"Congeners in spirit drinks are determined by direct injection of the spirit drink, or appropriately diluted spirit drink, into a gas chromatography (GC) system. A suitable internal standard is added to the spirit drink prior to injection. The congeners are separated by temperature programming on a suitable column and are detected using a flame ionisation detector (FID). The concentration of each congener is determined with respect to the internal standard from response factors, which are obtained during calibration under the same chromatographic conditions as those of the spirit drink analysis." Acceptable internal standards are listed as pentan-3-ol, pentan-1-ol, 4-methylpentan-1-ol, or methyl nonanoate.

Prior to official acceptance, this method was validated through an inter-laboratory study involving 31 laboratories in eight countries (Kelly et al. 1999); hence, all published datasets acquired using the prescribed methodology should be suitable for comparison with questioned samples.

Congener profiling is now recognized as a powerful method for establishing whisk(e)y authenticity but, as a comparative technique, relies on the availability of authentic samples to establish a database of reference compositions. In this respect, the approach is very much akin to the Joint Research Centre of the European Community (EC-JRC) database of Stable Natural Isotope Fractionation (SNIF)-NMR results for the authentication of wine. Although no equivalent "global" database exists for spirits, such a database may not need to be as comprehensive as the EC-JRC wine database. Research (Aylott et al. 1994) has shown that the congener profiles of five brands of blended Scotch whisky were consistent over production batches, unlike the yearly variability of wines.

Much as congeners were the subject of early studies by GC, they became the subject of a proof-of-concept study during the very early development of hyphenated isotopic analysis, specifically GC-C-IRMS (Parker et al. 1998). This study recognized that congener profiling could not detect adulteration with neutral alcohol such as cane molasses spirits, especially if flavors were added to restore the congener profile. As a means to detect this method of adulteration, the authors measured the carbon isotopic composition of five congeners (acetaldehyde, ethyl acetate, *n*-propanol, iso-butanol, amyl alcohol) by GC-C-IRMS. This publication demonstrated the potential of combining congener profiling with isotopic measurements; however, the authors noted that the use of an internal standard with well-characterized isotopic composition would improve the technique.

Some researchers have applied similar profiling techniques in an un-targeted manner. For example, Headley and Hardy (1989) analysed concentrated extracts from 48 brands of whiskey by GC without any knowledge of the identity of the separated components. Statistical analysis of these data was able to group the samples according to five categories. Using more modern instrumentation, researchers have simply infused diluted spirits into a mass spectrometer via an electrospray interface (ESI-MS). The resulting spectrum contains information not only about congeners, but also less volatile and polar components such as saccharides and phenolics. Once again, using sophisticated statistical tools this method has been described as a "fingerprint" for US and Scotch whiskies (Moller et al. 2005) and is claimed to be characteristic of Brazilian cachaça to the extent that it can distinguish between different woods used for maturation (de Souza et al. 2007; de Souza et al. 2014).

9.1.6 Stable isotopes

The relationship between photosynthetic pathways and the carbon isotopic composition of plant materials has long been studied (see Chapter 3) and Table 9.6 summarizes the photosynthetic pathway and typical isotopic compositions of some common fermentable substrates.

Figure 9.2 shows the $\delta^{13}C$ composition of 76 spirit drinks measured by the author as part of an investigation into the sale of counterfeit products. (Note that although every care was taken to ensure that the results shown here are from genuine samples, it is always possible that some products were not what was claimed.) The data shown in the figure reflect the patterns seen in Table 9.6—spirits derived from grapes and barley

Table 9.6 Typical carbon isotopic compositions of common fermentable substrates. Adapted from Simpkins and Rigby 1982, with permission from John Wiley and Sons.

Substrate	Typical $\delta^{13}C$ composition (‰)[a]	Photosynthetic pathway
Maize (corn)	−10.3	C_4
Sorghum	−11.4	C_4
Molasses/sugar cane	−11.8	C_4
Oats	−23.4	C_3
Barley	−24.0	C_3
Potato	−24.9	C_3
Wheat	−25.3	C_3
Rice	−25.9	C_3
Grape must	−26.0	C_3

[a] The values in the original manuscript are reported *vs.* PDB, not VPDB.

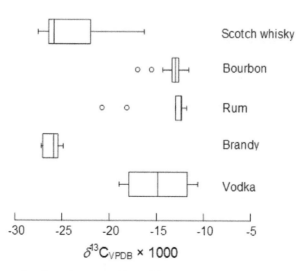

Figure 9.2 The carbon isotopic composition of 76 spirit drinks presented as box-plots. Open circles denote values that are deemed to be extreme outliers (i.e., more than three times the inter-quartile range from the median value).

(C_3 plants) typically had $\delta^{13}C$ values less than −21‰ whereas spirits derived from maize (corn) and sugar cane (C_4 plants) typically had values greater than −14‰. A single sample of tequila (not shown), derived from agave that fixes carbon via Crassulacean Acid Metabolism, had an intermediate value of −19.76‰. The range of isotopic compositions for any given class of spirits was wide, approximately 11‰ for whisky, 9‰ for rum, 8‰ for vodka, 5‰ for bourbon, and 2‰ for brandy. The magnitude of these ranges no doubt reflects the fact that brandy must be manufactured from grape juice whereas whisky and bourbon can be manufactured from a variety of grains, rum from different sources of molasses, and vodka from molasses, potatoes, or grains. Figure 9.2 also demonstrates the usefulness of box-plots to summarize large, non-parametric data sets—the range of compositions for each class of spirits are readily compared and outliers are easily identified.

Based on the isotopic differences between alcohol derived from C_3 and C_4 plants, the majority of published research addresses the detection of adulteration of expensive C_3 plant-derived spirits (primarily premium whisky and brandy) with low-value, molasses-derived alcohol—often intended for industrial use and therefore free from excise duty. An early study of the isotopic composition of spirits demonstrated that when grape spirit (C_3 origin) was adulterated with molasses spirit (C_4 origin) there was a linear relationship between the isotopic composition and the degree of adulteration (Simpkins and Rigby 1982).

Much like congener profiling, this method of detecting extraneous alcohol depends on analysis of an authentic sample. With the aid of reference sample(s), adulteration can be detected at levels as good as 3% whereas, without a reference sample, adulteration detection can be as poor as 40%—for example in blended Scotch whisky (Simpkins and Rigby 1982). Despite potential limitations, the carbon isotopic composition of alcohol has been used to detect adulteration with alcohol derived from cane, corn, beet, or potatoes in Brazilian brandies (Pissinato et al. 1999), tequila (Aguilar-Cisneros et al. 2002; Lopez 2005), rice spirits (Rau et al. 2005; Lai et al. 2009; Lai et al. 2014; Zhong et al. 2014), and fruit brandies (Winterova et al. 2008).

Shortly after the work of Simpkins and Rigby, Martin et al. (1983) demonstrated that SNIF-NMR (see Chapter 8) could also distinguish alcohol derived from C_3 and C_4 plants—specifically Scotch malt whisky and US bourbon. A decade later, a combination of SNIF-NMR and ^{13}C IRMS (Bauer-Christoph et al. 1997) was shown to distinguish alcoholic beverages manufactured from closely related varieties of fruit: cherry, plum, mirabelle, and apple.

Despite such promising research, both $\delta^{13}C$ and SNIF-NMR measurements are typically only used as adjuncts to congener profiling and interpreted as:

- if the congener, $\delta^{13}C$, and SNIF-NMR profile match an authentic sample —the sample is likely to be genuine,
- if the congener, $\delta^{13}C$, and SNIF-NMR profile do not match an authentic sample—the sample is unlikely to be genuine.

Isotopic analysis is typically expensive when compared to congener profiling and is likely to be used as a second line of proof rather than the primary means of analysis.

Although many areas of food forensics are moving towards geo-spatial models (isoscapes) as a means to determine country-of-origin, only one group appears to have tested a relationship between source water and finished product. Whisky is akin to beer in that most of the liquid content is derived from the distillery water supply rather than plant juice and in 2012 a Scottish-based research institute (Meier-Augentstein et al. 2012) studied the isotopic relationship between the water supply to 11 highland, islands, and lowlands distilleries and the bottled products. The authors found that both δ^2H and $\delta^{18}O$ values of production waters and the corresponding whisky samples were well correlated. The value for R^2 for solution of the linear regression analysis for δ^2H whisky/water was 0.71 and for $\delta^{18}O$ whisky/ water was 0.88. The observed ranges for δ^2H and $\delta^{18}O$ values of source waters were 28.0‰ and 4.87‰, respectively, while the corresponding ranges for whiskies were 26.5‰ and 1.39‰, respectively. The authors stressed that the dataset represented a small snapshot in time.

This isotopic analysis was also successful in identifying counterfeit whisky; not only were counterfeit samples separated from the authentic samples in well defined groups, they were also separated in two different groups, matching what was known by the Scottish Whisky Research Institute (SWRI) about their provenance.

9.2 Beer

"... proof that God loves us and wants us to be happy."

—*attributed to Benjamin Franklin (one of the Founding Fathers of the United States, 1706–1790)*

Beer is typically defined as:

"the product prepared by the yeast fermentation of an aqueous extract of malted and/or un-malted cereal, which is characterized by the presence of hops or preparations of hops."

Various regulations also allow for addition of flavorings, natural colors, thickening agents, and preservatives such as sulfur dioxide to the finished product.

It will be evident from earlier sections that numerous studies have addressed the authentication of wines and spirits through isotopic, elemental and chemical analysis. In contrast, only a handful of researchers have considered methods for the authentication of beer, although beer has the largest volume share of the global alcoholic beverage market and total sales valued at US$105 million in 2010 and was expected to reach US$137 million in 2015 [https://industrytoday.co.uk/market-research-industry-today/global-beer-market; accessed June 2016].

9.2.1 Brewing

Figure 9.3 shows an outline of the brewing process; for a detailed discussion the reader is referred to Hornsey 2013.

The primary ingredient of beer is barley, and less often, wheat, rice, maize, sorghum, oats, and rye, either solely or in combination.

The first step in brewing is *malting*, which is itself a three-stage process: *steeping, germination,* and *kilning*. The main function of malting is often said to be starch hydrolysis but, in truth, the primary function is to break down the endosperm cell wall, exposing the starch grains. Steeping increases the water content of the grain from 15–20% to 42–46%, initiating germination. The germinated grains are then dried to a water content between 5–8%.

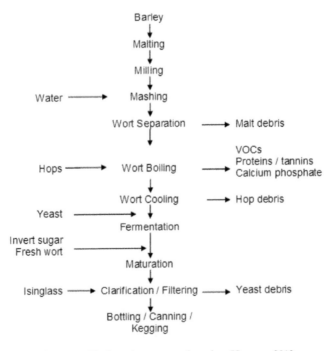

Figure 9.3 The brewing process, based on Hornsey 2013.

The malted grain is then mashed, with hot water, to produce a sugary liquid, called *wort*. Hops, or hop extracts, are added and the wort is boiled, primarily to sterilize and terminate the enzymatic reactions that began with malting. The wort is typically boiled for between 45 minutes and two hours, which can reduce the volume by up to 15% and which will have an effect on the isotopic composition (Brettell et al. 2012).

Although the processes of beer and wine making share some similarity, the fundamental difference is that wine-water is derived exclusively from grape-water, whereas beer-water is derived largely from the brewery water supply. Famous brewing regions, such as Burton-on-Trent in the UK and Munich in Germany owe the characteristic style of their beers to the natural chemical composition of local waters, especially pH and hardness due to calcium and magnesium salts. Before water chemistry was well understood, breweries depended on the local water source whereas today almost any water can be chemically modified to produce a specific style of beer. Barley (and other grains) and hops may be transported considerable distances to the brewery, as the brewer seeks out quality ingredients from other regions or countries. In contrast, there is little economic or quality advantage in using anything other than local water, which can provide a fingerprint for authentication purposes.

During wort boiling, numerous chemical processes occur, which develop the flavor of the beer, but also precipitate proteins, tannins, and calcium phosphate. Volatile organic compounds are also lost. The vessels for wort boiling were traditionally made of copper and traditionally known as *coppers*. Although commercial breweries have largely replaced copper with stainless steel, metallic copper must still be present to catalyze color development and to bind undesirable sulfur compounds. After boiling, the wort is cooled, hop debris removed, and yeast added to initiate the primary fermentation, the product of which is known as *green beer*. The nature of the following stages—*maturation, conditioning,* or *lagering*—will depend largely on the style of beer, but typically involves secondary fermentation, induced by the addition of invert sugar or fresh wort.

In contrast to wine, beer is typically clarified by filtration, rather than by the use of fining agents (when finings are used, isinglass is the preferred agent), and stabilized by filtration and/or pasteurization rather than by the addition of anti-microbial agents such as sulfur dioxide. Cask conditioned beers are the domain of the real ale enthusiast and are neither filtered nor stabilized. Unlike highly crafted wines or spirits, these products are relatively inexpensive and are unlikely to be the subject of counterfeit, not least due to the very short shelf life.

Finally, once beers have been packaged for sale, copper and zinc can be introduced into beer from welded cans; for example, canned beers are reported to contain three times the zinc concentration of the equivalent bottled beer (Hornsey 2013).

9.2.2 Beer authentication

In comparison to wines and spirits, only a limited number of studies have examined the isotopic, elemental, and chemical compositions of beers; these studies have primarily attempted to discriminate between styles of beer (lager, ale, stout, etc.) based on elemental composition (Bellido-Milla et al. 2000; Alcázar et al. 2002). This might be viewed as a somewhat academic exercise, as the style of beer is usually apparent and is, moreover, a subjective assessment. One study has suggested that a combination of iron, phosphorus, poly-phenol, and potassium concentrations could distinguish beers from Germany, Portugal, and Spain (Alcázar et al. 2012). In order to achieve this level of discrimination, sophisticated statistical tools such as support vector machines (SVM) were employed.

A recent study of the isotopic composition of beers (Chesson et al. 2010) examined 33 beers as a potential source of liquid input to the human body water pool, and concluded that there was a correlation, albeit not strong, between local drinking water $\delta^2H/\delta^{18}O$ composition and isotopic composition of water from beers purchased at the same location.

An earlier study of 160 beers focused on the carbon isotopic composition of the dry residue (comprised mostly of sugars) as a means to determine the proportion of added C_4 plant material, specifically high fructose corn syrup (HFCS) or cane sugar (Brooks et al. 2002), based on a mixing model approach, Equation 9.1.

$$\%C_{4CARBON} = \frac{(\delta^{13}C_{BEVERAGE} - \delta^{13}C_3)}{(\delta^{13}C_4 - \delta^{13}C_3)} \times 100 \qquad \text{Equation 9.1}$$

In which $\delta^{13}C_3$ is the measured isotopic composition of a C_3 plant + 1 standard deviation (in this case barley, −25.2‰), and $\delta^{13}C_4$ is the typical isotopic composition of C_4 plant sugars (−12.5‰).

The study concluded that inexpensive beers and those from large production facilities were most likely to contain detectable quantities of C_4 plant derived carbon, possibly added as a cheap alternative to cereals such as barley or wheat. This finding was confirmed by a recent study of Brazilian beers (Mardegan et al. 2013), which concluded that beers from large breweries typically had $\delta^{13}C$ values of −20‰ compared to artisan breweries with $\delta^{13}C$ values of −25‰. These researchers also commented that beer prices internationally were directly related to $\delta^{13}C$ composition!

Other researchers have successfully used this mixing model approach to identify adulteration of a variety of beverages.

In the most comprehensive study to date (Carter et al. 2015a), the authors analyzed a total of 162 bottled or canned beers from around the globe. The authors determined the $\delta^2H/\delta^{18}O$ composition of the whole beer, the $\delta^{13}C$ composition of the dry residue (mostly sugars), and the concentrations of five anions (F, Cl, NO$_3$, SO$_4$, PO$_4$) and seven cations (Ca, K, Mg, SiO$_2$, V, Mn, Sr). The initial intention was that these data could serve as a reference dataset against which beers suspected of being substitute or counterfeit could be compared. Strong correlations between the concentrations of certain anions and cations (Ca, Sr, Cl, SO$_4$) appeared to confirm the widespread use of two common brewing salts, calcium chloride and calcium sulfate (gypsum), to adjust the character of the brewing water.

The relationship between $\delta^2H/\delta^{18}O$ of the beer-water was found to have a slope similar to the GMWL but with a slight offset, Equation 9.2.

$$\delta^2H = \delta^{18}O \times 8.1 + 6.3, R^2 = 0.95. \hspace{2cm} \text{Equation 9.2}$$

This offset was consistent with a study of the $\delta^{18}O$ composition of brewing water (Brettell et al. 2012), which demonstrated fractionation at every stage of the process and an overall increase in $\delta^{18}O$ of +1.3‰.

The $\delta^2H/\delta^{18}O$ composition of the beers correlated strongly with the longitude and latitude of the stated brewing location and, when these data were geo-spatially mapped (coined an *alcoscape*), the relationship between the isotopic composition of beers and local precipitation was visually apparent (Figure 9.4).

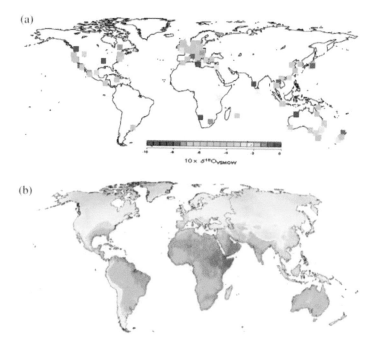

Figure 9.4 $\delta^{18}O$ isoscapes of (a) beer samples (an "alcoscape") and (b) global precipitation (www.waterisotopes.org).
Reproduced with permission from Carter et al. 2015a. Copyright 2015 Elsevier.

Based on the $\delta^{13}C$ composition of the dry residues, it is possible to generate a different kind of alcoscape, which shows the geographical distribution of added C_4 plant materials (Figure 9.5). The majority of beers brewed in northern Europe were color-coded blue, corresponding to the $\delta^{13}C$ signature of barley or similar grains. In contrast, beers brewed in Mediterranean countries were isotopically heavier, due to the inclusion of C_4 plant materials, such as maize or rice, which were (mostly) declared as ingredients. A single red data point in central Europe corresponds to a *gluten-free beer*, which appeared to be manufactured from cane sugar or HFCS to avoid proteins from cereal crops. The use of mapped $\delta^{13}C$ and $\delta^{18}O$ data could be of great benefit in identifying substitute and counterfeit beers, as this technique has the potential to predict the composition of samples when no authentic sample can be obtained.

-28   -26   -24   -22   -20

$\delta^{13}C_{VPDB}$

Figure 9.5 Alcoscape showing the geographic distribution of carbon isotopic compositions of European beers.

9.3 Sake or Saké

"It is the man who drinks the first bottle of saké; then the second bottle drinks the first, and finally it is the saké that drinks the man."

Japanese proverb

Sake or saké is universally described as a Japanese *rice wine* although the production method has far more in common with beer. The alcoholic strength of sake is, however, more akin to wine—typically 15 to 20% ABV.

9.3.1 Sake making

The rice used to produce sake is grown specifically for that purpose and contains less protein and lipid than rice typically eaten in Japan. In the first stage of sake production the rice is *polished* to remove most of the outer surface, which contains the majority of protein and lipid—leaving mostly starch. (Note that as a result of removing most of the protein, sake contains very low concentrations of higher alcohols.) The polished rice is then washed and steeped in water for several hours before being carefully steamed. Like beer, water is a critical ingredient throughout the whole process of sake making and the mineral content of the water plays a significant role in the character of the finished product. For this reason certain regions of Japan are renowned for distinctive and high quality products.

Unlike the malt used to brew beer, rice does not contain amylase, which converts starch to fermentable sugar, and a two-stage fermentation must be employed. The initial fermentation is brought about by the action of an *Aspergillus* mold, followed by the addition of water and yeast to promote alcoholic fermentation. Following fermentation, the sake is extracted by filtration at which stage alcohol (known as *brewer's alcohol*) can be added to improve the extraction of flavor compounds—although cheaper products are extended by the addition of more alcohol. Finally, the sake is diluted to approximately 15% ABV before being bottled and matured for between nine and twelve months.

There are several styles of sake, distinguished primarily by the degree to which the rice has been polished and by the percentage of brewer's alcohol added during production: *Honjozo-shu, Ginjo-shu,* and *Daiginjo-shu* denote sake made from rice which has 30%, 40%, and 50%, respectively, polished away. The prefix *Junmai* denotes a pure rice wine with no added alcohol—e.g., *Junmai Daiginjo*.

9.3.2 Sake authentication

Federal regulations in the U.S., which prohibit the addition of sugar to rice prior to fermentation or the addition of distilled alcohol to brewed sake, gave rise to one of the very few applications of ^{14}C measurement in food forensics (Martin et al. 1983). These researchers reasoned that ethanol derived from petroleum would be *carbon dead* (an activity of approximately 2.8 mBq) compared to freshly fermented sake (approximately 259 mBq). The authors first distilled the ethanol from sake as an azeotropic mix, combusted this to CO_2, which was then converted to acetylene and finally trimerized to benzene. The ^{14}C of the resultant benzene was measured by liquid scintillation counting. Of 22 samples tested, the authors found no evidence for the addition of petroleum-derived alcohol. The complexity of this experimental procedure may give some indication as to why radiocarbon techniques are not widely used in food forensics (but see Chapter 13). The authors also measured the stable carbon isotopic composition of natural sake and found that the typical $\delta^{13}C$ value for ethanol was −26.2‰ and −25.2‰ for sugar. They concluded that ethanol $\delta^{13}C$ values greater than −22.7‰ were indicative of the addition of cane or corn-derived alcohol.

Recent, preliminary research has investigated δ^2H, $\delta^{13}C$, and $\delta^{18}O$ measurements of sake as a means to determine adulteration with brewer's alcohol derived from the fermentation of rice (Hashiguchi et al. 2015). The authors concluded that the addition of rice-derived brewer's alcohol was characterized by a decrease in δ^2H values, but caused little effect on $\delta^{13}C$ or $\delta^{18}O$ values. The authors agreed with the earlier research that it was easy to detect the addition of alcohol derived from corn or cane sugars.

9.4 Cider, Cyder, or Cidré

> "There are three sorts of cider; singing cider, fighting cider, and sleeping cider."
>
> —*Paul Chant, Somerset cider maker (@ChantCider)*

Traditional cider (or *hard cider* as it is known in the U.S.) can be defined as:

> "the fruit wine prepared from the juice of apples and no more than 25% of the juice of pears."

And conversely, pear cider as:

> "the fruit wine prepared from the juice of pears and no more than 25% of the juice of apples."

Pear cider is sometimes branded as *perry*, although EU regulations reserve that name for alcoholic beverages made from specific cultivars of European pears, with high quantities of tannins in the juice—e.g., *perry pears* or *snow pears*.

In the EU more than 800,000 tons of fruit are fermented every year to produce over 14 million hectoliters of cider and perry and, in the U.S., cider sales for 2011 were estimated at US$90 million.

9.4.1 Cider making

After harvest, cider apples (or pears) are typically stored to allow them to ripen further, before being crushed in a mill to form pulp or *pomace*, which is then pressed (traditionally between layers of straw or muslin cloth) to release the *must*. Fermentation traditionally occurs as a result of yeasts naturally present on the fruit, carried out at relatively low temperatures (14 to 16°C) in open vats or barrels. In contrast, commercial cider makers will treat the must with sulfur dioxide, to inhibit natural wild yeasts, and then add pure yeast cultures for fast fermentation at around 20°C, in order to obtain a faster turnover.

Following initial fermentation the cider is siphoned into new vats, leaving the sediment behind, and then left to mature for six to twelve months before the cider is blended with new and old ciders, to maintain a consistent character. Commercial cider blends are nearly always cleared by centrifugation or kieselguhr filtration (a form of diatomaceous earth) and sterilized by filtration or flash pasteurization. Finally, the product is artificially carbonated in the bottle and sulfur dioxide is added to inhibit microbial activity. In contrast, traditional ciders are typically served flat and cloudy and may also be available as cask-conditioned cider, analogous to real ale. In France, cider is generally produced by the Charmat process (see Chapter 8) and is highly carbonated and more like an apple wine than traditional English cider.

Although Europe remains the world's major consumer, cider has recently experienced a significant global increase in popularity, with Australian and U.S. sales growing by more than 50% in 2011–2012. Although cider captures only 1% of the volume market for beer, it is often sold to the high end of the market: reports claim that prestigious English ciders sell in stylish U.S. restaurants for US$26 a bottle, compared to the UK retail price of £2.59 (equivalent to about US$3.70).

9.4.2 Cider authentication

On a commercial scale there are considerable cost advantages to be gained by adding sugar and/or water to the must, as these are much cheaper than apple/pear juice. Online sources claim that "many commercial ciders are now made from around 35% juice and 65% glucose syrup" (http://www. cider.org.uk/part3.htm, accessed 30th May 2016). The author cannot support or refute such claims and there appears to be no published methodology to identify the addition of water or sugar to ciders, although, intuitively, methods developed to detect the adulteration of fruit juice (Chapter 7) or wine (Chapter 8) could easily be adapted. The research that exists for the authentication of cider has focused on methods to determine geographical origins because, like all alcoholic beverages, this has a major influence on the retail price.

Some researchers have considered the poly-phenolic composition of ciders as a potential guide to origin, but attributed variations more to apple variety and maturity at harvest, rather than geography (Alonso-Salces et al. 2005; Alonso-Salces et al. 2006). One study has concluded that a combination of multiple elemental concentrations (Al, As, Ba, Ca, K, Mg, Mn, Mo, Na, Rb, Si, Sr, Ti, V, and Zn) and strontium isotope ratio analysis provided a means to distinguish between ciders from England, France, Spain, and Switzerland (García-Ruiz et al. 2007).

Until recently, the isotopic analysis carbon present in ciders had been limited to identifying endogenous carbon dioxide introduced by carbonation with industrial gases (Calderone et al. 2007; Cabañero et al. 2012; Gaillard et al. 2013). These studies reported that the $\delta^{13}C$ compositions of endogenous CO_2 from ciders ranged from −24.80 to −20.89‰, which was significantly depleted (−37.13 to −26.00‰) if industrial CO_2 was added. The method was also reported to be capable of identifying the addition of C_4 sugar prior to fermentation.

The most recent study of ciders reported the elemental and stable isotopic compositions of seven Australian, three New Zealand, and 12 European (Belgium, France, Ireland, Spain, Sweden, and the UK) ciders (Carter et al. 2015b). None of the analytical data acquired during this study found differences between ciders produced from apples or pears, which was in agreement with the European Fruit Juice Association (AIJN) reference guides for apple and pear juice (AIJN COP 6.3, AIJN COP 6.8). The majority of samples had $\delta^{13}C$ compositions typical of unadulterated apple or pear juice (*ca.* −26‰), whereas seven of the nine Australian/New Zealand samples and a single EU sample (produced in Spain) were found to be significantly enriched in ^{13}C, suggesting the inclusion of C_4 sugar (based on Equation 9.1). Analysis of these samples by HPLC provided additional evidence for the addition of sucrose, glucose, and/or HFCS.

Figure 9.6 shows the relationship determined between $\delta^2H/\delta^{18}O$ of cider-waters. These data can be modelled with the linear relationship shown in Equation 9.3, the slope of which approximates the GMWL with an enrichment in ^{18}O. As noted in Chapters 7 and 8, fruit water would be expected to have a slope of approximately four and a slope of 8 might indicate the addition of tap water.

$$\delta^2H = \delta^{18}O \times 7.6 + 4.3\ (R^2 = 0.98) \qquad \text{Equation 9.3}$$

$$\delta^2H = \delta^{18}O^2 \times -0.2 + 3.7 \times \delta^{18}O - 10\ (R^2 = 0.99) \qquad \text{Equation 9.4}$$

It was also possible to model the data shown in Figure 9.4 using the quadratic Equation 9.4. The linear term in this model has a slope of 3.7, which is the same as determined for grape-water and was attributed to evapotranspiration (see Chapter 8), suggesting that the isotopic composition

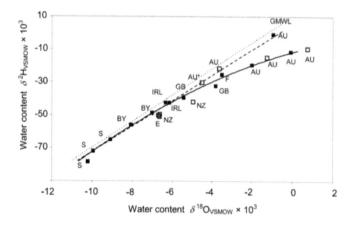

Figure 9.6 The relationship between δ^2H and δ^{18}O of the water content of ciders. The dashed line shows the data modelled as a linear relationship (Equation 9.3) while the solid line shows the data modelled as a quadratic relationship (Equation 9.4).

Reproduced with permission from Carter et al. 2015b. Copyright 2015 American Chemical Society.

of ciders is derived from multiple sources. Further, recent research has highlighted the complexity of the relationship between the δ^2H composition of plant tissue and the δ^2H/δ^{18}O of tissue water as a result of water loss during post-harvest storage of apples (Greule et al. 2015).

Despite the obvious complexity within these data it was still possible to make some inferences about the geographical origin of cider based on Figure 9.6. Simplistically, the most isotopically depleted samples originated from Sweden (S) or Belgium (BY), whereas the most isotopically enriched samples originated from Australia (AU). Samples from similar geographical regions such as the United Kingdom (GB), France (F), or Ireland (IRL) had similar isotopic compositions, as did samples from regions with similar climates; the EU, NZ, and Tasmania (AU*), especially Spanish and New Zealand samples. The carbon isotopic composition of many of the ciders was indicative of the addition of C_4 derived sugars and HPLC analysis identified the presence of cane sucrose, HFCS, or glucose syrup. Further investigation will be needed to determine whether these were added as legitimate, post-fermentation sweetening agents or illegally added prior to alcoholic fermentation.

The sources of isotopic composition variations within ciders appear to be more complex than beers (and possibly wines) and more research is needed for accurate predictions of geographic origin of all forms of alcoholic beverages. This must include (but not be limited to):

- the relationship between ground and atmospheric waters and fruit-water,

- the mixed contributions of waters from precipitation and irrigation,
- the relationship between underlying geology and elemental composition,
- the possible long distance transport of ingredients,
- the effects of boiling, brewing, distillation, and storage, and
- the relationships between water, sugar, alcohol, and CO_2.

References

Aguilar-Cisneros, B. O., M. G. López, E. Richling, F. Heckel and P. Schreier. 2002. Tequila authenticity assessment by headspace SPME-HRGC-IRMS analysis of $^{13}C/^{12}C$ and $^{18}O/^{16}O$ ratios of ethanol. *Journal of Agricultural and Food Chemistry* 50: 7520–7523.

Alcázar, A., F. Pablos, M. J. Martin and A. G. Gonzalez. 2002. Multivariate characterisation of beers according to their mineral content. *Talanta* 57: 45–52.

Alcázar, A., J. M. Jurado, A. Palacios-Morillo, F. de Pablos and M. J. Martin. 2012. Recognition of the geographical origin of beer based on support vector machines applied to chemical descriptors. *Food Control* 23: 258–262.

Alonso-Salces, R. M., S. Guyot, C. Herrero, L. A. Berrueta, J. -F. Drilleau, B. Gallo and F. Vicente. 2005. Chemometric classification of Basque and French ciders based on their total polyphenol contents and Cielab parameters. *Food Chemistry* 91: 91–98.

Alonso-Salces, R. M., C. Herrero, A. Barranco, D. M. López-Márquez, L. A. Berrueta, B. Gallo and F. Vicente. 2006. Polyphenolic compositions of Basque natural ciders: A chemometric study. *Food Chemistry* 97: 438–446.

Arbuzov, V. N. and S. A. Savchuk. 2002. Identification of vodkas by ion chromatography and gas chromatography. *Journal of Analytical Chemistry* 57: 428–433.

Association of the Industry of Juices and Nectars from fruits and vegetables of the EEC. COP 6.3 Reference guidelines for Apple—revision April 2014: AIJN: Brussels, Belgium, 2014.

Association of the Industry of Juices and Nectars from fruits and vegetables of the EEC. COP 6.8 Reference guidelines for Pear—revision April 2014: AIJN: Brussels, Belgium, 2014.

Aylott, R. I., A. H. Clyne, A. P. Fox and D. A. Walker. 1994. Analytical strategies to confirm Scotch whisky authenticity. *Analyst* 119: 1741–1746.

Baudler, R., L. Adam, A. Rossmann, G. Versini and K. -H. Engel. 2006. Influence of the distillation step on the ratios of stable isotopes of ethanol in cherry brandies. *Journal of Agricultural and Food Chemistry* 54: 864–869.

Bauer-Christoph, C., H. Wachter, N. Christoph, A. Roßmann and L. Adam. 1997. Assignment of raw material and authentication of spirits by gas chromatography, hydrogen- and carbon-isotope ratio measurements I. Analytical methods and results of a study of commercial products. *Zeitschrift für Lebensmittel-Untersuchung und –Forschung A* 204: 445–452.

Bellido-Milla, D., J. M. Moreno-Perez and M. P. Hernandez-Artiga. 2000. Differentiation and classification of beers with flame atomic spectrometry and molecular absorption spectrometry and sample preparation assisted by microwaves. *Spectrochimica Acta Part B: Atomic Spectroscopy* 55: 855–864.

Brettell, R., J. Montgomery and J. Evans. 2012. Brewing and stewing: The effect of culturally mediated behaviour on the oxygen isotope composition of ingested fluids and the implications for human provenance studies. *Journal of Analytical Atomic Spectrometry* 27: 778–785.

Brooks, J. R., N. Buchmann, S. Phillips, B. Ehleringer, R. D. Evans, M. Lott, L. A. Martinelli et al. 2002. Heavy and light beer: A carbon isotope approach to detect C_4 carbon in beers of different origins, styles, and prices. *Journal of Agricultural and Food Chemistry* 50: 6413–6418.

Cabañero, A. I. and M. Rupérez. 2012. Carbon isotopic characterization of cider CO_2 by isotope ratio mass spectrometry: A tool for quality and authenticity assessment. *Rapid Communications in Mass Spectrometry* 26: 1753–1760.

Calderone, G., C. Guillou, F. Reniero and N. Naulet. 2007. Helping to authenticate sparkling drinks with $^{13}C/^{12}C$ of CO_2 by gas chromatography-isotope ratio mass spectrometry. *Food Research International* 40: 324–331.

Carter, J. F., H. S. A. Yates and U. Tinggi. 2015a. A global survey of the stable isotope and chemical compositions of bottled and canned beers as a guide to authenticity. *Science & Justice* 55: 18–26.

Carter, J. F., H. S. A. Yates and U. Tinggi. 2015b. Stable isotope and chemical compositions of European and Australasian ciders as a guide to authenticity. *Journal of Agricultural and Food Chemistry* 63: 975–982.

Chesson, L. A., L. O. Valenzuela, S. P. O'Grady, T. E. Cerling and J. R. Ehleringer. 2010. Links between purchase location and stable isotope ratios of bottled water, soda, and beer in the United States. *Journal of Agricultural and Food Chemistry* 58: 7311–7316.

Cunningham, J. 2011. The drinks international millionaires club: The definitive ranking of the World's biggest spirits brands. *Drinks International* July: 5–41.

de Souza, P. P., A. M. M. Resende, D. V. Augusti, F. Badotti, F. de C. O. Gomes, R. R. Catharino, M. N. Eberlin and R. Augusti. 2014. Artificially-aged cachaça samples characterised by direct infusion electrospray ionisation mass spectrometry. *Food Chemistry* 143: 77–81.

de Souza, P. P., H. G. L. Siebald, D. V. Augusti, W. B. Neto, V. M. Amorim, R. R. Catharino, M. N. Eberlin and R. Augusti. 2007. Electrospray ionization mass spectrometry fingerprinting of Brazilian artisan cachaça aged in different wood casks. *Journal of Agricultural and Food Chemistry* 55: 2094–2102.

Fotakis, C. and M. Zervou. 2016. NMR metabolic fingerprinting and chemometrics driven authentication of Greek grape marc spirits. *Food Chemistry* 196: 760–768.

Gaillard, L., F. Guyon, M. -H. Salagoity and B. Medina. 2013. Authenticity of carbon dioxide bubbles in French ciders through multiflow-isotope ratio mass spectrometry measurements. *Food Chemistry* 141: 2103–2107.

García-Ruiz, S., M. Moldovan, G. Fortunato, S. Wunderli and J. I. García Alonso. 2007. Evaluation of strontium isotope abundance ratios in combination with multi-elemental analysis as a possible tool to study the geographical origin of ciders. *Analytica Chimica Acta* 590: 55–66.

Greule, M., A. Rossmann, H. -L. Schmidt, A. Mosandl and F. Keppler. 2015. A stable isotope approach to assessing water loss in fruits and vegetables during storage. *Journal of Agricultural and Food Chemistry* 63: 1974–1981.

Hashiguchi, T., F. Akamatsi, H. Izu and T. Fujii. 2015. Preliminary detection method for added rice- and sugarcane-derived brewer's alcohol in bulk samples of sake by measurement of hydrogen, oxygen, and carbon isotopes. *Bioscience, Biotechnology, and Biochemistry* 79: 1018–1020.

Headley, L. M. and J. K. Hardy. 1989. Classification of whiskies by principal component analysis. *Journal of Food Science* 54: 1351–1358.

Hornsey, L. 2013. Brewing 2nd Ed. Cambridge, Royal Society of Chemistry.

Iglesias Rodriguez, R., M. F. Delgado, J. B. Garcia, R. M. P. Crecente, S. G. Martin and C. H. Latorre. 2010. Comparison of several chemometric techniques for the classification of orujo distillate alcoholic samples from Galicia (Northwest Spain) according to their certified brand of origin. *Analytical and Bioanalytical Chemistry* 397: 2603–2614.

Jansco, G. L. L., P. N. Rebelo and W. A. Van Hook. 1994. A nonideality in isotopic mixtures. *Chemical Society Reviews* 23: 257–264.

Kelly, J., S. Chapman, P. Brereton, A. Bertrand, C. Guillou and R. Witkowski. 1999. Gas chromatographic determination of volatile congeners in spirit drinks: Interlaboratory study. *Journal of AOAC International* 82: 1375–1388.

Kokkinofta, R., P. V. Petrakis, T. Mavromoustakos and C. R. Theocharis. 2003. Authenticity of the traditional Cypriot spirit "Zivania" on the basis of metal content using a combination of coupled plasma spectroscopy and statistical analysis. *Journal of Agricultural and Food Chemistry* 51: 6233–6239.

Lai, C. -H., C. -W. Hsieh and W.-C. Ko. 2014. Detection limit of molasses spirits mixed in rice spirits using the SNIF-NMR method. *Journal of Food and Drug Analysis* 22: 197–201.

Lai, C. -H., W. -C. Ko, E. C. -F. Chen and C. -W. Hsieh. 2009. Detection of adulteration in Taiwanese rice-spirits by $^{13}C/^{12}C$ stable isotope ratio analysis and SNIF-NMR methods. Paper presented at the 237th ACS National Meeting, March 22–26, 2009, Salt Lake City, UT, United States.

Lopez, M. G. 2005. Authenticity: The case of Tequila. Paper presented at the 229th ACS National Meeting, March 13–17, 2005, San Diego, CA, United States.

MacNamara, K. 2002. Flavour components of whiskey. Ph.D. Dissertation, University of Stellenbosch. https://scholar.sun.ac.za/handle/10019.1/52849.

Mardegan, S. F., T. M. B. Andrade, E. R. de Sousa Neto, E. B. de Castro Vasconcellos, L. F. B. Martins, T. M. Mendonca and L. A. Martinelli. 2013. Stable carbon isotopic composition of Brazilian beers—A comparison between large- and small-scale breweries. *Journal of Food Composition and Analysis* 29: 52–57.

Martin, G. J., M. L. Martin, F. Mabon and M. J. Michon. 1983. A new method for the identification of the origin of ethanols in grain and fruit spirits: High-field quantitative deuterium nuclear magnetic resonance at the natural abundance level. *Journal of Agricultural and Food Chemistry* 31: 311–315.

Martin, G. E., J. M. Burggraff, F. C. Alfonso and D. M. Figert. 1983. Determination of authentic sake by carbon isotope ratio analysis. *Journal of AOAC International* 66: 1405–1408.

Meier-Augenstein, W., H. F. Kemp and S. M. L. Hardie. 2012. Detection of counterfeit Scotch whisky by ^2H and ^{18}O stable isotope analysis. *Food Chemistry* 133: 1070–1074.

Moller, J. K. S., R. R. Catharino and M. N. Eberlin. 2005. Electrospray ionization mass spectrometry fingerprinting of whisky: Immediate proof of origin and authenticity. *Analyst* 130: 890–897.

Parker, I. G., S. D. Kelly, M. Sharman, M. J. Dennis and D. Howie. 1998. Investigation into the use of carbon isotope ratios ($^{13}C/^{12}C$) of Scotch whisky congeners to establish brand authenticity using gas chromatography-combustion-isotope ratio mass spectrometry. *Food Chemistry* 63: 423–428.

Pissinatto, L., L. A. Martinelli, R. L. Victoria and P. B. D. Camargo. 1999. Stable carbon isotopic analysis and the botanical origin of ethanol in Brazilian brandies. *Food Research International* 32: 665–668.

Rau, Y. -H., G. -P. Lin, W. -S. Chang, S. -S. Wen and W. Fu. 2005. Using $^{13}C/^{12}C$ isotopic ratio analysis to differentiate between rice spirits made from rice and cane molasses [Chinese]. *Yaowu Shipin Fenxi* 13: 159–162.

Samarajeewa, U., M. R. Adams and J. M. Robinson. 1981. Major volatiles in Sri Lankan Arrack, a palm wine distillate. *International Journal of Food Science & Technology* 16: 437–444.

Simpkins, W. A. and D. Rigby. 1982. Detection of the illicit extension of potable spirituous liquors using $^{13}C/^{12}C$ ratios. *Journal of Agricultural and Food Chemistry* 33: 898–903.

Stevens, R. 1960. Beer flavour I. Volatile products of fermentation: A review. *Journal of the Institute of Brewing* 66: 453–471.

Wright, S. 2011. Trends in global spirits production. *Brewer and Distiller International* September: 25–27.

Yucesoy, D. and B. Ozen. 2013. Authentication of a Turkish traditional aniseed flavored distilled spirit, Raki. *Food Chemistry* 141: 1461–1465.

Zhong, Q., D. Wang and Z. Xiong. 2014. Application of stable isotope technique on distinguish between Chinese spirit by traditional fermentation and Chinese spirit made from traditional and liquid fermentation [Chinese]. *Zhipu Xuebao* 35: 66–71.

References from Table

Table 9.4

Aguilar-Cisneros, B. O., M. G. López, E. Richling, F. Heckel and P. Schreier. 2002. Tequila authenticity assessment by headspace SPME-HRGC-IRMS analysis of $^{13}C/^{12}C$ and $^{18}O/^{16}O$ ratios of ethanol. *Journal of Agricultural and Food Chemistry* 50: 7520–7523.

Angel Cantarelli, M., S. M. Azcarate, M. Savio, E. J. Marchevsky and J. M. Camia. 2015. Authentication and discrimination of whiskies of high commercial value by pattern recognition. *Food Analytical Methods* 8: 790–798.

Arbuzov, V. N. and S. A. Savchuk. 2002. Identification of vodkas by ion chromatography and gas chromatography. *Journal of Analytical Chemistry* 57: 428–433.

Aylott, R. I., A. H. Clyne, A. P. Fox and D. A. Walker. 1994. Analytical strategies to confirm Scotch whisky authenticity. *Analyst* 119: 1741–1746.

Barbosa-García, O., G. Ramos-Ortiz, J. L. Maldonado, J. L. Pichardo-Molina, M. A. Meneses-Nava, J. E. A. Landgrave and J. Cervantes-Martínez. 2007. UV–vis absorption spectroscopy and multivariate analysis as a method to discriminate tequila. *Spectrochimica Acta, Part A* 66: 129–134.

Bauer-Christoph, C., N. Christoph, B. O. Aguilar-Cisneros, M. G. López, E. Richling, A. Rossmann and P. Schreier. 2003. Authentication of tequila by gas chromatography and stable isotope ratio analyses. *European Food Research and Technology* 217: 438–443.

Bauer-Christoph, C., H. Wachter, N. Christoph, A. Roβmann and L. Adam. 1997. Assignment of raw material and authentication of spirits by gas chromatography, hydrogen- and carbon-isotope ratio measurements I. Analytical methods and results of a study of commercial products. *Zeitschrift für Lebensmittel-Untersuchung und –Forschung A* 204: 445–452.

Ceballos-Magaña, S. G., J. M. Jurado, M. J. Martín and F. Pablos. 2009. Quantitation of twelve metals in tequila and mezcal spirits as authenticity parameters. *Journal of Agricultural and Food Chemistry* 57: 1372–1376.

Chen, H., C. Tan, T. Wu, L. Wang and W. Zhu. 2014. Discrimination between authentic and adulterated liquors by near-infrared spectroscopy and ensemble classification. *Spectrochimica Acta, Part A* 130: 245–249.

Contreras, U., O. Barbosa-Garcia, J. L. Pichardo-Molina, G. Ramos-Ortiz, J. L. Maldonado, M. A. Meneses-Nava, N. E. Ornelas-Soto and P. L. Lopez-De-Alba. 2010. Screening method for identification of adulterate and fake tequilas by using UV-vis spectroscopy and chemometrics. *Food Research International* 43: 2356–2362.

Costa, R. S. D., S. R. B. Santos, L. F. Almeida, E. C. L. Nascimento, M. J. C. Pontes, R. a. C. Lima, S. S. Simoes and M. C. U. Araujo. 2004. A novel strategy to verification of adulteration in alcoholic beverages based on Schlieren effect measurements and chemometric techniques. *Microchemical Journal* 78: 27–33.

De León-Rodríguez, A. and L. González-Hernández. 2006. Characterization of volatile compounds of mezcal, an ethnic alcoholic beverage obtained from agave salmiana. *Journal of Agricultural and Food Chemistry* 54: 1337–1341.

de Souza, P. P., A. M. M. Resende, D. V. Augusti, F. Badotti, F. D. C. O. Gomes, R. R. Catharino, M. N. Eberlin and R. Augusti. 2014. Artificially-aged cachaça samples characterised by direct infusion electrospray ionisation mass spectrometry. *Food Chemistry* 143: 77–81.

de Souza, P. P., H. G. L. Siebald, D. V. Augusti, W. B. Neto, V. M. Amorim, R. R. Catharino, M. N. Eberlin and R. Augusti. 2007. Electrospray ionization mass spectrometry fingerprinting of Brazilian artisan cachaca aged in different wood casks. *Journal of Agricultural and Food Chemistry* 55: 2094–2102.

Espinosa-Sanchez, Y. M., D. Luna-Moreno and D. Monzon-Hernandez. 2015. Detection of aromatic compounds in tequila through the use of surface plasmon resonance. *Applied Optics* 54: 4439–4446.

Fernández, A. P., M. C. Santos, S. G. Lemos, M. M. C. Ferreira, A. R. Nogueria and J. A. Nobrega. 2005. Pattern recognition applied to mineral characterization of Brazilian coffees and sugar-cane spirits. *Spectrochimica Acta, Part B* 60: 717–724.

Filajdić, M. and J. Djuković. 1973. Gas-chromatographic determination of volatile constituents in Yugoslav plum brandies. *Journal of the Science of Food and Agriculture* 24: 835–842.

Fotakis, C. and M. Zervou. 2016. NMR metabolic fingerprinting and chemometrics driven authentication of Greek grape marc spirits. *Food Chemistry* 196: 760–768.

González-Arjona, D., V. González-Gallero, F. Pablos and A. G. González. 1999. Authentication and differentiation of Irish whiskeys by higher-alcohol congener analysis. *Analytica Chimica Acta* 381: 257–264.

González-Arjona, D., G. López-Pérez, V. González-Gallero and A. G. González. 2006. Supervised pattern recognition procedures for discrimination of whiskeys from gas chromatography/mass spectrometry congener analysis. *Journal of Agricultural and Food Chemistry* 54: 1982–1989.

Headley, L. M. and J. K. Hardy. 1989. Classification of whiskies by principal component analysis. *Journal of Food Science* 54: 1351–1358.

Heller, M., L. Vitali, M. a. L. Oliveira, A. C. O. Costa and G. A. Micke. 2011. A rapid sample screening method for authenticity control of whiskey using capillary electrophoresis with online preconcentration. *Journal of Agricultural and Food Chemistry* 59: 6882–6888.

Iglesias Rodriguez, R., M. Fernandez Delgado, J. Barciela Garcia, R. M. Pena Crecente, S. Garcia Martin and C. Herrero Latorre. 2010. Comparison of several chemometric techniques for the classification of orujo distillate alcoholic samples from Galicia (Northwest Spain) according to their certified brand of origin. *Analytical and Bioanalytical Chemistry* 397: 2603–2614.

Jelen, H. H., A. Ziolkowska and A. Kaczmarek. 2010. Identification of the botanical origin of raw spirits produced from rye, potato, and corn based on volatile compounds analysis using a SPME-MS method. *Journal of Agricultural and Food Chemistry* 58: 12585–12591.

Jochmann, M. A., D. Steinmann, M. Stephen and T. C. Schmidt. 2009. Flow injection analysis-isotope ratio mass spectrometry for bulk carbon stable isotope analysis of alcoholic beverages. *Journal of Agricultural and Food Chemistry* 57: 10489–10496.

Kokkinofta, R., P. V. Petrakis, T. Mavromoustakos and C. R. Theocharis. 2003. Authenticity of the traditional Cypriot spirit "Zivania" on the basis of metal content using a combination of coupled plasma spectroscopy and statistical analysis. *Journal of Agricultural and Food Chemistry* 51: 6233–6239.

Lachenmeier, D. W. 2007. Rapid quality control of spirit drinks and beer using multivariate data analysis of Fourier transform infrared spectra. *Food Chemistry* 101: 825–832.

Lachenmeier, D. W., R. Attig, W. Frank and C. Athanasakis. 2003. The use of ion chromatography to detect adulteration of vodka and rum. *European Food Research and Technology* 218: 105–110.

Lachenmeier, D. W., E. Richling, M. G. López, W. Frank and P. Schreier. 2005. Multivariate analysis of FTIR and ion chromatographic data for the quality control of tequila. *Journal of Agricultural and Food Chemistry* 53: 2151–2157.

Lachenmeier, D. W., E. M. Sohnius, R. Attig and M. G. López. 2006. Quantification of selected volatile constituents and anions in Mexican agave spirits (tequila, mezcal, sotol, bacanora). *Journal of Agricultural and Food Chemistry* 54: 3911–3915.

Lai, C. -H., C. -W. Hsieh and W. -C. Ko. 2014. Detection limit of molasses spirits mixed in rice spirits using the SNIF-NMR method. *Journal of Food and Drug Analysis* 22: 197–201.

Lai, C. -H., W. -C. Ko, E. C. -F. Chen and C. -W. Hsieh. 2009. Detection of adulteration in Taiwanese rice-spirits by $^{13}C/^{12}C$ stable isotope ratio analysis and SNIF-NMR methods. Paper presented at the 237th ACS National Meeting, March 22–26, 2009, Salt Lake City, UT, United States.

Lopez, M. G. 2005. Authenticity: The case of tequila. Paper presented at the 229th ACS National Meeting, March 13–17, 2005, San Diego, CA, United States.

Mackenzie, W. M. and R. I. Aylott. 2004. Analytical strategies to confirm Scotch whisky authenticity. Part II: Mobile brand authentication. *Analyst* 129: 607–612.

Martin, G. J., M. L. Martin, F. Mabon and M. J. Michon. 1983. A new method for the identification of the origin of ethanols in grain and fruit spirits: High-field quantitative deuterium nuclear magnetic resonance at the natural abundance level. *Journal of Agricultural and Food Chemistry* 31: 311–315.

Martin-Alvarez, P. J., M. D. Cabezudo, J. Sanz, A. Herranz, P. De La Serna and C. Barro. 1988. Application of several statistical classification techniques to the differentiation of whisky brands. *Journal of the Science of Food and Agriculture* 45: 347–358.

McIntyre, A. C., M. L. Bilyk, A. Nordon, G. Colquhoun and D. Littlejohn. 2011. Detection of counterfeit Scotch whisky samples using mid-infrared spectrometry with an attenuated total reflectance probe incorporating polycrystalline silver halide fibres. *Analytica Chimica Acta* 690: 228–233.

Meier-Augenstein, W., H. F. Kemp and S. M. L. Hardie. 2012. Detection of counterfeit Scotch whisky by ^2H and ^{18}O stable isotope analysis. *Food Chemistry* 133: 1070–1074.

Moller, J. K. S., R. R. Catharino and M. N. Eberlin. 2005. Electrospray ionization mass spectrometry fingerprinting of whisky: Immediate proof of origin and authenticity. *Analyst* 130: 890–897.

Ng, T. L. and S. O. Woo. 1980. Characterisation of Chinese alcoholic beverages by their congener contents. *Journal of the Science of Food and Agriculture* 31: 503–509.

Nietner, T., S. A. Haughey, N. Ogle, C. Fauhl-Hassek and C. T. Elliott. 2014. Determination of geographical origin of distillers dried grains and solubles using isotope ratio mass spectrometry. *Food Research International* 60: 146–153.

Palma, M. and C. G. Barroso. 2002. Application of FT-IR spectroscopy to the characterisation and classification of wines, brandies and other distilled drinks. *Talanta* 58: 265–271.

Panossian, A., G. Mamikonyan, M. Torosyan, E. Gabrielyan and S. Mkhitaryan. 2001. Analysis of aromatic aldehydes in brandy and wine by high-performance capillary electrophoresis. *Analytical Chemistry* 73: 4379–4383.

Parker, I. G., S. D. Kelly, M. Sharman, M. J. Dennis and D. Howie. 1998. Investigation into the use of carbon isotope ratios ($^{13}C/^{12}C$) of Scotch whisky congeners to establish brand authenticity using gas chromatography-combustion-isotope ratio mass. *Food Chemistry* 63: 423–428.

Peng, Q., R. Tian, F. Chen, B. Li and H. Gao. 2015. Discrimination of producing area of Chinese Tongshan Kaoliang spirit using electronic nose sensing characteristics combined with the chemometrics methods. *Food Chemistry* 178: 301–305.

Petrakis, P., I. Touris, M. Liouni, M. Zervou, I. Kyrikou, R. Kokkinofta, C. R. Theocharis and T. M. Mavromoustakos. 2005. Authenticity of the traditional Cypriot spirit "Zivania" on the basis of ^1H NMR spectroscopy diagnostic parameters and statistical analysis. *Journal of Agricultural and Food Chemistry* 53: 5293–303.

Picque, D., P. Lieben, G. Corrieu, R. Cantagrel, O. Lablanquie and G. Snakkers. 2006. Discrimination of cognacs and other distilled drinks by mid-infrared spectroscopy. *Journal of Agricultural and Food Chemistry* 54: 5220–5226.

Pilar-Marti, M., J. Pino, R. Boque, O. Busto and J. Guasch. 2005. Determination of ageing time of spirits in oak barrels using a headspace-mass spectrometry (HS-MS) electronic nose system and multivariate calibration. *Analytical and Bioanalytical Chemistry* 382: 440–443.

Pissinatto, L., L. A. Martinelli, R. L. Victoria and P. B. D. Camargo. 1999. Stable carbon isotopic analysis and the botanical origin of ethanol in Brazilian brandies. *Food Research International* 32: 665–668.

Pontes, M. J. C., S. R. B. Santos, M. C. U. Araújo, L. F. Almeida, R. a. C. Lima, E. N. Gaião and U. T. C. P. Souto. 2006. Classification of distilled alcoholic beverages and verification of adulteration by near infrared spectrometry. *Food Research International* 39: 182–189.

Rau, Y. -H., G. -P. Lin, W. -S. Chang, S. -S. Wen and W. Fu. 2005. Using $^{13}C/^{12}C$ isotopic ratio analysis to differentiate between rice spirits made from rice and cane molasses [Chinese]. *Yaowu Shipin Fenxi* 13: 159–162.

Sadecka, J., J. Tothova and P. Majek. 2009. Classification of brandies and wine distillates using front face fluorescence spectroscopy. *Food Chemistry* 117: 491–498.

Samarajeewa, U., M. R. Adams and J. M. Robinson. 1981. Major volatiles in Sri Lankan arrack, a palm wine distillate. *International Journal of Food Science & Technology* 16: 437–444.

Savchuk, S. A. and G. M. Kolesov. 2007. Chromatographic determination of phthalic acid esters as an indicator of adulterated cognacs and cognac spirits. *Journal of Analytical Chemistry* 62: 761–772.

Savchuk, S. A., V. N. Vlasov, S. A. Appolonova, V. N. Arbuzov, A. N. Vedenin, A. B. Mezinov and B. R. Grigor'yan. 2001. Application of chromatography and spectrometry to the authentication of alcoholic beverages. *Journal of Analytical Chemistry* 56: 214–231.

Schipilliti, L., I. Bonaccorsi, A. Cotroneo, P. Dugo and L. Mondello. 2012. Evaluation of gas chromatography-combustion-isotope ratio mass spectrometry (GC-C-IRMS) for the quality assessment of citrus liqueurs. *Journal of Agricultural and Food Chemistry* 61: 1661–1670.

Simpkins, W. A. and D. Rigby. 1982. Detection of the illicit extension of potable spirituous liquors using $^{13}C/^{12}C$ ratios. *Journal of the Science of Food and Agriculture* 33: 898–903.

Spanik, I., L. Cirka and P. Majek. 2015. Classification of wine distillates using multivariate statistical methods based on their direct GC-MS analysis. *Chemical Papers* 69: 395–401.

Thomas, F., C. Randet, A. Gilbert, V. Silvestre, E. Jamin, S. Akoka, G. Remaud, N. Segebarth and C. Guillou. 2010. Improved characterization of the botanical origin of sugar by carbon-13 SNIF-NMR applied to ethanol. *Journal of Agricultural and Food Chemistry* 58: 11580–11585.

Urickova, V., J. Sadecka and P. Majek. 2013. Right-angle fluorescence spectroscopy for differentiation of distilled alcoholic beverages. *Nova Biotechnologica et Chimica* 12: 83–92.

Vichi, S., M. Riu-Aumatell, S. Buxaderas and E. Lopez-Tamames. 2008. Assessment of some diterpenoids in commercial distilled gin. *Analytica Chimica Acta* 628: 222–229.

Villanueva, S., H. Escalona, M. Estarron, S. T. Martin Del Campo, E. Cantor and K. Aguilera. 2005. Classification of tequilas from different categories and regions using gas chromatography-mass spectrometry and sensory evaluation. Paper presented at the 229th ACS National Meeting, March 13–17, 2005, San Diego, CA, United States.

Winterova, R., R. Mikulikova, J. Mazac and P. Havelec. 2008. Assessment of the authenticity of fruit spirits by gas chromatography and stable isotope ratio analyses. *Czech Journal of Food Sciences* 26: 368–375.

Yucesoy, D. and B. Ozen. 2013. Authentication of a Turkish traditional aniseed flavored distilled spirit, Raki. *Food Chemistry* 141: 1461–1465.

Zhong, Q. -D., D. -B. Wang and Z. -H. Xiong. 2014. Application of stable isotope technique on distinguish between Chinese spirit by traditional fermentation and Chinese spirit made from traditional and liquid fermentation [Chinese]. *Zhipu Xuebao* 35: 66–71.

Zhou, X., Z. Yang, S. A. Haughey, P. Galvin-King, L. Han and C. T. Elliott. 2015. Classification the geographical origin of corn distillers dried grains with solubles by near infrared reflectance spectroscopy combined with chemometrics: A feasibility study. *Food Chemistry* 189: 13–18.

CHAPTER 10

Stable Isotope Measurements and Modeling to Verify the Authenticity of Dairy Products

Emad Ehtesham,[1] Federica Camin,[2] Luana Bontempo[2] and Russell D. Frew[3],*

10.1 Introduction

"Poets have been mysteriously silent on the subject of cheese."

G. K. Chesterton, English writer (1874–1936)

Globalization and the increasing complexity of the trade in food provide opportunities and risks to producers and consumers alike. Dairy is a priority commodity area due to its simple processing procedures, high level of trade, and use as a major or minor ingredient in many processed and sensitive products such as infant formula.

This chapter reviews the application of stable isotope ratio measurements to verify the authenticity of dairy products such as milk and cheese using a conventional *batch* approach. We then present an alternative using a *predictive model* to reduce the data burden and, hence lower some of the barriers to those wishing to use isotope ratio analysis as an authentication tool for dairy products.

[1] Department of Chemistry and Biotechnology, Swedish University of Agricultural Sciences (SLU), Uppsala, SWEDEN.
  Email: emad.ehtesham@slu.se
[2] Department of Food Quality and Nutrition, Research and Innovation Centre, Fondazione Edmund Mach (FEM), Via E. Mach 1, 38010 San Michele all'Adige, ITALY.
  Email: federica.camin@fmach.it; luana.bontempo@fmach.it
[3] Department of Chemistry, University of Otago, PO Box 56, Dunedin, NEW ZEALAND.
* Corresponding author: rfrew@chemistry.otago.ac.nz

10.2 Current trends in food fraud and milk adulteration

As the global food supply chain becomes more integrated, the magnitude and hazards associated with food fraud are creating increased worldwide concern. A potentially contaminated ingredient, such as milk powder, could be incorporated into a food product such as infant formula that would be traded in many countries other than the country of origin. In such a case finding the source of the adulterated ingredient becomes a priority, especially when the investigation has to be made through industrial organizations, or legal actions have to be taken by governmental bodies to halt the distribution of the adulterated food product and to prosecute the perpetrator. Apart from negative impacts on regional milk producers, this may also present serious consequences to public health. A high-profile international case was the addition of melamine to milk powder, which was later incorporated into infant formula (Figure 10.1). As a consequence, China lost confidence with consumers around the globe and, more seriously, thousands of children were hospitalized while six died (Sharma and Paradakar 2010; Wei and Liu 2012). Such incidents demonstrate the necessity for a prompt response to cases of potentially harmful adulterated foods (Jia et al. 2012; Lyu 2012).

There are several ways of trading sub-standard food that would all be considered *adulteration*, such as the addition of water to milk or the abstraction of fat from milk (Karoui 2012). The relative simplicity of adulterating milk with low cost ingredients for economic gain has often made milk a target for such fraudulent actions. A recent study that analyzed peer-reviewed academic databases between 1980 and 2010 revealed that milk was the food ingredient with the second highest levels of adulteration incidents (Moore et al. 2012). Despite this, milk fraud received little media attention compared to other food ingredients. The study authors categorized this fraud as (1) false declaration of milk's geographical or species origin, or (2) dilution of milk or addition of lower cost adulterant mixtures to increase the value of a lower quality milk. They concluded that if the data had been

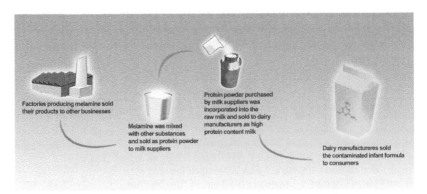

Factories producing melamine sold their products to other businesses

Melamine was mixed with other substances and sold as protein powder to milk suppliers

Protein powder purchased by milk suppliers was incorporated into the raw milk and sold to dairy manufacturers as high protein content milk

Dairy manufactureres sold the contaminated infant formula to consumers

Figure 10.1 Schematic illustration of melamine adulteration in milk powder.

available to risk assessors at the time of the milk powder-melamine incident in China, there would have been more motivation to develop methods to track such fraudulent actions.

10.3 Milk composition

The major constituents of milk are water, lactose (a disaccharide sugar), protein, and fat, which typically comprise 87%, 4.6%, 3.3% and 4.0% of the total milk composition, respectively (Walstra et al. 2005). Bovine milk is a complex oil-in-water emulsion produced from the animal's metabolism. Volatile and conjugated fatty acids are produced through biosynthesis in the rumen and some of these will be incorporated into milk without further modification (Bauman et al. 2000; Siciliano-Jones and Murphy 1989).

The fat content of milk is one quality assessment for the milk trade (Ohtani et al. 2005). Milk fat, which is an important component in various dairy and food products, is a complex mixture of approximately 400 different fatty acids (Pesek et al. 2005) predominantly in the esterified form of triacylglycerides (~ 98.0 g/100 g) and minor forms such as phospholipids (1.0 g/100 g), sterols (0.45 g/100 g), and free fatty acids (0.4 g/100 g) (MacGibbon and Taylor 2006; Tunick 2010). Only a limited number of fatty acids constitute the larger portion of milk fat, as shown in Table 10.1.

Table 10.1 Major bovine milk fatty acids. Compiled from Kaylegian and Lindsay (1995).

Fatty acid carbon number	Fatty acid common name	Average range (wt%)[1]
4:0	butyric	2–5
6:0	caproic	1–5
8:0	caprylic	1–3
10:0	capric	2–4
12:0	lauric	2–5
14:0	myristic	8–14
15:0	pentadecanoic	1–2
16:0	palmitic	22–35
16:1	palmoleic	1–3
17:0	margaric	0.5–1.5
18:0	stearic	9–14
18:1	oleic	20–30
18:2	linoleic	1–3
18:3	linolenic	0.5–2

[1] Total fatty acids.

Triacylglycerides (TAGs) consist of three fatty acid moieties with the carboxyl groups attached to a glycerol backbone via an ester linkage, as depicted in Figure 10.2. Fatty acids are either saturated or unsaturated and the degree of unsaturation is defined by the number of double bonds present in the molecule.

Resolving the large number of fatty acids present in milk requires specialized chromatographic approaches, such as two-dimensional chromatographic techniques (Akoto et al. 2008) or specialized columns for fatty acid isomer analysis (Ecker et al. 2012).

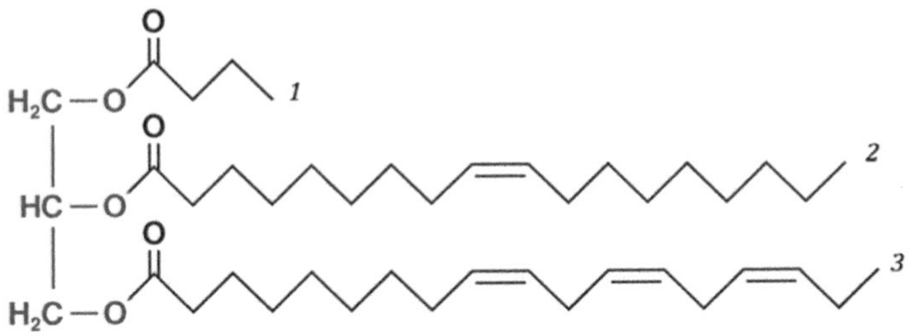

Figure 10.2 Typical structure of a triacylglyceride molecule with (blue color) butyric acid (C4:0) in position 1, mono-unsaturated linoleic acid (C18:1) in position 2, and poly-unsaturated linolenic acid (C18:3) in position 3, attached to glycerol backbone (red color).

10.4 Production and processing of milk

Milk is mostly traded in a dehydrated (spray-dried) form that can be reconstituted as an ingredient in different milk-based products, such as infant formula, yogurt, and chocolate. Dehydration facilitates transportation, reduces corresponding shipping costs, and preserves the milk for a longer time (Augustin and Margetts 2003). Milk byproducts, such as whey proteins and lactose from cheese manufacturing (González-Martínez et al. 2002; Zadow 2003), or casein and milk fat from milk powder manufacturing, are also used in specific applications (e.g., adding nutritional and physical functionality to a final product). Most of these processed milk products contain some milk lipid—with the notable exception of lactose—and thus potentially contain fatty acids as biomarkers.

Milk powder is the most common form of milk traded globally, with major markets in North America, South America, and the European Union (Baldwin and Pearce 2005) and has been analyzed to assess its provenance. For this reason we first focus on milk powder—as opposed to other dairy products—in this chapter because there is an increasing demand for tools for dairy authentication from consumers and the food industry sector (Romdhane 2010).

New Zealand exports 95% of its dairy products (MFTE New Zealand Government 2011) and is the largest dairy exporter in the world, accounting for one-third of cross-border dairy trade. In 2012, New Zealand's dairy export revenue was approximately NZ$14.6 billion and increasing (MPI New Zealand Government, Statistics New Zealand, DairyNZ and Fonterra Co-operative Group 2012). The global demand for New Zealand milk powder can be attributed to its quality, which has made it a prime target for fraud. This is of particular importance for milk powder used in the preparation of infant formula that claims to be sourced from New Zealand and tools are required to complement the traditional traceability systems and provide assurance of origin to customers.

Determining a range of fatty acid isotopic data from milk, in addition to the isotopic analysis of bulk milk solid, may reveal distinctive differences between milk production regions using multivariate statistical analysis. In order to determine which of the major fatty acids in milk are characteristic and representative geographical tracers, a range of fatty acids has been investigated.

10.5 Milk isotopic fingerprint, a result of biogeochemical influence

The synthesis of milk components in mammary glands follows defined biochemical pathways (Larson 1979) that should reflect characteristic isotopic values of the local environment (Kornexl et al. 1997). Sun (2008) proposed that the isotopic distribution of molecules from a biological organism is affected by its environmental and biological factors and can thus be used to trace back to the origin of a specific compound.

Kornexl et al. (1997) reported the first application of isotope ratio mass spectrometry (IRMS) to determine the geographical origin of milk using bulk $\delta^{15}N$ and $\delta^{13}C$ values to authenticate origin. This study, however, concluded that the combination of $\delta^{13}C$ and $\delta^{15}N$ values reflected only the cattle dietary habit and that nitrogen and carbon isotopes were not useful to assign the origin of the milk. The $\delta^{13}C$ and $\delta^{15}N$ values of total milk casein (a protein) were found to remain reasonably constant throughout a year of production with a *ca.* 1‰ variance for both elements. Furthermore, the authors suggested that the $\delta^{18}O$ composition of milk water might be a more useful tool to determine milk's geographical origin than analysis of carbon and nitrogen isotopes. Crittenden et al. (2007) determined stable isotope ratios of C, N, O, S, and Sr in Australasian milk and found that C and O isotope ratios conformed to predicted isotope fractionation patterns based on latitude and climate. The study found that casein was relatively enriched in both ^{13}C and ^{15}N, but depleted in ^{34}S, compared to whole milk. Australasian milk was significantly enriched in both ^{18}O and ^{34}S compared to European dairy products (Crittenden et al. 2007).

In another attempt to assign the origin of the milk, Rossmann et al. (1998) proposed the use of $\delta^{18}O$ signatures of milk water in combination with $\delta^{13}C$ and $\delta^{15}N$ values of casein and $\delta^{34}S$ signatures of an amino acid (methionine) derived from milk casein. It was found that $\delta^{18}O$ values of milk collected from across alpine Europe varied by 3‰, with a 2‰ enrichment compared to cows' local drinking water. No correlation was observed between $\delta^{34}S$ and $\delta^{15}N$ values in methionine or bulk milk. A small variation in $\delta^{15}N$ values was observed and was attributed to the fertilizers applied to pastures utilized by the cows.

Recently, Ehtesham et al. (2015) found that the δ^2H composition of bovine milk was influenced by both the ^2H composition of grass fatty acids and drinking water and that a significant correlation existed between the ^2H isotopic composition of milk and the fatty acids in cattle feed. It was proposed that δ^2H values of milk could help to determine the provenance of feed and the milk, as illustrated in Figure 10.3.

The influence of processing of milk has been examined (Scampicchio et al. 2012) and heat treatment was found to induce changes in $\delta^{13}C$ and $\delta^{15}N$ signatures of milk, particularly in the fat and whey fractions. These authors did not de-fat the casein fraction and so residues of fat are likely to have influenced their results. A further study of milk samples sourced from Tyrol (Austria) demonstrated that a combination of stable isotope data with vibrational spectroscopy and GC analysis could be used to determine

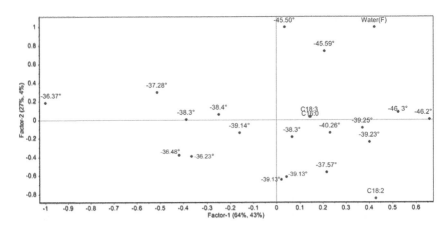

Figure 10.3 Partial Least Square (PLS) bi-plot of ^2H data of milk components from 18 dairy farms (latitudes, blue coded) and the discriminating influence of *x*-loadings (farm water and grass fatty acid ^2H composition, red coded) on milk ^2H composition. Reprinted with permission from Ehtesham et al. 2015. Copyright 2015 Elsevier.

the geographic origin of milk. The data from the various techniques were combined using multivariate classification methods and an error rate of 5% was achieved in assigning geographic origin to milk (Scampicchio et al. 2016).

10.6 Preparation of milk components for isotopic analysis

An understanding of the production and processing of raw milk[1] is necessary to determine whether these processes affect the chemical and/or physical composition of the milk matrix and the measurement of isotopic composition, and to avoid erroneous interpretation. As an example, if the isotopic measurements of raw milk components are to be considered, a record of milk pre-harvesting (e.g., microbial contamination) and post-harvesting (e.g., storage conditions) treatments for each sample would help to assess whether the composition of the compounds of interest in that sample may have been altered due to mishandling. Fortunately, due to the strict quality auditing procedures in milk powder production plants (Augustin et al. 2003), the probability of receiving milk powder of sub-standard quality is minimized. Prior to the analysis of milk powder and subsequent data interpretation, other potentially influential factors should be taken into account such as: the addition of lecithin to whole milk powder during spray drying to improve solubility (in-process factors) (Augustin and Margetts 2003; Kim et al. 2009; Schuck 2011); exposure of raw milk to elevated temperatures during pasteurization and spray drying (Farkye 2006); and re-activation of milk's indigenous enzymes as a result of heat treatment of the milk powder samples that may arise during procedures required for isotopic analysis under certain temperatures (post-process factors) (Deeth and Fitz-Gerald 1983; Hayes et al. 2005).

10.7 Application of stable isotope analysis to dairy products

10.7.1 Infant formula

Because most of the milk powder traded internationally is used as an ingredient in other products (e.g., infant formula, confectionaries), it is essential to have a system that can provide information on the source of the dairy component of that mixture. Conventional stable isotopes methods have, however, proved unable to distinguish the origin of adulterated dairy products once they are incorporated into mixtures. This is principally because ingredients from different origins will mask the product's original isotopic composition. Infant formula has been a target for fraud, resulting in large financial losses to the industry of the producing countries

---

[1] Raw milk here means an unheated and unprocessed milk.

(Xiu and Klein 2010) and, therefore, a robust technique for authentication is in demand. Fatty acid compounds in infant formula comprise approximately 27% of the dry weight (Cesa et al. 2012). Isotopic fractionation in the reactions involved in biochemical and physicochemical processes can lead to wide variations in the isotopic values of individual fatty acids (Chikaraishi et al. 2004) and can potentially be used to determine milk's region of origin.

In order to adopt an appropriate analytical strategy it is important that the analyst understands the nature of the infant formula being examined. For example, to enhance similarity with human breast milk TAG profile, synthetic TAG formulations have been developed of which Betapol® was the first commonly used formula (Wells 1996; Spurgeon et al. 2003; Álvarez and Akoh 2015). The TAG composition in Betapol® formula is enriched (70%) with palmitic acid (C16:0) at the sn-2 position. Most infant formulas are manufactured using a mixture of vegetable oil, animal oil (butterfat), or synthesized oil (Berger et al. 2000; López-López et al. 2002; Maduko et al. 2007). A comprehensive study of fatty acids from vegetable- and animal-based lipids incorporated into infant formula may provide evidence for this approach in infant formula traceability. An extension of this research might apply already developed compound specific isotope analysis (CSIA) methodology to the provenance of complex milk powder based products such as infant formula.

The addition of various fat sources such as vegetable fats to complex milk-based foods such as infant formula cannot be simply determined by this approach. More sophisticated separation techniques must be employed in order to separate the incorporated fat prior to any CSIA assay. A simple preliminary screening step for adulterants is desirable prior to more labor intensive, sophisticated, and expensive tests.

Because the isotopic values recorded in organic material are the product of the physical, chemical, and biological interactions with their immediate environment (see Chapters 3 and 4), it is necessary to decouple the various sources in a complex matrix such as infant formula before isotopic analysis. In the case of fatty acid analysis, animal or vegetable sources need to be separated before ^{2}H and ^{13}C CSIA analysis. The fatty acid in vegetable oil, TAGs, are not randomly distributed (Vander Wal 1960). As an example, saturated fatty acids such as palmitic acid (C16:0) occur infrequently at the sn-2 position in vegetable oil (Martínez-Force et al. 2009), but are key functional fatty acids, used in the manufacture of infant formula, when esterified at the sn-2 position (Shahidi and Senanayake 2006). The fatty acids in most vegetable oils are also not randomly distributed at the sn-1 and sn-3 positions of TAGs (Vander Wal 1960).

Because of this non-random distribution, a position-specific enzymatic hydrolysis can separate similar fatty acids present in different TAGs—for example, single seed vegetable oil fatty acids that are incorporated into infant formula (Kelly et al. 1997). Goat milk infant formula contains unmodified vegetable oils such as sunflower and/or canola (Prosser et al. 2010) and CSIA of the fatty acids can be used as an indicator of the source of plant oil that is incorporated into the formula. Lipolytic enzymes that hydrolyze specific position at the TAG molecule (such as pancreatic lipase family) (Rogalska et al. 1993) are among those used for such assays (Jensen 1971; Jensen and Jensen 1992; Martínez-Force et al. 2009). By employing such enzymatic techniques, it is possible to determine the incorporation of fat into infant formula based on specific compounds (fatty acids from various sources) which, to date, has not been yet implemented in forensic application of dairy products. This approach may, however, be helpful for provenancing infant formula and its constituent ingredients.

10.7.2 Cheese

"Cheese is the ripened or unripened soft, semi-hard, hard, or extra-hard product ...obtained by coagulating wholly or partly the protein of milk ...through the action of rennet, coagulating agents or other suitable processing techniques and by partially draining the whey resulting from the coagulation" (Codex Standard 283-1978). On a global scale, cheese is an important agricultural product, according to the Food and Agricultural Organization of the United Nations, with over 21 million tons of cheese produced worldwide in 2013 (FAO, http://faostat3.fao.org/). The largest cheese producer is the United States, accounting for approximately 25% of world production, followed by Germany, France, and Italy. Many cheeses are synoymous with specific regions and have Protected Designation of Origin (PDO) or Protected Geographical Indication (PGI) status. European law provides that, in the case of fresh milk and PDO/PGI cheeses (Regulation 510/2006/EC and the following EU Regulation 1151/2012), an indication of the origin of the milk and other raw materials used for manufacture has to be provided on the label. The U.S. Food and Drug Adminstration also lays down the legally required characteristics of particular types of cheese—for example, it regulates what can legally be called *Parmesan* according to standards established in the 1950s and updated in 2015.

There have been many food fraud scandals in the last few years involving cheese. In the latest (at the time of writing), Parmesan adulterated with wood pulp was distributed to some of the biggest grocery chains in the U.S. (Mulvany 2016). Furthermore, a recent study organized by an independent consumer watchdog group examined the provenance of the UK' goat cheese supply and found that approximately 12% of the samples were adulterated with milk from animals other than goats (Charlton 2014).

This form of adulteration has the potential to cause intolerance or allergy as well as legal, religious, ethical, or cultural objections.

Stable isotope ratios of bio-elements (H, C, N, O, and/or S) have been applied in the last twenty years to cheese, in particular to determine its geographical origin as well as the production method (C_4 plants or fresh herbage in animal's diet, organic *vs.* conventional production, etc.). $\delta^{13}C$ and $\delta^{15}N$ signatures have been determined for casein and combined with the isotopic ratios of other elements (^2H, ^{34}S, ^{87}Sr) and/or other cheese components ($\delta^{13}C$ and $\delta^{18}O$ signatures of glycerol) to characterize the geographical provenance of particularly high-priced cheeses (Pillonel et al. 2003; Camin et al. 2004; Brescia et al. 2005; Manca et al. 2006; Stevenson et al. 2015; Necemer et al. 2016). Throughout these studies some specific features have been observed. Alpine cheeses, produced at high altitudes, have $\delta^{15}N$ values significantly lower than products from lower elevation areas (Pillonel et al. 2003; Bontempo et al. 2011; Bontempo et al. 2012). Both $\delta^{13}C_{CASEIN}$ and $\delta^{13}C_{GLYCEROL}$ allowed researchers to verify if the maize (corn) uptake in animal diets was in compliance with the maximum level established in some PDO production protocols. Furthermore, $\delta^{18}O$ signatures were found to be useful in determining if animals ate silage or fresh herbage, as the latter contains water enriched in ^{18}O due to evapotranspiration (Bontempo et al. 2011).

Some studies have combined stable isotope ratios and elemental data (Bontempo et al. 2011; Camin et al. 2012; Camin et al. 2015), improving the characterization of Alpine PDO cheeses and the discrimination of Parmigiano Reggiano PDO cheese from non-PDO hard cheeses. The reliability and efficiency of this approach has also been recognized by the Italian consortia of PDO Grana Padano and Parmigiano Reggiano cheeses. Since 2000, a large reference database for PDO and non-PDO hard cheeses has been created and in 2011 stable isotope ratio analysis was officially adopted to verify the authenticity of grated and shredded PDO Grana Padano (production protocol, EU Regulation 584/2011). With regards to the authentication of Parmigiano Reggiano, using a combination of δ^2H, $\delta^{13}C$, $\delta^{15}N$, $\delta^{34}S$, and select elemental concentrations two Random Forests models were created to trace the origin of cheese in grated and shredded forms, for which it is not possible to check the logo on the rind. One model was able to predict the origin of seven types of European hard cheeses (in a validation step, 236 samples out of 240 were correctly classified). The other model, which was designed specifically to discriminate the PDO Parmigiano Reggiano cheese from nine European and two non-European imitators, correctly classified 260 out of 264 samples.

Molkentin (2013) proposed the use of $\delta^{13}C$ signatures of fat together with C18:3w3 fatty acid concentration to determine if a dairy product was organic or conventional, as already demonstrated in a previous study of German milk (Molkentin and Giesemann 2007; Molkentin 2013). The established $\delta^{13}C$ threshold of −26.5‰ was found to be widely applicable to processed dairy products. In the case of products such as evaporated milk, $\delta^{13}C$ values have to be for both fat and de-fatted dry matter. A difference of < 1‰ between these two fractions could provide evidence of a fraud, or of conventional rather than organic production. Furthermore, the $\delta^{15}N$ values of de-fatted dry matter were typically ≤5.5‰ for organic dairy products. Finally, a recent paper (Capici et al. 2015) verified that it was possible to distinguish between cheeses produced from raw milk and those prepared from pasteurized milk based on differences in the $\delta^{13}C$ and $\delta^{15}N$ isotope values between the fat and de-fatted fractions. These authors also demonstrated that it was possible to detect the fraudulent addition of extraneous matter, having different isotope values, such as reconstituted milk obtained from powdered milk samples.

10.7.3 Other dairy products

Butter has been the subject of a series of food scandals, including the substitution of the product with animal fats or synthetic products (Merrett 2008) as well as the mislabelling with incorrect geographical origin to increase EU subsidies (Balling and Rossmann 2004). To date, only one study has applied stable isotope ratio analysis to butter (Rossmann et al. 2000) in which a combination of $\delta^{13}C$, $\delta^{15}N$, $\delta^{18}O$, $\delta^{34}S$, and $^{87}Sr/^{86}Sr$ data could reliably determine the regional provenance of butter from several European and non-European countries.

McLeod et al. (2016) determined $\delta^{13}C$ and $\delta^{15}N$ values and the concentration of trace elements in goat milk powder, which could be used to verify the origin of the product even, in some cases, at the manufacturer or factory level.

10.8 Predictive modelling

The approaches presented above and generally used in food traceability utilize measurement of selected parameters of authentic samples to define specifications that will distinguish the *genuine article* and products from other sources. This can be a robust approach provided the training set used has sufficient samples to define the specifications for the genuine product and encompasses the variability inherent in that product. It can be an expensive exercise to obtain and measure all the necessary samples and the training set may only be relevant to specific products—i.e., inter-batch or inter-seasonal differences may require regular resampling. The cost of this approach may present a barrier for producers to implement this technology in their authentication and traceability systems.

One of the main advantages of using stable isotope systems is that the controlling factors in variation may be well understood and the patterns observed in isotopic compositon can be mapped as 'isoscapes' (e.g., West et al. 2014). Ehtesham et al. (2013) have applied a close correlation between patterns of rainfall δ^2H composition and milk powder to produce a map of the predicted composition of milk δ^2H for New Zealand (Figure 10.4).

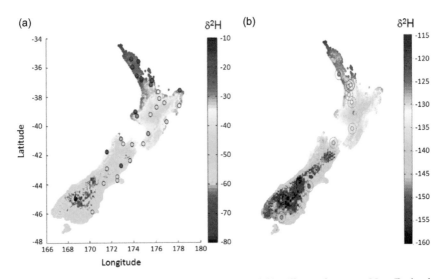

Figure 10.4 Maps of δ^2H composition of (a) rainfall and (b) milk powder across New Zealand. The milk composition map is derived from linear regression of authentic sample composition (inner circles) with annually averaged rainfall from 25 km radius (outer circles); redrawn with permission from Ehtesham et al. 2013.

10.9 A case study of deliberate contamination

The importance of milk in the human diet and especially its use as an ingredient in infant formula makes it a target for deliberate contamination for pecuniary gain. In New Zealand in 2015 Jeremy Kerr mixed highly concentrated monofluoroacetate (MFA, or pesticide "1080") with infant formula and posted envelopes containing the contaminated powder to the Fonterra Dairy Cooperative and to Federated Farmers. Together with the powders were letters demanding that the country stop using 1080 for pest control (mainly invasive possums) or 1080-contaminated milk would be released onto the international market. The cost of the subsequent police investigation and damage to the New Zealand milk market was estimated to be in excess of NZ$37 million (NZ Herald 2016).

Police obtained 43 samples of MFA, which were considered to be representative of the supply of MFA available in New Zealand at the time of the incident. These samples would have been produced in an unknown number of batches although each batch would have a characteristic history in terms of reagents use in production, reaction conditions, post-production processing such as purification, mixing, etc. It was assumed that samples with identical history (i.e., the same production batch) would have indistinguishable stable isotope ratios whereas samples from different batches might have different stable isotope ratios and therefore be distinguishable. The stable isotope ratios of H, C, and O were measured on the reference samples and MFA extracted from the case samples to:

- Determine that there was sufficient variation in the isotope ratios to allow differentiation of MFA from different batches.
- Identify the likely supply of the MFA found in the contaminated milk powder contained with the envelopes.

The reference samples were all powders and so the isotopic ratios could be determined by bulk analysis. The traces of MFA recovered from the case samples were converted to the dicyclohexylcarbodiimide derivative (Figure 10.5) for identification and quantification. The isotopic compositions of these derivatives were then determined by GC-C-IRMS. Raw data were corrected for the contribution of carbon (6/8) and hydrogen (4/6) atoms from the derivatizing agent by mass balance.

A wide range of stable isotope ratios were found within the reference samples (δ^2H: 190‰; δ^{13}C: 22‰; δ^{18}O: 25‰), allowing discrimination between the majority of the reference samples and the two case samples (Figure 10.6). Statistical analysis enabled all but five of the reference samples to be eliminated as potential sources of the case MFA and it was later revealed that two of these five were found in the suspect's possession and a third was from an associated company. The stable isotope evidence provided significant support to the police case and a confession and conviction were subsequently obtained. Mr. Kerr was subsequently sentenced to eight and a half years in prison for attempting to blackmail the New Zealand Government.

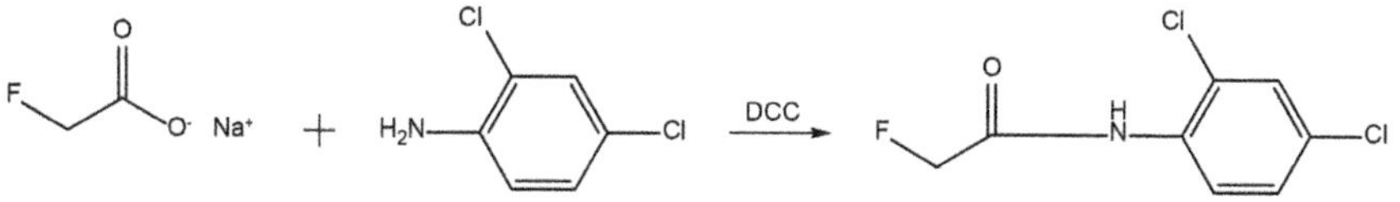

Figure 10.5 Derivatization of monofluoroacetate (MFA) with 2,4-dichloroaniline (DCA) in the presence of N,N′-dicyclohexylcarbodiimide (DCC) to form MFA-DCA.

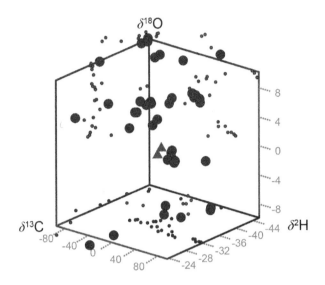

Figure 10.6 Three dimensional plot of the reference MFA samples (dots) and the case samples (triangles). The larger circles are the sample data and the smaller dots are the projections onto the 2D plane—i.e., the scatter plot between two parameters.

References

Akoto, L., F. Stellaard, H. Irth, R. J. J. Vreuls and R. Pel. 2008. Improved fatty acid detection in micro-algae and aquatic meiofauna species using a direct thermal desorption interface combined with comprehensive gas chromatography–time-of-flight mass spectrometry. *Journal of Chromatography A* 1186: 254–261.

Álvarez, C. A. and C. C. Akoh. 2015. Enzymatic synthesis of infant formula fat analog enriched with capric acid. *Journal of the American Oil Chemists' Society* 92: 1003–1014.

Augustin, M. A., P. T. Clarke and H. Craven. 2003. Powdered milk—characteristics of milk powders. pp. 4703–4711. *In*: C. Benjamin (ed.). *Encyclopedia of Food Sciences and Nutrition (Second Edition)*, Oxford: Academic Press.

Augustin, M. A. and C. L. Margetts. 2003. Powdered milk—milk powders in the marketplace. pp. 4694–4702. *In*: C. Benjamin (ed.). *Encyclopedia of Food Sciences and Nutrition (Second Edition)*. Oxford: Academic Press.

Baldwin, A. and D. Pearce. 2005. Milk powder. pp. 387–433. *In*: C. Onwulata (ed.). *Encapsulated and Powdered Foods*. CRC Press.

Balling, H. P. and A. Rossmann. 2004. Countering fraud via isotope analysis—Case report. *Kriminalistik* 58: 44–47.

Bauman, D. E., L. H. Baumgard, B. A. Corl and J. M. Griinari. 2000. Biosynthesis of conjugated linoleic acid in ruminants. *Journal of Animal Science* 77(E-Suppl.): 1–15.

Berger, A., M. Fleith and G. Crozier. 2000. Nutritional implications of replacing bovine milk fat with vegetable oil in infant formulas. *Journal of Pediatric Gastroenterology and Nutrition* 30: 115–130.

Bontempo, L., R. Larcher, F. Camin, S. Holzl, A. Rossmann, P. Horn and G. Nicolini. 2011. Elemental and isotopic characterisation of typical Italian alpine cheeses. *International Dairy Journal* 21: 441–446.

Bontempo, L., G. Lombardi, R. Paoletti, L. Ziller and F. Camin. 2012. H, C, N and O stable isotope characteristics of alpine forage, milk and cheese. *International Dairy Journal* 23: 99–104.

Brescia, M. A., M. Monfreda, A. Buccolieri and C. Carrino. 2005. Characterisation of the geographical origin of buffalo milk and mozzarella cheese by means of analytical and spectroscopic determinations. *Food Chemistry* 89: 139–147.

Camin, F., D. Bertoldi, A. Santato, L. Bontempo, M. Perini, L. Ziller, A. Stroppa and R. Larcher. 2015. Validation of methods for H, C, N and S stable isotopes and elemental analysis of cheese: Results of an international collaborative study. *Rapid Communications in Mass Spectrometry* 29: 415–423.

Camin, F., R. Wehrens, D. Bertoldi, L. Bontempo, L. Ziller, M. Perini, G. Nicolini, M. Nocetti and R. Larcher. 2012. H, C, N and S stable isotopes and mineral profiles to objectively guarantee the authenticity of grated hard cheeses. *Analytica Chimica Acta* 711: 54–59.

Camin, F., K. Wietzerbin, A. B. Cortes, G. Haberhauer, M. Lees and G. Versini. 2004. Application of multielement stable isotope ratio analysis to the characterization of French, Italian, and Spanish cheeses. *Journal of Agricultural and Food Chemistry* 52: 6592–6601.

Capici, C., T. Mimmo, L. Kerschbaumer, S. Cesco and M. Scampicchio. 2015. Determination of cheese authenticity by carbon and nitrogen isotope analysis: Stelvio cheese as a case study. *Food Analytical Methods* 8: 2157–2162.

Cesa, S., M. A. Casadei, F. Cerreto and P. Paolicelli. 2012. Influence of fat extraction methods on the peroxide value in infant formulas. *Food Research International* 48: 584–591.

Charlton, C. 2014. Baaa-humbug! Investigation reveals goat's cheese made from sheep milk is being sold in British supermarkets. *Daily Mail*. Published 22 Oct 2014; Retrieved 29 Aug. 2016.

Chikaraishi, Y., Y. Suzuki and H. Naraoka. 2004. Hydrogen isotopic fractionations during desaturation and elongation associated with polyunsaturated fatty acid biosynthesis in marine macroalgae. *Phytochemistry* 65: 2293–2300.

Crittenden, R. G., A. S. Andrew, M. LeFournour, M. D. Young, H. Middleton and R. Stockmann. 2007. Determining the geographic origin of milk in Australasia using multi-element stable isotope ratio analysis. *International Dairy Journal* 17: 421–428.

Deeth, H. C. and C. H. Fitz-Gerald. 1983. Lipolytic enzymes and hydrolytic rancidity in milk and milk products. pp. 195–239. *In*: P. F. Fox (ed.). *Developments in Dairy Chemistry—2*. Springer Netherlands.

Ecker, J., M. Scherer, G. Schmitz and G. Liebisch. 2012. A rapid GC–MS method for quantification of positional and geometric isomers of fatty acid methyl esters. *Journal of Chromatography B* 897: 98–104.

Ehtesham, E., W. T. Baisden, E. D. Keller, A. R. Hayman, R. Van Hale and R. D. Frew. 2013. Correlation between precipitation and geographical location of the δ^2H values of the fatty acids in milk and bulk milk powder. *Geochimica et Cosmochimica Acta* 111: 105–116.

Ehtesham, E., A. Hayman, R. Van Hale and R. Frew 2015. Influence of feed and water on the stable isotopic composition of dairy milk. *International Dairy Journal* 47: 37–45.

Farkye, N. Y. 2006. Significance of milk fat in milk powder. pp. 451–465. *In*: P. F. Fox and P. L. H. McSweeney (eds.). *Advanced Dairy Chemistry* (Vol. 2). Springer US.

González-Martínez, C., M. Becerra, M. Cháfer, A. Albors, J. M. Carot and A. Chiralt. 2002. Influence of substituting milk powder for whey powder on yoghurt quality. *Trends in Food Science & Technology* 13: 334–340.

Hayes, M. G., P. F. Fox and A. L. Kelly. 2005. Potential applications of high pressure homogenisation in processing of liquid milk. *Journal of Dairy Research* 72: 25–33.

Jensen, R. G. and G. L. Jensen. 1992. Specialty lipids for infant nutrition. I. Milks and formulas. *Journal of Pediatric Gastroenterology and Nutrition* 15: 232–245.

Jensen, R. G. 1971. Lipolytic enzymes. *Progress in the Chemistry of Fats and other Lipids* 11: 347–394.

Jia, X. P., J. K. Huang, H. Luan, S. Rozelle and J. Swinnen. 2012. China's Milk Scandal, government policy and production decisions of dairy farmers: The case of Greater Beijing. *Food Policy* 37: 390–400.

Karoui, R. 2012. Chapter 15—Food Authenticity and Fraud. pp. 499–517. *In:* Y. Picó (ed.). *Chemical Analysis of Food: Techniques and Applications*. Boston: Academic Press.

Kaylegian, K. E. and R. C. Lindsay. 1995. Milk fat usage and modification. pp. 1–18. *Handbook of Milk fat Fractionation Technology and Application*. American Oil Chemists Society Press: Champaign, IL.

Kelly, S., I. Parker, M. Sharman, J. Dennis and I. Goodall. 1997. Assessing the authenticity of single seed vegetable oils using fatty acid stable carbon isotope ratios ($^{13}C/^{12}C$). *Food Chemistry* 59: 181–186.

Kim, E. H. J., X. D. Chen and D. Pearce. 2009. Surface composition of industrial spray-dried milk powders. 1. Development of surface composition during manufacture. *Journal of Food Engineering* 94: 163–168.

Kornexl, B. E., T. Werner, A. Roßmann and H. -L. Schmidt. 1997. Measurement of stable isotope abundances in milk and milk ingredients—a possible tool for origin assignment and quality control. *Zeitschrift für Lebensmitteluntersuchung und-Forschung A* 205: 19–24.

Larson, B. L. 1979. Biosynthesis and secretion of milk proteins: A review. *Journal of Dairy Research* 46: 161–174.

López-López, A., M. C. López-Sabater, C. Campoy-Folgoso, M. Rivero-Urgell and A. I. Castellote-Bargalló. 2002. Fatty acid and sn-2 fatty acid composition in human milk from Granada (Spain) and in infant formulas. *European Journal of Clinical Nutrition* 56: 1242–1254.

Lyu, J. C. 2012. A comparative study of crisis communication strategies between Mainland China and Taiwan: The melamine-tainted milk powder crisis in the Chinese context. *Public Relations Review* 38: 779–791.

MacGibbon, A. K. H. and M. W. Taylor. 2006. Composition and structure of bovine milk lipids. pp. 1–42. *In:* P. F. Fox and P. L. H. McSweeney (eds.). *Advanced Dairy Chemistry Volume 2 Lipids*. Springer US.

Maduko, C. O., C. C. Akoh and Y. W. Park. 2007. Enzymatic production of infant milk fat analogs containing palmitic acid: Optimization of reactions by response surface methodology. *Journal of Dairy Science* 90: 2147–2154.

Manca, G., M. A. Franco, G. Versini, F. Camin, A. Rossmann and A. Tola. 2006. Correlation between multielement stable isotope ratio and geographical origin in Peretta cows' milk cheese. *Journal of Dairy Science* 89: 831–839.

Martínez-Force, E., N. Ruiz-López and R. Garcés. 2009. Influence of specific fatty acids on the asymmetric distribution of saturated fatty acids in sunflower (*Helianthus annuus* L.) triacylglycerols. *Journal of Agricultural and Food Chemistry* 57: 1595–1599.

McLeod, R. J., C. G. Prosser and J. W. Wakefield. 2016. Identification of goat milk powder by manufacturer using multiple chemical parameters. *Journal of Dairy Science* 99: 982–993.

Merrett, N. 2008. Mixed verdict in managers' butter adultery case. *DairyReporter.com*. Published 16 Jan. 2008; Retrieved 29 Aug. 2016.

Molkentin, J. 2013. The effect of cheese ripening on milkfat composition and the detection of fat from non-dairy origin. *International Dairy Journal* 33: 16–21.

Molkentin, J. and A. Giesemann. 2007. Differentiation of organically and conventionally produced milk by stable isotope and fatty acid analysis. *Analytical and Bioanalytical Chemistry* 388: 297–305.

Moore, J. C., J. Spink and M. Lipp. 2012. Development and application of a database of food ingredient fraud and economically motivated adulteration from 1980 to 2010. *Journal of Food Science* 77: R118–R126.

Mulvany, L. 2016. The Parmesan cheese you sprinkle on your penne could be wood. *Bloomberg. com*. Published 16 Feb. 2016; Retrieved 29 August 2016.

Necemer, M., D. Potocnik and N. Ogrinc. 2016. Discrimination between Slovenian cow, goat and sheep milk and cheese according to geographical origin using a combination of elemental content and stable isotope data. *Journal of Food Composition and Analysis* 52: 16–23.

New Zealand Government, ministry for primary industries, Statistics New Zealand, DairyNZ and Fonterra Co-operative Group. (2012). Dairy, facts and figures. Retrieved 04.06.2013, from http://www.mpi.govt.nz/agriculture/pastoral/dairy.aspx.

New Zealand Government, Ministry for the Environment. 2011. Valuing New Zealand's clean green image. Retrieved 27.07.2013, from http://www.mfe.govt.nz/publications/sus-dev/clean-green-image-value-aug01/chapter-2-aug01.pdf.

NZ Herald. 2016. 1080 blackmailer Jeremy Kerr jailed for eight and a half years. Retrieved 26.10.2016, from http://www.nzherald.co.nz/nz/news/article.cfm?c_id=1&objectid=11610428.

Ohtani, S., T. Wang, K. Nishimura and M. Irie. 2005. Milk fat analysis by fiber-optic spectroscopy. *Asian-Australasian Journal of Animal Sciences* 18: 580–583.

Pesek, M., J. Spicka and E. Samkova. 2005. Comparison of fatty acid composition in milk fat of Czech Pied cattle and Holstein cattle. *Czech Journal of Animal Science* 50: 122–128.

Pillonel, L., R. Badertscher, P. Froidevaux, G. Haberhauer, S. Holzl, P. Horn, A. Jakob, E. Pfammatter, U. Piantini, A. Rossmann, R. Tabacchi and J. O. Bosset. 2003. Stable isotope ratios, major, trace and radioactive elements in emmental cheeses of different origins. *Lebensmittel-Wissenschaft Und-Technologie-Food Science and Technology* 36: 615–623.

Prosser, C. G., V. I. Svetashev, M. V. Vyssotski and D. J. Lowry. 2010. Composition and distribution of fatty acids in triglycerides from goat infant formulas with milk fat. *Journal of Dairy Science* 93: 2857–2862.

Rogalska, E., C. Cudrey, F. Ferrato and R. Verger. 1993. Stereoselective hydrolysis of triglycerides by animal and microbial lipases. *Chirality* 5: 24–30.

Romdhane, K. 2010. Determination of identity and quality of dairy products. pp. 413–434. *In*: L.M.L. Nollet and F. Toldra (eds.). *Sensory Analysis of Foods of Animal Origin*. CRC Press.

Rossmann, A., G. Haberhauer, S. Holzl, P. Horn, F. Pichlmayer and S. Voerkelius. 2000. The potential of multielement stable isotope analysis for regional origin assignment of butter. *European Food Research and Technology* 211: 32–40.

Rossmann, A., B. E. Kornexl, G. Versini, F. Pichlmayer and G. Lamprecht. 1998. Origin assignment of milk from alpine regions by multielement stable isotope ratio analysis (SIRA). pp. 37–50. *In*: *Atti del primo convegno internazionale Gli alimenti montani*. Istituto agrario di San Michele all'Adige. http://hdl.handle.net/10449/16381.

Scampicchio, M., D. Eisenstecken, L. De Benedictis, C. Capici, D. Ballabio, T. Mimmo, P. Robatscher, L. Kerschbaumer, M. Oberhuber, A. Kaser, C. W. Huck and S. Cesco. 2016. Multi-method approach to trace the geographical origin of alpine milk: A case study of Tyrol region. *Food Analytical Methods* 9: 1262–1273.

Scampicchio, M., T. Mimmo, C. Capici, C. Huck, N. Innocente, S. Drusch and S. Cesco. 2012. Identification of milk origin and process-induced changes in milk by stable isotope ratio mass spectrometry. *Journal of Agricultural and Food Chemistry* 60: 11268–11273.

Schuck, P. 2011. Dehydrated dairy products—milk powder: Types and manufacture. pp. 108–116. *In*: W. F. John (ed.). *Encyclopedia of Dairy Sciences (Second Edition)*. San Diego: Academic Press.

Shahidi, F. and S. Senanayake. 2006. Nutraceutical and specialty lipids. pp. 1–25. *In*: F. Shahidi (ed.). *Nutraceutical and Specialty Lipids and their Co-Products*. CRC Press.

Sharma, K. and M. Paradakar. 2010. The melamine adulteration scandal. *Food Security* 2: 97–107.

Siciliano-Jones, J. and M. R. Murphy. 1989. Production of volatile fatty acids in the rumen and cecum-colon of steers as affected by forage: Concentrate and forage physical form. *Journal of Dairy Science* 72: 485–492.

Spurgeon, M. J., A. K. Palmer and P. A. Hepburn. 2003. An investigation of the general, reproductive and postnatal developmental toxicity of Betapol™, a human milk fat equivalent. *Food and Chemical Toxicology* 41: 1355–1366.

Stevenson, R., S. Desrochers and J. F. Helie. 2015. Stable and radiogenic isotopes as indicators of agri-food provenance: Insights from artisanal cheeses from Quebec, Canada. *International Dairy Journal* 49: 37–45.

Sun, D. W. 2008. *Modern Techniques for Food Authentication* (1st ed.). Academic Press.

Tunick, M. H. 2010. Milk lipids. pp. 313–325. *In*: A. Kolakowska (ed.). *Chemical, Biological, and Functional Aspects of Food Lipids, Second Edition*. CRC Press.

Vander Wal, R. J. 1960. Calculation of the distribution of the saturated and unsaturated acyl groups in fats, from pancreatic lipase hydrolysis data. *Journal of the American Oil Chemists Society* 37: 18–20.

Walstra, P., J. T. M. Wouters and T. J. Geurts. 2005. Milk. pp. 3–16. *In*: *Dairy Science and Technology*. CRC Press.

Wei, Y. and D. Liu. 2012. Review of melamine scandal: Still a long way ahead. *Toxicology and Industrial Health* 28: 579–582.

Wells, J. C. K. 1996. Nutritional considerations in infant formula design. *Seminars in Neonatology* 1: 19–26.

West, A. G., E. C. February and G. J. Bowen. 2014. Spatial analysis of hydrogen and oxygen stable isotopes ("isoscapes") in ground water and tap water across South Africa. *Journal of Geochemical Exploration* 145: 213–222.

Xiu, C. and K. K. Klein. 2010. Melamine in milk products in China: Examining the factors that led to deliberate use of the contaminant. *Food Policy* 35: 463–470.

Zadow, J. G. 2003. Whey and whey powders—production and uses. pp. 6147–6152. *In*: C. Benjamin (ed.). *Encyclopedia of Food Sciences and Nutrition (Second Edition)*. Oxford: Academic Press.

CHAPTER 11

Edible Vegetable Oils

Federica Camin* and Luana Bontempo

11.1 Introduction

'When I grew up in Italy in the 1950s, it was still very agricultural. Food was very important; produce was very important. Everyone made their own olive oil.'

Isabella Rossellini (b. 1952), Italian model, actress, author, and philanthropist.

Fats and oils form part of a healthy diet, but the type of fat—as well as the total amount of fat consumed daily—makes a difference. A high intake of saturated fats, trans fats, and cholesterol has been linked with an increased risk of coronary heart disease. On the other hand, many components naturally present in vegetable oils (e.g., linoleic and linolenic acids, phytosterols, vitamin E) have been shown to have beneficial health effects. The increase in the quantity and change in the quality of fats and oils in the diet have important consequences for health. Of particular interest is the content of essential fatty acids—for example, soybean oil contains less linoleic acid than sunflower seed oil. Both soybean and sunflower oils, however, provide adequate amounts of linoleic acid. Essential fatty acids from the omega-3 (ω-3 or n-3) family may be supplied from vegetable oils containing α-linolenic acid, such as canola and soybean oils. For these reasons the *Dietary Guidelines for Americans*, 2015–2020 (http://health.gov/dietaryguidelines/2015/guidelines/) state that "oils should replace solid fats rather than being added to the diet." A list of the most common fatty acids found in vegetable oils is presented in Table 11.1.

Department of Food Quality and Nutrition, Research and Innovation Centre, Fondazione Edmund Mach (FEM), Via E. Mach 1, 38010 San Michele all'Adige, ITALY.
Email: luana.bontempo@fmach.it
* Corresponding author: federica.camin@fmach.it

Table 11.1 List of the most common fatty acids found in vegetable oils.

Common name	N of carbon atoms	Double bonds	Scientific name	Vegetable oil sources
Caprylic Acid	8	0	octanoic acid	coconut oil
Capric Acid	10	0	decanoic acid	coconut oil
Lauric Acid	12	0	dodecanoic acid	coconut oil
Myristic Acid	14	0	tetradecanoic acid	palm kernel oil
Palmitic Acid	16	0	hexadecanoic acid	palm and coconut oils
Stearic Acid	18	0	octadecanoic acid	palm and coconut oils
Oleic Acid	18	1	9-octadecenoic acid	olive, rapeseed, sesame oils
Ricinoleic acid	18	1	12-hydroxy-9-octadecenoic acid	castor oil
Linoleic Acid	18	2	9,12-octadecadienoic acid	corn, sunflower, grape seed oils
Alpha-Linolenic Acid	18	3	9,12,15-octadecatrienoic acid	flaxseed, canola, soy oils
Gamma-Linolenic Acid	18	3	6,9,12-octadecatrienoic acid	borage oil
Arachidic Acid	20	0	eicosanoic acid	peanut oil
Behenic acid	22	0	docosanoic acid	rapeseed oil
Erucic acid	22	1	13-docosenoic acid	rapeseed oil

Edible oils are lipid products that are typically liquids at room temperature whereas animal-derived fats tend to be more saturated and are therefore solid at room temperature. The chemical composition of vegetable edible oils is predominantly triacylglycerides of fatty acids mixed with minor quantities of other lipids (e.g., phosphatides), unsaponifiable constituents (e.g., alcohols, sterols, polyphenols), and free fatty acids (Codex Alimentraius, CODEX STAN 19-1981, Rev. 2-1999). The differences between the various types of vegetable oils mainly concern the fatty acid composition. With some exceptions, and in contrast to animal fats, vegetable oils contain predominantly unsaturated fatty acids of two kinds: monounsaturated (oleic acid, primarily found in extra virgin olive oil) and polyunsaturated (linoleic acid and linolenic acid, found in oils extracted from oilseeds) (FAO data, Definition and classification of commodities: 14. Vegetable and animal oils and fats, http://www.fao.org/es/faodef/fdef14e.htm). Tropical oils (e.g., oils derived from coconut palm) are an exception and contain a higher percentage of saturated lipids than the other vegetable oils. The profile of oil fatty acids not only varies according to the plant species, but also due to climatic conditions and the type of soil. It can also vary due to genetic modifications of the plants.

Edible vegetable oils are mainly derived from oily nuts and seeds but can also be obtained from other parts of a plant, such as flowering tops, leaves, flowers, fruits, roots, and rhizomes. Commercial oil extraction is typically performed in one of two ways:

Through a chemical extraction, using solvents—a process which produces a high yield and is fast and inexpensive. Oil extracted with solvents generally needs to be refined to make it suitable for human consumption and the production chain involves several steps. Initially, plant parts used for the extraction are placed in a bath of hexane or heptane and after separating the oil-solvent together from what remains of the plant, the solvent is removed by evaporation at a temperature of around 150°C. Subsequently, a degumming process is applied to remove fibers and complex carbohydrates. The oil is then de-acidified by the addition of an extremely corrosive base (e.g., sodium hydroxide or a mix of sodium hydroxide and sodium carbonate) and de-colorized via suitable filtration. During this process, the essential fatty acids are altered (to form peroxides and conjugated fatty acids). Finally, the oil is de-odorized by steam distillation at high temperature under vacuum, removing aromatic substances, free fatty acids, and molecules generated from previous processes that can give an unpleasant odor/taste to the oil. During this process, up to 5% of trans fatty acids may be formed.

Alternatively, physical oil extraction, without the use of solvents, may be employed. This extraction takes place by pressing plant parts and is often used to produce cooking and edible oil, because it is preferred by consumers (at least in Europe). From these mechanical procedures *virgin* and *cold pressed* oils are obtained, without altering the nature of the oil. In these physically-extracted oils, additives are not allowed.

Edible vegetable oils have a wide variety of food uses. Some of these oils are sold individually as commodities and are used as condiments for salads or for cooking, whereas others, such as palm oil, are essential ingredients in many processed food products such as margarines, cookies, and pastries.

According to the Food and Agriculture Organization of the United Nations (FAO), the worldwide production of vegetable edible oils was more than 150 million tons in 2011, with two oils—soybean oil and palm oil—accounting for roughly 60% of the total world production (Table 11.2). The vegetable oil sector encompasses products of differing value and sales volume. For example, at wholesale market, soybean oil has a price of about US$677 per ton whereas extra virgin olive oil has a price of about US$4,200 per ton, approximately six times higher (data from IndexMundi, December 2015). This significant price difference between vegetable oils provides a significant incentive for fraud, in particular, the dilution of expensive oils with cheaper ones. Some examples previously observed include: the addition of lower grades of olive oil to extra virgin olive oil, blending palm stearin and olein with palm oil, the addition of undeclared small quantities of cheaper oils (e.g., soybean oil) to groundnut oils, the addition of rapeseed oil to the more expensive soybean oil and maize (corn) oil, and the dilution of cottonseed oil with palm olein (Bell and Gillatt 2011).

There are several official regulations worldwide governing edible vegetable oils. In general, EU Regulations have been the most restrictive and, since 1966, the EU has established a common organization of the

Table 11.2 World 2011 production of edible vegetable oils (FAOSTAT data, http://faostat3. fao.org/home/E).

Type of oil	Million tons
Palm oil	48.520
Soybean oil	41.831
Rape and mustard oil (canola)	23.264
Sunflower oil	13.360
Groundnut oil	5.373
Cottonseed oil	5.145
Other oil crops	4.340
Olive oil	3.353
Coconut oil	3.111
Maize (corn) oil	2.330
Sesame oil	1.094

market in oils and fats. This allowed prices to be stabilized and ensured a fair level of income for farmers by laying down arrangements for trade with non-EU countries. Since 1992, the EU Regulation has related almost exclusively to olives and olive oil (see below). Recently, EU Regulation 1169/2011, enacted on 14 December 2014, stated that different types of oil have to be specifically identified in the list of ingredients, rather than as the generic "vegetable oil," which means that it is now compulsory to specify the type of vegetable oil—or oil blend—used in food products.

Among other regulations, U.S. requirements refer to the characteristics stated by the Codex Alimentarius (Codex Standard for Named Vegetable Oils—CODEX-STAN 210-1999, and Codex Standard for Olive Oil, Virgin and Refined and for Refined Olive-Pomace Oil—CODEX STAN 33-1981, Rev. 1-1989). The Codex defines the vegetable oils usable for human consumption and their essential composition and quality factors; it also defines the ranges of fatty acid composition, the maximum contents of food additives such as antioxidants or anti-foaming agents, as well as contaminants such as heavy metals and pesticides. Recently, China has indicated that the country's standards of edible vegetable oils are under revision (GAIN report CH15047, USDA, 21st October 2015).

In general, the parameters considered by the official regulations for edible vegetable oils concern the product's quality (e.g., moisture content, impurities, free fatty acid profiles, and peroxide value) to verify, for example, whether the oil is crude or refined and the extent of any rancidity. At present the officially recognized methods do not provide a means for the detection of dilution of expensive oils with cheaper ones or verification of claims of origin. For this reason, there is an increasing demand for analytical methods and statistical tools capable of effectively detecting the various types of vegetable oil fraud.

Stable isotope analysis of the H, C, and O compositions of bulk oils using IRMS or oil sub-components using GC-C-IRMS has become the technique of choice for oil authentication, because the isotopic ratios of H, C, and O change with the botanical origin (type of plant/oil) and with the climatic and geographical characteristics of the location where the plant was grown.

11.2 Methods

The carbon isotope ratio ($^{13}C/^{12}C$) of bulk oil can be determined by EA-IRMS directly from the raw sample without any need for pre-treatment (Banerjee et al. 2015; Kim et al. 2015; Jeon et al. 2015; Camin et al. 2010a; Camin et al. 2010b).

The analysis of hydrogen ($^{2}H/^{1}H$) and oxygen ($^{18}O/^{16}O$) isotope ratios of bulk oil is performed via TC/EA-IRMS. For analysis of hydrogen and oxygen isotopic compositions, dehydration is essential, in order to remove any contribution from sample water and/or absorbed moisture (Carter and

Fry 2013). Dehydration can be performed by keeping the samples under vacuum for 7 days before analysis and analyzing dried samples without delay after that period, using a suitable autosampler maintained under a flow of helium or nitrogen (Camin et al. 2010b; Bontempo et al. 2009). To remove moisture, Banerjee and colleagues (2015) heated filled capsules in an oven at 100°C for 15 min. Some authors have performed the analysis of δ^2H and $\delta^{18}O$ directly on the raw sample (Kim et al. 2015; Jeon et al. 2015; Richter et al. 2010), without any dehydration, but this choice is not recommended.

At present only one international matrix-matched standard for oils is available: NBS 22, a mineral oil (as opposed to a vegetable oil), which has internationally agreed (International Atomic Energy Agency, IAEA) isotopic compositions of $\delta^{13}C$ (−30.031 ± 0.043‰) and δ^2H (−120 ± 1‰).

In more than twenty years of applications of stable isotope analysis on vegetable oils, the isotopic values of many specific oil components have been investigated. The components investigated include glycerol, sterols, aliphatic acids, fatty acids, and *normal*-alkanes (*n*-alkanes) (e.g., Angerosa et al. 1997; Angerosa et al. 1999; Faberi et al. 2014; Mihailova et al. 2015; Camin et al. 2010b). For the majority of these compounds, the $\delta^{13}C$ composition was determined by EA-IRMS or GC-C-IRMS. The analysis of specific oil components requires a suitable pre-treatment of the oil sample; for example, $\delta^{13}C$ and δ^2H compositions were recently determined for *n*-alkanes (Mihailova et al. 2015) previously isolated from olive oils using column chromatography.

11.3 Applications of IRMS for detecting the authenticity of edible vegetable oils

A summary of publications concerning the application of H, C, and O stable isotope ratio analysis to edible vegetable oil is given in Table 11.3. In the following sections, details of some of this research, grouped by the type of oil, are presented.

11.3.1 Olive oil

Most publications have focused on the characterization of olive oil, which is obtained from the fruit of the olive tree (*Olea europaea*). The oil is produced by pressing whole olives (producing extra virgin and virgin olive oils) or by solvent extraction and refinement, yielding oils of much lower quality (e.g., refined olive oil and olive pomace oil).

Extra virgin olive oil is a fundamental component of the *Mediterranean diet* and different international regulations have been established to define olive oil authenticity and quality [EC Reg. 182/2009, Implementing Regulation (EU) 29/2012 and the subsequent Commission Implementing Regulation (EU) 1335/2013]. Extra virgin olive oil commands a high retail

Table 11.3 List of some publications concerning the application of the H, C, and/or O stable isotope ratio analysis to edible vegetable oils.

Authors	Isotopes	IRMS technique	Type of oil	Aim
Angerosa et al. 1997	$\delta^{13}C$	GC-C Aliphatic alcohols, sterols	Olive oil	Adulteration (pomace oil)
Angerosa et al. 1999	$\delta^{13}C$, $\delta^{18}O$	EA, TC/EA, GC-C Aliphatic alcohols, sterols	Olive oil	Geographical origin
Aramendia et al. 2007	$\delta^{18}O$	TC/EA	Olive oil	Geographical origin
Banerjee et al. 2015	$\delta^{13}C$, δ^2H, $\delta^{18}O$	EA, TC/EA	Olive, sesame, sunflower, walnut, coconut, pumpkin seed, mustard, hemp seed, soybean, palm, canola	Botanical and geographical origin
Bianchi et al. 1993	$\delta^{13}C$	EA	Olive oil	
Bontempo et al. 2009	$\delta^{13}C$, δ^2H, $\delta^{18}O$	EA, TC/EA	Olive oil	Geographical origin
Bréas et al. 1998	$\delta^{18}O$	TC/EA	Almond, coconut, grapestone, maize, pine, ricinus (castorbean), sesame, soybean, sunflower, wheat	Botanical and geographical origin
Camin et al. 2010a	$\delta^{13}C$, δ^2H, $\delta^{18}O$	EA, TC/EA	Olive oil	Geographical origin
Camin et al. 2010b	$\delta^{13}C$, δ^2H, $\delta^{18}O$	EA, TC/EA Bulk, glycerol	Olive oil	Geographical origin
Faberi et al. 2014	$\delta^{13}C$	EA, GC-C Fatty acids (C16:0, C18:0, C18:1, C18:2)	Olive oil	Geographical origin
Guo et al. 2010	$\delta^{13}C$	EA, GC-C Fatty acids (C16:0, C18:0, C18:1, C18:2)	Camellia seed, perilla seed, flax seed, maize, peanut, sesame, pine nut, pumpkin seed, sunflower seed, soybean, walnut	Botanical origin
Lu et al. 2015	$\delta^{13}C$	EA	Soybean, maize, rapeseed, peanut, sesame, sunflower, olive, mixed recycled cooking oil	Recycling of cooking oil

Table 11.3 contd. ...

...Table 11.3 contd.

Authors	Isotopes	IRMS technique	Type of oil	Aim
Horacek et al. 2015	$\delta^{13}C$, $\delta^{2}H$	EA, TC/EA	Sesame oil	Geographical origin
Hrastar et al. 2009	$\delta^{13}C$	EA, GC-C Fatty acids (C16:0, C18:0, C18:1n7, C18:1n9, C18:2n6, C18:3n3, C20:0, C20:1n9, C20:3n3, C22:1n9)	*Camelina sativa* (false flax)	Authentication
Iacumin et al. 2009	$\delta^{13}C$, $\delta^{2}H$	EA, TC/EA	Olive oil	Geographical origin
Jeon et al. 2015	$\delta^{13}C$, $\delta^{2}H$, $\delta^{18}O$	EA, TC/EA	Sesame oil	Geographical origin
Kelly et al. 1997	$\delta^{13}C$	GC-C Fatty acids (C16:0, C18:0, C18:1, C18:2)	Sunflower, rapeseed, groundnut, palm	Botanical origin
Kim et al. 2015	$\delta^{13}C$, $\delta^{2}H$, $\delta^{18}O$	EA, TC/EA	Sesame oil	Authentication
Medini et al. 2015	$^{87}Sr/^{86}Sr$	TIMS	Olive oil	Geographical origin
Mihailova et al. 2015	$\delta^{13}C$, $\delta^{2}H$	GC-C n-alkanes (n-C_{29})	Olive oil	Geographical origin
Portarena et al. 2015	$\delta^{13}C$, $\delta^{2}H$	EA, TC/EA	Olive oil	Geographical origin
Portarena et al. 2014	$\delta^{13}C$, $\delta^{2}H$	EA, TC/EA	Olive oil	Geographical origin
Richter et al. 2010	$\delta^{13}C$, $\delta^{2}H$, $\delta^{18}O$	EA, GC-C Fatty acids (C16:0, C18:1, C18:2, C18:3)	Rapeseed oils *vs.* flax oil and false flax (*Camelina sativa*) oil, poppy, sunflower, and safflower	Authentication
Royer et al. 1999	$\delta^{13}C$	GC-C Fatty acids (C16:0, C18:1, C18:2)	Olive oil	Geographical origin
Seo et al. 2010	$\delta^{13}C$	EA	Sesame oil	Authentication (maize oil addition)

			Sesame oil	Authentication (maize oil addition)
Seol et al. 2012	$\delta^{13}C$	EA		
Spangenberg et al. 1998	$\delta^{13}C$	EA, GC-C Fatty acids (C16:0, C18:1)	Olive oil	Adulteration (pomace oil)
Spangenberg and Ogrinc 2001	$\delta^{13}C$	EA, GC-C Fatty acids (C16:0, C16:1, C18:0, C18:1, C18:2, C18:3)	Olive, pumpkin, sunflower, maize, rape, soybean, sesame	Botanical and geographical origin
Woodbury et al. 1995	$\delta^{13}C$	GC-C Fatty acids (C16:0, C18:1, C18:2)	Maize, groundnut	Authentication

value (two to four times that of other oils) because of its sensorial qualities and purported health benefits (Salas-Salvado et al. 2014). Moreover, unlike other categories of edible olive oil, the characteristics of a virgin olive oil are related to the geographical origin of the olives in addition to the specific extraction techniques used during the production. Starting in 1992, the European Union created the Protected Designation of Origin (PDO) and Protected Geographical Indication (PGI) systems to promote and protect foodstuffs of particular quality (Regulation 2081/92/EEC and then Regulation 510/2006/EEC subsequently amended by Regulation 1151/2012/EU). On 4 February 2009, the EU Member States agreed to compulsory origin labelling for virgin and extra virgin olive oil (EC Reg. No. 182/2009).

As a high-priced product, olive oil is frequently the subject of counterfeiting or adulteration, as evident from recent olive oil scandals (e.g., *New York Times*, "Extra Virgin suicide: The adulteration of Italian olive oil" Jan. 26th 2014; *The Telegraph*, "Italian olive oil scandal: Seven top brands 'sold fake extra-virgin'" Nov. 11th 2015). For this reason, recent attention has been focused more and more on analytical methods for oil authentication, including verification of the geographical origin of olive oils.

Together with other methods, stable isotope ratio analysis has been applied to oil authentication since 1993, when $\delta^{13}C$ values were determined for whole oil, sterols, long-chain alcohols, and glycerol of Italian olive oil (Bianchi et al. 1993). In this study, glycerol showed the most depleted $\delta^{13}C$ values, sterols the most enriched, and the alcohols had intermediate values. The ripening stage of the fruit was reported not to affect the carbon isotopic composition. A few years later, the $\delta^{13}C$ composition of bulk olive oil or of some sub-components (individual fatty acids or aliphatic alcohols and sterols) proved to be useful for detecting the adulteration of olive oil with cheaper pomace olive oil or with other vegetable oils (Angerosa et al. 1997; Spangenberg et al. 1998). The $\delta^{13}C$ values of aliphatic alcohols of pomace oils were found to be more depleted than values recorded for virgin and refined oils (Angerosa et al. 1997).

The $^{13}C/^{12}C$ composition, especially in combination with $^{2}H/^{1}H$ and/or $^{18}O/^{16}O$ measurements, of bulk oil as well as in specific components has proved to be a good tool for characterizing geographical origin. The carbon isotopic compositions of palmitic, oleic, and linoleic fatty acids of olive oils were found to differentiate French, Italian, and Greek olive oils—providing a regional classification for Greek olive oils (Royer et al. 1999) and for different Italian regions (Faberi et al. 2014). Some authors (Bréas et al. 1998; Angerosa et al. 1999) have found that both the $^{13}C/^{12}C$ and $^{18}O/^{16}O$ ratios of olive oils from Italy, Greece, Spain, Tunisia, Morocco, and Turkey changed according to latitude, distance from the sea, and environmental conditions during plant growth (through changes in water stress, atmospheric moisture, and temperature). The $\delta^{13}C$ and δ^2H values of n-C_{29} alkane were found to be positively correlated with the mean temperature during August–December and negatively correlated with mean relative humidity during these months (Mihailova et al. 2015). Carbon and hydrogen isotopic compositions were significantly more enriched in olive oils from the southern (Greece, southern Italy, Spain, Portugal, and Morocco) compared with the northern (Slovenia, Croatia, France, northern and central Italy) Mediterranean countries.

A number of other studies have focused on different combinations of δ^2H, $\delta^{13}C$, and $\delta^{18}O$ measurements of bulk oil to characterize Italian extra virgin olive oil in particular (Bontempo et al. 2009; Iacumin et al. 2009; Camin et al. 2010b; Portarena et al. 2014), as Italy has the highest number of PDO and PGI registered extra virgin olive oils in Europe (38 recognized brands; last update 22/07/2015, http://ec.europa.eu/agriculture/quality/schemes/index_en.htm). A recent paper by Chiocchini and colleagues (2016) related $\delta^{13}C$ and $\delta^{18}O$ values determined in Italian extra virgin olive oils to GIS (Geographic Information System) layers of $\delta^{18}O$ values of source water and also climate data. These authors determined a geospatial model of $\delta^{18}O$ and $\delta^{13}C$ signatures of olive oil that identified four main Italian geographical areas (Figure 11.1). It was proposed that this model could be used for the authentication and verification of the geographical origin of Italian extra virgin olive oils.

11.3.2 Sesame oil

Sesame oil is obtained from sesame seeds (*Sesamum indicum* L.) that are roasted before pressing. The oil is frequently used to impart a specific flavor to a variety of Asian dishes and foods due to the characteristic aroma that develops during roasting. Sesame seeds are available in three different colors—black, brown, and white. White sesame seeds are mostly used in the manufacturing of roasted sesame oil in Korea (Jeon et al. 2015). Because the domestic production of sesame seeds does not supply the national need, a large amount of sesame seeds are imported into Korea from China and India (Kim et al. 2015). The retail price of Korean sesame seeds is typically two to three times higher than that of imported sesame seeds and can be 10 or 20 times higher than other vegetable oils (Seol et al. 2012). Again, there is a temptation for dishonest producers to label foreign sesame oils as Korean and/or to adulterate sesame oil by blending with low-cost edible oils (e.g., maize and soybean oils) or other low quality ingredients (e.g., extracts from sesame cakes intended as animal feed), sometimes in unsanitary conditions (Kim et al. 2015).

By combining the analysis of $\delta^{13}C$ composition of whole oil with fatty acid profiles, in particular the proportion of stearic acid, it was possible to detect the addition of the cheaper maize oil to the sesame oil. This was demonstrated in 2010: as the proportion of maize oil increased, $\delta^{13}C$ values

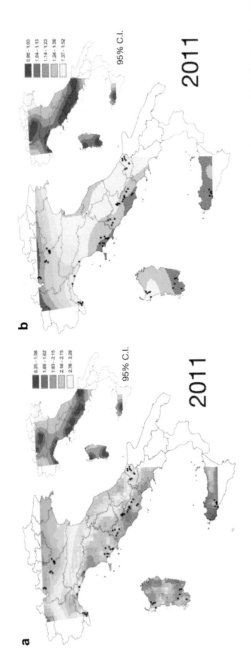

Figure 11.1 $\delta^{18}O$ (a) and $\delta^{13}C$ (b) isoscapes and relative 95% confidence intervals (C.I.) for Italian extra virgin olive oils collected in 2011. Reprinted from *Food Chemistry*, 202, Chiocchini, F., S. Portarena, M. Ciolfi, E. Brugnoli and M. Lauteri, Isoscapes of carbon and oxygen stable isotope compositions in tracing authenticity and geographical origin of Italian extra-virgin olive oils, 291–301 Copyright (2016), with permission from Elsevier.

increased linearly; the content of palmitic, linoleic, and linolenic acids increased gradually; and the content of stearic and oleic acids decreased (Seo et al. 2010). These results were subsequently confirmed (Seol et al. 2012) by the analysis of $\delta^{13}C$ values in addition to color, fluorescence intensity, and fatty acid profiles of sesame oils prepared from sesame seeds with different roasting conditions (175, 200, 225, or 250°C for 15 or 30 min), mixing sesame oil with maize oil in different proportions. Roasting conditions were found not to influence the $\delta^{13}C$ values or fatty acid composition, but did influence the color and fluorescence properties of the oil. The $\delta^{13}C$ values, as well as some fatty acid ratios, were found to provide good discrimination between sesame oils blended with maize oil.

Another recent study (Horacek et al. 2015) considered δ^2H in addition to $\delta^{13}C$ composition in an investigation focused on identifying the geographical origins of sesame oil. Sesame oils from tropical and subtropical/moderate climatic provenances could be discriminated on the basis of $\delta^{13}C$ and δ^2H values, which were significantly, positively correlated (Horacek et al. 2015). The $\delta^{13}C$ values of these oils also showed differences between extracts from black and white sesame seeds grown in the same location, potentially indicating higher water use efficiency of plants producing black seeds.

11.3.3 Other edible vegetable oils

Rapeseed or *canola* oil, produced from the crops *Brassica napus, Brassica rapa,* and other *Brassica* species (Cruciferae, also known as the mustard family), has become an important oilseed crop in cool, temperate, northern climates where most other oilseed crops do not grow. A study of the $\delta^{13}C$ values of fatty acids of rapeseed oil found that palmitic acid was relatively enriched compared to oleic acid (like groundnut and palm oils), whereas stearic acid was relatively depleted in ^{13}C with respect to oleic acid in all the oils considered (Kelly et al. 1997).

Camelina sativa is another member of the mustard family that is traditionally cultivated as an oilseed crop to produce vegetable oils and has been the subject of a study of the $\delta^{13}C$ values of individual fatty acids as well as of bulk oil from different geographical origins (Hrastar et al. 2009). The authors found a correlation between linoleic and linolenic acids that could highlight cases in which impurity or adulteration was suspected. Principal component analysis of the $\delta^{13}C$ values of bulk oil—combined with oleic, linoleic, linolenic, and eicosenoic acids—was able to separate oils from different continents.

A recent paper (Banerjee et al. 2015) summarized the H, C, and O isotopic compositions of several types of vegetable edible oils from C_3 plants (olive, sesame, sunflower, walnut, coconut, pumpkin seed, mustard, hemp seed, soybean, palm, and canola) from different geographical origins. The authors reported an average $\delta^{13}C$ value of $-29.5‰$ for C_3 plant oils, and confirmed that it was possible to detect the addition of low amount of C_4 plant oil [in this case, maize oil, with average $\delta^{13}C$ value of $-15‰$ (Kelly and Rhodes 2002)].

The δ^2H and $\delta^{18}O$ compositions of oils have been shown to be correlated with and dependent on geographical origin. In 1998 the first paper describing ^{18}O analysis of oil via TC/EA-IRMS (Bréas et al. 1998) reported the $\delta^{18}O$ values of almond, coconut, grapeseed, olive, maize, pine, castor oil, sesame, soybean, sunflower, and wheat oils, and concluded that oxygen isotope ratio is more closely linked to the geographical origin of the oil than to its botanical origin. Both hydrogen and oxygen isotope ratios are now generally recognized as those most closely related to the geo-climatic characteristics of plant growth location. In general, the hydrogen and oxygen isotope ratios of edible vegetable oils have been shown to be defined by a relationship expressed by $\delta^2H_{edibleoil} = 6.7 \times \delta^{18}O_{edibleoil} - 304$ (Reeves and Weihrauch 1979), the slope of which closely parallels the global meteoric water line, suggesting that the oils record the signatures of local meteoric waters from where seeds or fruits of edible oils are harvested. Later research found a near constant H and O isotopic fractionations (R^2 approximately 0.9) between extra virgin olive oils and local meteoric waters from different locations (Banerjee et al. 2015). This research supported the idea that the isotopic composition of local meteoric water of an oil-producing location may be used to estimate the δ^2H and/or $\delta^{18}O$ values of authentic edible vegetable oil from that location—calculated using a relationship between the net hydrogen and oxygen isotopic fractionation factors between edible oil and local meteoric water.

11.4 Conclusions

The stable isotope ratio analysis of bulk oil allows the detection of the addition of C_4 plant oil, such as maize oil, to other edible vegetable oils. Using compound specific carbon stable isotope ratio analysis, it is possible to identify other types of adulteration, such as the addition of pomace oil to virgin olive oil, or to distinguish different types of edible vegetable oil.

By combining the stable isotope ratios of C, H, and O it is possible to characterize oils on the basis of the geographical origin, because these ratios in oil are related to those of local meteoric water and with the climatic characteristics.

References

Angerosa, F., O. Bréas, S. Contento, C. Guillou, F. Reniero and E. Sada. 1999. Application of stable isotope ratio analysis to the characterization of the geographical origin of olive oils. *Journal of Agricultural and Food Chemistry* 47: 1013–1017.

Angerosa, F., L. Camera, S. Cumitini, G. Gleixner and F. Reniero. 1997. Carbon stable isotopes and olive oil adulteration with pomace oil. *Journal of Agricultural and Food Chemistry* 45: 3044–3048.

Aramendia, M.A., A. Marinas, J. M. Marinas et al. 2007. Oxygen-18 measurement of Andalusian olive oils by continuous flow pyrolysis/isotope ratio mass spectrometry. *Rapid Communications in Mass Spectrometry* 21: 487–496.

Bianchi, G., F. Angerosa, L. Camera, F. Reniero and C. Anglani. 1993. Stable carbon isotope ratios (carbon-13/carbon-12) of olive oil components. *Journal of Agricultural and Food Chemistry* 41: 1936–1940.

Bontempo, L., F. Camin, R. Larcher, G. Nicolini, M. Perini and A. Rossmann. 2009. Coast and year effect on H, O and C stable isotope ratios of Tyrrhenian and Adriatic Italian olive oils. *Rapid Communications in Mass Spectrometry* 23: 1043–1048.

Banerjee, S., T. K. Kyser, A. Vuletich and E. Leduc. 2015. Elemental and stable isotopic study of sweeteners and edible oils: Constraints on food authentication. *Journal of Food Composition and Analysis* 42: 98–116.

Bell, J. R. and P. N. Gillatt. 2011. Standards to ensure the authenticity of edible oils and fats. FAO Corporate Document Repository, 1–9. www.fao.org/docrep/T4660t/t46600e. htm#TopOfPage [accessed 29/07/2016].

Bréas, O., C. Guillou, F. Reniero, E. Sada and F. Angerosa. 1998. Oxygen-18 measurement by continuous flow pyrolysis/isotope ratio mass spectrometry of vegetable oils. *Rapid Communications in Mass Spectrometry* 12: 188–192.

Camin, F., R. Larcher, G. Nicolini et al. 2010a. Isotopic and elemental data for tracing the origin of European olive oils. *Journal of Agricultural and Food Chemistry* 58: 570–577.

Camin, F., R. Larcher, M. Perini et al. 2010b. Characterisation of authentic Italian extra-virgin olive oils by stable isotope ratios of C, O and H and mineral composition. *Food Chemistry* 118: 901–909.

Carter, J. F. and B. Fry. 2013. Ensuring the reliability of stable isotope ratio data—Beyond the principle of identical treatment. *Analytical and Bioanalytical Chemistry* 405: 2799–2814.

Chiocchini, F., S. Portarena, M. Ciolfi, E. Brugnoli and M. Lauteri. 2016. Isoscapes of carbon and oxygen stable isotope compositions in tracing authenticity and geographical origin of Italian extra-virgin olive oils. *Food Chemistry* 202: 291–301.

Faberi, A., R. M. Marianella, F. Fuselli et al. 2014. Fatty acid composition and $\delta^{13}C$ of bulk and individual fatty acids as marker for authenticating Italian PDO/PGI extra virgin olive oils by means of isotopic ratio mass spectrometry. *Journal of Mass Spectrometry* 49: 840–849.

Guo, L. X., X. M. Xu, J. P. Yuan, C. F. Wu and J. H. Wang. 2010. Characterization and authentication of significant Chinese edible oilseed oils by stable carbon isotope analysis. *JAOCS Journal of the American Oil Chemists' Society* 87: 839–848.

Horacek, M., K. Hansel-Hohl, K. Burg, G. Soja, W. Okello-Anyanga and S. Fluch. 2015. Control of origin of sesame oil from various countries by stable isotope analysis and DNA based markers—A pilot study. *PLoS ONE* 10: e0123020.

Hrastar, R., M. G. Petrišič, N. Ogrinc and I. J. Košir. 2009. Fatty acid and stable carbon isotope characterization of *Camelina sativa* oil: Implications for authentication. *Journal of Agricultural and Food Chemistry* 57: 579–585.

Iacumin, P., L. Bernini and T. Boschetti. 2009. Climatic factors influencing the isotope composition of Italian olive oils and geographic characterization. *Rapid Communications in Mass Spectrometry* 23: 448–454.

Jeon, H., S. C. Lee, Y. J. Cho, J. H. Oh, K. Kwon and B. H. Kim. 2015. A triple-isotope approach for discriminating the geographic origin of Asian sesame oils. *Food Chemistry* 167: 363–369.

Kelly, S. D. and C. Rhodes. 2002. Emerging techniques in vegetable oil analysis using stable isotope ratio mass spectrometry. *Grasas Aceites* 53: 34–44.

Kelly, S., I. Parker, M. Sharman, J. Dennis and I. Goodall. 1997. Assessing the authenticity of single seed vegetable oils using fatty acid stable carbon isotope ratios ($^{13}C/^{12}C$). *Food Chemistry* 59: 181–186.

Kim, J., G. Jin, Y. Lee, H. S. Chun, S. Ahn and B. H. Kim. 2015. Combined analysis of stable isotope, ¹H NMR, and fatty acid to verify sesame oil authenticity. *Journal of Agricultural and Food Chemistry* 63: 8955–8965.

Lu, H., C. Wei, H. Fu, X. Li, Q. Zhang and J. Wang. 2015. Identification of recycled cooking oil and edible oils by iodine determination and carbon isotopic analysis. *JAOCS Journal of the American Oil Chemists' Society* 92: 1549–1553.

Medini, S., M. Janin, P. Verdoux and I. Techer. 2015. Methodological development for $^{87}Sr/^{86}Sr$ measurement in olive oil and preliminary discussion of its use for geographical traceability of PDO Nîmes (France). *Food Chemistry* 171: 78–83.

Mihailova, A., D. Abbado, S. D. Kelly and N. Pedentchouk. 2015. The impact of environmental factors on molecular and stable isotope compositions of *n*-alkanes in Mediterranean extra virgin olive oils. *Food Chemistry* 173: 114–121.

Portarena, S., D. Farinelli, M. Lauteri, F. Famiani, M. Esti and E. Brugnoli. 2015. Stable isotope and fatty acid compositions of monovarietal olive oils: Implications of ripening stage and climate effects as determinants in traceability studies. *Food Control* 57: 129–135.

Portarena, S., O. Gavrichkova, M. Lauteri and E. Brugnoli. 2014. Authentication and traceability of Italian extra-virgin olive oils by means of stable isotopes techniques. *Food Chemistry* 164: 12–16.

Reeves, J. B. and J. L. Weihrauch. 1979. *Composition of foods: fats and oils. Agriculture handbook n. 8–4.* Washington D.C.: U.S. Dept. of Agriculture, Science and Education Administration.

Richter, K. E., J. E. Spangenberg, M. Kreuzer and F. Leiber. 2010. Characterization of rapeseed (*Brassica napus*) oils by bulk C, O, H, and fatty acid C stable isotope analyses. *Journal of Agricultural and Food Chemistry* 58: 8048–8055.

Royer, A., C. Gerard, N. Naulet, M. Lees and G. J. Martin. 1999. Stable isotope characterization of olive oils. I—Compositional and carbon-13 profiles of fatty acids. *JAOCS Journal of the American Oil Chemists' Society* 76: 357–363.

Salas-Salvado, J., M. Bullo, R. Estruch, E. Ros, M. I. Covas, N. Ibarrola-Jurado et al. 2014. Prevention of diabetes with Mediterranean diets: A subgroup analysis of a randomized trial. *Annals of Internal Medicine* 160: 1–10.

Seo, H. Y., J. Ha, D. B. Shin et al. 2010. Detection of corn oil in adulterated sesame oil by chromatography and carbon isotope analysis. *JAOCS Journal of the American Oil Chemists' Society* 87: 621–626.

Seol, N. G., E. Y. Jang, M. -J. Kim and J. Lee. 2012. Effects of roasting conditions on the changes of stable carbon isotope ratios ($\delta^{13}C$) in sesame oil and usefulness of $\delta^{13}C$ to differentiate blended sesame oil from corn oil. *Journal of Food Science* 77: 1263–1268.

Spangenberg, J. E., S. A. Macko and J. Hunziker. 1998. Characterization of olive oil by carbon isotope analysis of individual fatty acids: Implications for authentication. *Journal of Agricultural and Food Chemistry* 46: 4179–4184.

Spangenberg, J. E. and N. Ogrinc. 2001. Authentication of vegetable oils by bulk and molecular carbon isotope analyses with emphasis on olive oil and pumpkin seed oil. *Journal of Agricultural and Food Chemistry* 49: 1534–1540.

Woodbury, S. E., R. P. Evershed, J. B. Rossell, R. E. Griffith and P. Farnell. 1995. Detection of vegetable oil adulteration using gas chromatography combustion/isotope ratio mass spectrometry. *Analytical Chemistry* 67: 2685–2690.

CHAPTER 12

Organic Food Authenticity

Simon D. Kelly[1],* and Alina Mihailova[2]

12.1 Introduction

"To eat is a necessity, but to eat intelligently is an art."

François de La Rochefoucauld, French author (1613–1680)

There has been a rapid growth in the *organic food* sector over recent years, driven largely by consumer demand. Organic foods are becoming widely available not only in organic food supermarkets and from specialist retailers, but in a wide variety of conventional supermarkets, from mass-market merchandizers and through online purchase/home delivery services. Suppliers are increasingly introducing new varieties, and retailers now offer many organic fresh-produce items all year-round (Oberholtzer et al. 2005).

This chapter highlights a small selection of fraud cases in the organic food sector and sets out the regulatory context against which organic food authenticity must be determined. It also summarizes the findings from research studies that have important implications for using stable isotope analysis to infer information about organic agriculture, to draw together any general findings from the research, and to highlight any potential pitfalls of using stable isotope analysis in this area of food forensics.

Organic produce typically retails at a higher price than the conventionally produced counterpart, partly because organic production is relatively labor

[1] Food and Environmental Protection Laboratory, Joint FAO/IAEA Division of Nuclear Applications in Food and Agriculture, Department of Nuclear Sciences and Applications, International Atomic Energy Agency, Vienna International Centre, PO Box 100, 1400 Vienna, AUSTRIA.

[2] Science Analytical Facility, Faculty of Science, University of East Anglia, Norwich Research Park, Norwich, NR4 7TJ, UNITED KINGDOM.
Email: a.mihailova@uea.ac.uk

* Corresponding author: S.Kelly@iaea.org
With contributions from Karyne M. Rogers

intensive and partly because of smaller economies of scale in production, distribution, and processing. More controversially, it has been suggested that some of the difference in price may be due to retailers, manufacturers, and/or farmers taking advantage of the consumers' willingness to pay extra for organic products (Pretty et al. 2005). Notwithstanding the reasons for the price differential, consumers paying a premium price for organic produce have the right to receive what they have actively chosen and paid for. Higher retail prices for organic goods mean that there is a financial enticement for dishonest traders, wholesalers, packing houses, and growers to mislabel and try to pass a cheaper conventional product as organic.

Assurance schemes for organic quality and authenticity are available from certification bodies mainly through farm or on-site inspections and documentation designed to allow traceability from the producer to the consumer. The traceability of globally traded organic produce has also become increasingly important as more than one quarter of the world's organically managed land (nine million hectares) is located in developing countries (Sahota 2009) of Latin America, Asia, and Africa. The countries with the largest areas under organic systems are Argentina, Brazil, China, India, and Uruguay.

There can be little doubt that detecting the substitution of organic produce with conventionally produced foods presents a significant analytical challenge because these foods are ostensibly the same chemically, a view supported by a systematic review of the scientific literature (Dangour et al. 2009). Even more challenging is the detection of subtle deviations from accepted organic practices, which may not impart a characteristic chemical or physical "tracer." Confirming the authenticity of organic food does not, therefore, simply address aspects such as the use/misuse of synthetic fertilizers, pesticides, and herbicides. For example, the lack of an appropriate and ethical on-farm wildlife policy would very likely prove impossible to detect in a retail product by analytical means. Consequently, it is important to acknowledge that none of the methods reviewed in this chapter has the capability to ascertain whether a farmer/grower is complying with all the legal requirements of organic production.

Currently, the authenticity of organic produce is based on inspections and the primary focus is on the production process rather than the products reaching the consumers. Despite strict production standards, certification, and inspection of organic producers and supply chains, the potential scale and breadth of organic food fraud, in the same way as conventional food fraud, is a major challenge—from the local farmer's market to international organized crime syndicates generating illegal profits through cereal production and distribution of €220 million (EWFC 2016; Independent 2016). Market prices of organic products are typically 30–75% higher than conventional counterparts (Brown and Sperow 2005; Winter and Davis 2006; Willer and Kilcher 2010; Jaenicke and Carlson 2015) and depending on the product type, the difference in price can be greater than 200% (Oberholtzer et al. 2005; Yu et al. 2014). Products labelled "organic" are distinguished from conventional foods by production and processing principles rather than attributes that are discernible in the products (McCluskey 2000) and consumers must rely on the information on the product label or simply trust the retailer—e.g., at farmers' markets. This makes the organic market arguably more susceptible to fraudulent behavior in the supply chain than most conventional food markets.

Recently, there have been a number of reported cases in the European Union (EU) in which conventional products were misrepresented as organic and priced accordingly (Beck 2012; Ferrante 2012; Huber 2012; Piva 2012). One example took place from 2007 to 2011, when tons of commodities of plant origin (mainly wheat), were imported from Romania with false certificates and exported to other countries of the EU where they were sold at premium prices (Beck 2012; Ferrante 2012; Huber 2012; Piva 2012).

In the United States, a Department of Agriculture (USDA) survey revealed that 43% of organic produce contained measurable quantities of pesticide residues (United States Department of Agriculture 2012) although only 4% exceeded the Environmental Protection Agency reporting limits and were therefore in violation of organic regulations. The USDA suggested that such residues could result from contamination from adjacent conventional farms, a view that proved controversial because cross-contamination will depend on a number of factors such as proximity, application rates, droplet size, wind-speed, etc. Organic fraud is not limited to the food products but also to products such as the fertilizers used in production. Between 2000 and 2006, a California-based fertilizer company sold a registered organic fertilizer *fortified* with synthetic ammonium chloride generating over US$6.5 million in gross sales (Federal Bureau of Investigation 2012a). In a similar fashion, between 2003 and 2009, Port Organic Products Ltd. sold an estimated US$40 million worth of organic fertilizer fortified with synthetic aqueous ammonia and ammonium sulfate (Federal Bureau of Investigation 2012b). The sale of adulterated organic fertilizer over several years meant that significant quantities of produce, cultivated on organic farms, were fertilized with synthetic nitrogen sources in contravention of USDA regulations.

Pre-packaged, processed foods, and ready-to-eat meals are also sold under the organic label, which means that the ingredients must be certified organic, presenting a challenge for traceability. In 2011, counterfeit organic certificates were brought to the attention of enforcement agencies in the US regarding organic hibiscus, jasmine, and beetroot extract powders produced in China (Food Safety News 2011).

Organic fraud also affected animal and dairy production—for example, an egg packing business in England generated £3m profit over a two-year period by systematically mis-describing eggs (Pidd 2010). At this time (2004 and 2006) farmers could expect a price of about 90 p per dozen for organic eggs, 70 p for free-range eggs, and 35 p for caged eggs. Such practices were described as widespread by those who were caught and prosecuted.

12.2 Regulations

Organic food operators must guarantee that all stages of production and processing comply with organic production rules and that all necessary steps are taken to prevent contamination, mixing, or comingling with non-organic products. It is these procedures and their effectiveness that will be the subject of a detailed inspection (Beck 2012). The development of reliable and robust analytical tools can only act to support and underpin their effectiveness.

Legislation defines the principles of organic production at farm, preparation, storage, transport, labelling, and marketing stages, and provides an indication of permitted soil fertilization and conditioning, pest and disease control, food additives, and processing aids. *Organic standards* are adopted in order to:

- protect consumers against deception of unsubstantiated product claims in the market place;
- protect organic producers against misrepresentation of other agricultural produce as being organic;
- ensure that all stages of production, preparation, storage, transport, and marketing of organic foods are subject to inspection and comply with set regulations; and
- harmonize the requirements for organic production, marketing, inspection, and labelling in order to facilitate recognition of national systems as equivalent for the purposes of trade (Codex Alimentarius Commission 2013).

In 1999 Codex Alimentarius Commission—the joint intergovernmental body of the Food and Agriculture Organization of the United Nations (FAO) and World Health Organization—produced "Guidelines for the Production, Processing, Labelling and Marketing of Organically Produced Foods" (CAC/GL32-1999). The International Federation of Organic Agriculture Movements (IFOAM), a worldwide umbrella organization for the organic agriculture movement, set out IFOAM Norms, which are guidelines for organic production worldwide. The Norms have been developed through a joint effort of IFOAM, FAO, and the UN Conference on Trade and Development (UNCTAD). The Norms include common objectives and requirements of organic standards, the IFOAM standard for organic production and processing, and accreditation requirements for bodies certifying organic production and processing (IFOAM 2014). Codex Alimentarius and IFOAM guidelines are minimum standards for organic agriculture, intended to guide governments and private certification bodies in standard setting.

A large number of organic standards currently exist worldwide. Some of these organic standards are governmental regulations, others are industry standards; some have only local scope (e.g., one country or a region) while others are used internationally. Table 12.1 lists countries that currently have organic regulations and those countries in the process of drafting legislation (adopted from Huber et al. 2015). Many countries (e.g., EU member states, Switzerland, United States, Canada, Japan, Korea) have comprehensive organic regulations and other countries (e.g., Hong Kong, Kuwait, Bahrain, Laos, United Arab Emirates, Vietnam, Ghana, Kenya, Zambia, Zimbabwe) have national production standards, which provide a national definition of organic production system and serve as a reference for certification (Huber et al. 2015).

In the EU, the minimum standards for organic production, processing, certification, and labelling are currently set by the Council Regulation

Table 12.1 Countries with regulations on organic agriculture (adopted from Huber et al. 2015).

Region	Countries with implemented regulations	Countries drafting regulations
Europe	EU: Austria, Belgium, Bulgaria, Croatia, Cyprus, Czech Republic, Denmark, Estonia, Finland, France, Germany, Greece, Hungary, Ireland, Italy, Latvia, Lithuania, Luxemburg, Malta, Poland, Portugal, Romania, Slovak Republic, Slovenia, Spain, Sweden, Netherlands, United Kingdom	Belarus, Bosnia and Herzegovina, Russia
	Non-EU: Albania, Iceland, Kosovo, Macedonia, Moldova, Montenegro, Norway, Serbia, Switzerland, Turkey, Ukraine	
The Americas and Caribbean	Argentina, Bolivia, Brazil, Canada, Chile, Colombia, Costa Rica, Cuba, Dominican Republic, Ecuador, El Salvador, Guatemala, Honduras, Mexico, Nicaragua, Panama, Paraguay, Peru, Uruguay, USA, Venezuela	Jamaica, St. Lucia
Asia and Pacific	Armenia, Australia, Azerbaijan, China, Georgia, India, Indonesia, Iran, Israel, Japan, Jordan, South Korea, Lebanon, Malaysia, New Caledonia, New Zealand, Philippines, Saudi Arabia, Taiwan, Thailand	Jordan, Kyrgyzstan, Nepal, Pakistan
Africa	Morocco, Tunisia	Algeria, Egypt, Kenya, Namibia, Senegal, South Africa, Sudan

(EC) No. 834/2007 and Commission Regulations (EC) No. 889/2008 and (EC) No. 1235/2008. Once a product has been certified by an accredited certification body, it is automatically recognized EU-wide. A mandatory EU organic logo (introduced in 2010) must be displayed on all pre-packed organic food produced in the EU, accompanied by an indication of the origin of the raw materials: "EU Agriculture," "non-EU Agriculture," or "EU/non-EU Agriculture" [Council Regulation (EC) No. 834/2007]. To be labeled "organic" the product must contain a minimum of 95% organic agricultural ingredients by weight. In the case where all agricultural raw materials have been farmed in the same country, the terms "EU" and "non-EU" can be replaced or supplemented by the name of that country [Council Regulation (EC) No. 834/2007].

Most EU countries employ a system of private control bodies that are accredited by government authorities; few countries (e.g., Denmark and Finland) have governmental control authorities (Janssen and Hamm 2014). The most common private organic labels are from farmers and organic sector associations (e.g., Organic Farmers and Growers, Soil Association, Demeter) and control bodies (e.g., Ecocert, Institute for Marketecology—IMO). Some of these organizations (e.g., Demeter) have production standards that exceed and supplement the EU Regulations whereas other control bodies (e.g., Ecocert) indicate that the product has been certified according to EU Regulations by the respective certification body (Janssen and Hamm 2014).

The EU permits organic goods to be imported from non-EU countries [Commission Regulation (EC) No. 1235/2008] if the rules on organic production and control are equivalent to the EU's, such as members of the European Economic Area (Norway and Iceland), Switzerland, USA, Canada, Australia, Japan, Republic of Korea, New Zealand, Argentina, Costa Rica, India, Israel, and Tunisia [Commission Regulation (EC) No. 1235/2008, Commission Implementing Regulation (EU) 2015/131]. Imports from non-EU countries can be certified by independent private control bodies approved by the European Commission.

In the United States, the National Organic Program (NOP) (US Code of Federal Regulations, 7CRF—Part 205) develops the laws that regulate the production, handling, labelling, trade, and enforcement of all organic products in accordance with USDA regulations. The NOP incorporates all USDA organic standards, including prohibited practices, requirements, and lists of allowed and prohibited materials. The NOP allows for four categories of products that are labeled as (i) "100% organic" (100% organically produced ingredients by weight or volume), (ii) "organic" (not less than 95% organically produced raw or processed agricultural ingredients), (iii) "made with organic ingredients or food groups" (at least 70% organically produced ingredients), and (iv) "products with less than 70% organically produced ingredients" (specific ingredients may be identified as organic) (US Code of Federal Regulations, 7CRF—Part 205). The name of the certifying agent of the final product must be displayed on the information panel and must be accredited by the USDA. In contrast to EU practice, producers have the option to display the USDA organic seal and/or the logo of the certification body on "100% organic" and "organic" products.

The NOP works with the Foreign Agricultural Service and Office of the United States Trade Representative to establish international trade arrangements for organic products. For example, the United States currently has an equivalence arrangement with the EU, Canada, Taiwan, Japan, and Korea and recognition agreements with other countries—e.g., Israel, India, and New Zealand. The latter allows a foreign government to accredit certifying agents to certify organic farms and processing facilities, ensuring that USDA organic products are met or exceeded. These products can then be exported for sale in the United States and other countries that accept USDA organic products (USDA 2015).

In Canada, food products labelled as "organic" or displaying the Canadian organic logo must meet the requirements set in the Organic Products Regulations (Canada Organic Products Regulations, SOR/2009-176). These are developed by the Governor General in Council, on the recommendation of the Minister of Agriculture and Agri-Food in accordance with the Canada Agricultural Products Act (2015). The regulations require mandatory certification to the revised Canadian Organic Standards (CAN/CGSB-32.310-2006, CAN/CGSB-32.311-2006) for agricultural products represented as organic in import, export, and inter-provincial trade.

The Canadian Food Inspection Agency enforces the Organic Products Regulations and is responsible for the accreditation of organic certification bodies. The name of the certification body must be displayed on the product label; however, the use of the Canada Organic logo is voluntary. The logo is only permitted on products that have 95% or more organic content and have been certified according to Canada Organic Products Regulations (SOR/2009-176). Multi-ingredient products with 70–95% organic content may declare "contains X% organic ingredients," but may not use the Canada Organic logo and/or the claim "organic." Multi-ingredient products with less than 70% organic content may only display organic claims in the product's ingredient list and may not use the Canada Organic logo. Organic products sold within the province of origin are subject to provincial regulations, the Consumer Packaging and Labelling Regulations, and the Food and Drug Regulations. The Canadian provinces of Quebec, Manitoba, and British Columbia currently have organic certification systems in place.

In Australia, the organic sector is a co-regulatory, as opposed to a mandated system (e.g., as in the EU and US). The National Standard for Organic and Biodynamic Produce (first implemented in 1992) provides the organic industry with nationally agreed standards for the export of products labelled "organic" or "biodynamic." The standard is reviewed periodically by the Organic Industry Standards and Certification Council. The Department of Agriculture is responsible for export policy, including maintaining the National Standard for Organic and Biodynamic Produce and certifying exports of organic food against that standard. Another organic standard (AS 6000-2009 Organic and Biodynamic Products) was released by Standards Australia in 2009 through a representative technical committee comprising organic stakeholders. This standard is voluntary but can be used to execute existing legislation. All foods produced or imported for sale in Australia or New Zealand, including organic food, must be labelled in accordance with the Food Standards Code developed by Food Standards Australia New Zealand, which develops food standards for composition, labelling, and contaminants. The domestic market for organic products is commonly certified by one of Australia's seven certification bodies (e.g., Australian Certified Organic, The National Association of Sustainable Agriculture Australia, Organic Food Chain) based on the National Standard for Organic and Biodynamic Produce (2009). The Organic Federation of Australia is not a certifying body but represents the interests of Australian organic and biodynamic producers to industry and governments at the local, State, and Federal level.

In African countries, most certified organic production is targeted at export markets and most production has been certified according to EU regulations. Exports also go to the United States or to closer markets in the Gulf and South Africa (IFOAM OSEA Report 2008). Within domestic markets, African countries are reliant upon a number of national standards—e.g., East African Organic Products Standard (EAS 456:2007), Afrisco Standards for Organic Production (2011). In East Africa, stakeholders have developed a regional organic standard facilitated by the IFOAM and the UNEP/UNCTAD Capacity Building Task Force on Trade, Environment and Development. The East African Organic Products Standard was adopted by the East African Community as an East African standard and has been recognized in Burundi, Kenya, Rwanda, Tanzania, and Uganda (IFOAM OSEA Report 2008).

In China, a national organic standard (GB/T19630) was established in 2005 monitored by the Chinese government at local, regional, and national levels. All foreign certifiers must be in partnership with a Chinese-based organization approved by the Chinese Foreign Investment Committee. Organic products exports are largely aimed at EU countries and need to meet EU organic regulations. Other certifications for organic export include the United States (National Organic Program) and Japan (Japanese Agricultural Standard) (USDA 2010).

In 2003 the International Task Force (ITF) on Harmonization and Equivalence in Organic Agriculture was established as a result of combined effort of the UNCTAD, FAO, and IFOAM. ITF serves as an open platform for dialogue between public and private institutions (intergovernmental, governmental, and civil society) involved in trade and regulatory activities in the organic agriculture sector.

12.3 Discrimination between organic and conventional crops

From an analytical perspective, the authentication of organic produce presents a challenge and numerous attempts have been made to find specific compounds that can act as markers for organic crops (Woese et al. 1997; Bourn and Prescott 2002; Hoefkens et al. 2009). Unsurprisingly, pesticide residue concentrations were found to be systematically lower in organic plants than conventionally-grown plants (Woese et al. 1997; Bourn and Prescott 2002; Hoefkens et al. 2009). However, random pesticide residue testing of organic plant products cannot ensure successful authentication of organic production practice as often only a limited number of pesticides are tested, compared to the large number of possible contaminants. Furthermore, even in conventional plant products pesticide residues are frequently below the limit of detection (Hoefkens et al. 2009). Studies of the nutrient content of organic and conventional crops have been inconsistent and contradictory. A systematic review by Dangour et al. (2009) indicated that there were no significant differences in the chemical compositions of foods derived from organic and conventional growing systems, with the exceptions that conventionally produced crops had higher concentration of nitrogen, and organically produced crops had higher concentration of phosphorus and higher titratable acidity. Similar findings had been reported earlier in reviews by Woese et al. (1997) and Bourn and Prescott (2002). Although there are reports that organic and conventional fruits and vegetables differ in a sensory qualities and vitamin content, these findings were inconsistent (Bourn and Prescott 2002).

Over the past decade, a number of studies have applied isotope ratio mass spectrometry (IRMS) in order to identify the most suitable markers for the discrimination between organic and conventionally grown vegetables, cereals, and fruits (Table 12.2). In the vast majority of these studies the nitrogen isotope composition (δ^{15}N) of bulk matter has been investigated as a marker for organic cultivation. The differences in the δ^{15}N compositions of synthetic and organic fertilizers have been reflected in the δ^{15}N values of the crops (Choi et al. 2003; Nakano et al. 2003; Bateman et al. 2005; Verenitch and Mazumder 2015). Synthetic nitrogen fertilizers are produced by the *Haber-Bosch process* from atmospheric nitrogen (δ^{15}N$_{AIR}$ = 0‰) and generally have δ^{15}N values within a few ‰ of 0 (Freyer and Aly 1974; Vitòria et al. 2004; Bateman and Kelly 2007). In contrast, the nitrogen isotopic compositions of animal manures, composts, and other fertilizers permitted in organic production systems are significantly enriched compared to synthetic

Table 12.2 Overview of the analytical techniques and studies on isotope ratio analysis for the discrimination between organically and conventionally grown crops.

Type of analysis	Analytical technique	Isotope ratios measured	Crops analysed	References
Bulk isotope analysis	IRMS	$\delta^{15}N$, $\delta^{13}C$, $\delta^{2}H$, $\delta^{18}O$, $\delta^{34}S$	vegetables	Nakano et al. 2003; Bateman et al. 2005; Georgi et al. 2005; Schmidt et al. 2005; Yun et al. 2006; Bateman et al. 2007; Camin et al. 2007; Flores et al. 2007; Lim et al. 2007; del Amor et al. 2008; Rogers 2008; Kelly and Bateman 2010; Flores et al. 2011; Šturm et al. 2011; Šturm and Lojen 2011; Zhou et al. 2012; Yuan et al. 2012; Laursen et al. 2013; Flores et al. 2013; Verenitch and Mazumder 2015
	IRMS	$\delta^{15}N$, $\delta^{13}C$, $\delta^{2}H$, $\delta^{18}O$, $\delta^{34}S$	cereals	Choi et al. 2002; Schmidt et al. 2005; Choi et al. 2006; Suzuki et al. 2009; Yun et al. 2011; Laursen et al. 2013; Chung et al. 2015
	MC-ICP-MS	$\delta^{25}Mg$, $\delta^{26}Mg$	cereals	Laursen et al. 2013
	IRMS	$\delta^{15}N$, $\delta^{13}C$, $\delta^{2}H$, $\delta^{18}O$, $\delta^{34}S$	fruits	Rapisarda et al. 2005; Rapisarda et al. 2010; Camin et al. 2011; Bizjak Bat et al. 2012
Compound-specific isotope analysis (CSIA)	GC-IRMS, denitrifier method	Oxygen and nitrogen isotope ratios of plant-derived nitrate $\delta^{18}O_{NO3}$, $\delta^{15}N_{NO3}$	vegetables	Laursen et al. 2013; Mihailova et al. 2014
	GC-C-IRMS	Nitrogen isotope ratios of amino acids and proteins, carbon isotope ratios of amino acids, proteins, sugars	fruits	Rapisarda et al. 2005; Rapisarda et al. 2010; Bizjak Bat et al. 2012

nitrogen fertilizers (Kendall 1998; Bateman and Kelly 2007). The mean $\delta^{15}N$ values of organic fertilizers cluster around +8‰ and some fertilizers (e.g., animal manures) can have values higher than +35‰, due to the preferential volatilization of ^{15}N-depleted ammonia in the field or during storage (Bateman and Kelly 2007).

Bateman et al. (2005) conducted a series of greenhouse experiments to investigate the effects of the type and amount of fertilizer, growing medium, and irrigation water on the $\delta^{15}N$ values of carrots, tomatoes, and lettuces. Fertilizer type was a significant factor influencing the nitrogen isotope composition of the crop; the application of organic pelleted chicken manure resulted in crops with significantly higher $\delta^{15}N$ values compared to the application of synthetic ammonium nitrate. Plants grown in peat-based compost were generally found to be isotopically lighter than those grown in composted bark. This study also discussed how nitrate content and the $\delta^{15}N$ value of the nitrate in irrigation water might influence the $\delta^{15}N$ values of the crop. Šturm et al. (2011) investigated the combined use of synthetic and organic fertilizers on the bulk nitrogen composition of pot-grown lettuces. The authors reported significant differences in $\delta^{15}N$ values of whole plants, which were treated with organic or conventional fertilizers, applied in a single treatment. However, subsequent fertilization (with isotopically similar or different fertilizer) did not result in a significant change in plant $\delta^{15}N$ values. The authors suggested that $\delta^{15}N$ composition could be used to approximate the history of nitrogen fertilization in the case of a single application.

Camin et al. (2007) evaluated a number of markers to discriminate between organic and conventional potatoes from four separate field trials. Crops were subject to the same soil and climatic condition, but differed in respect to the types (organic/synthetic) of fertilizers applied. Irrespective of cultivar or environment, organic tubers showed a significant enrichment in ^{15}N compared to their conventional counterparts. When crops from all four field trials were considered, a threshold $\delta^{15}N$ value of +4.3‰ allowed the correct classification of all organic potatoes (potatoes with values above the threshold were organic and below the threshold, conventional) and misclassification of 15% of conventional samples.

Bateman et al. (2007) analyzed the nitrogen isotope composition of commercially available organic and conventional tomatoes, lettuces, and carrots to determine if there were any systematic differences in nitrogen isotope signatures associated with the cultivation system. The study demonstrated that for certain crop types bulk nitrogen isotope analysis could infer if synthetic nitrogen fertilizers were likely to have been applied. Tomato and lettuce datasets showed significant differences between the mean $\delta^{15}N$ values of organic and conventional samples, while no significant differences were found between organic and conventional carrots. The authors hypothesized that this may be because of the lower nitrogen requirements of carrots and consequently generally lower levels of fertilizer application, or because carrots are field grown crops whereas tomatoes and lettuces are often grown under protected conditions (i.e., in greenhouses) with more frequent fertilization. The study concluded that $\delta^{15}N$ composition of a questioned sample could not provide unequivocal evidence of organic/conventional production, but could be used as supporting evidence when there was suspicion of substitution.

Rogers (2008) explored the feasibility of nitrogen and carbon isotope analysis as a screening tool to discriminate between commercial organic and conventional vegetables from New Zealand. The study included a wide range of fast-growing (<80 days) and slow-growing (>80 days) vegetables and found that nitrogen isotopes could be used to differentiate between organic and conventional produce as long as the vegetables were not nitrogen fixing (e.g., legumes). Discrimination was more successful for the fast-growing crops such as tomatoes, broccoli, and zucchini than for slow-growing crops such as pumpkin, eggplant, potatoes, and corn (maize). Rogers (2008) speculated that long cultivation times would result in ^{15}N being used by soil microorganisms *in situ*, which would diminish the differences in $\delta^{15}N$ values between nitrogen pools available for crops.

A survey conducted by Šturm and Lojen (2011) applied nitrogen isotope analysis to discriminate between commercial organic and conventional vegetables from the Slovenian market and found distinctive differences in the mean $\delta^{15}N$ values (up to 6.3‰) between organically and conventionally grown vegetables. However, complete discrimination was only possible for 5 out of 13 types of vegetables (endive, rocket, leek, potato, and chicory). For the remaining 8 types of vegetables (parsley, tomato, sweet pepper, garlic, onion, kohlrabi, cauliflower, and carrot), differentiation was not possible due to considerable overlap of $\delta^{15}N$ values between organic and conventional samples, despite a significant difference in the means. The authors concluded that nitrogen isotope analysis could be a reliable indicator of organic production, but has to be validated with additional background information. In a comprehensive study of crop/fertilizer correlation (Verenitch and Mazumder 2015), bulk nitrogen isotope analysis was performed on over 1,000 samples of 40 commercial crop types, organically and conventionally grown, along with the samples of nitrogen fertilizers used for their production. These samples were collected from various locations in western North America (Seattle, WA, USA and Victoria, BC, Canada). Twenty-five of the crop types (basil, broccoli, carrots, cauliflower, cabbage, celery, chard, cucumber, parsley, leeks, lettuce, kale, radish, pepper, tomatoes, spinach, zucchini, etc.) had distinctive differences between the $\delta^{15}N$ values of organic and conventional samples. Fifteen of the crop types (avocado, banana, beetroot, blueberries, corn/maize, garlic, green beans, ginger, lemon, lime, yellow onion, mushrooms, pineapple, sprouts, and strawberries) could not be differentiated due to extensive overlap of the $\delta^{15}N$ values of organic and conventional samples.

Although the nitrogen isotope composition of certain crop types can differentiate between synthetic and organic fertilization regimes, the overlapping $\delta^{15}N$ values observed for other crops highlights the limitation of this approach. The problem becomes more complex when organic cultivation practices use nitrogen-fixing plants in order to increase the nitrogen content of soil. Nitrogen-fixing plants derive most of their nitrogen from air and have $\delta^{15}N$ values close to that of air nitrogen ($\delta^{15}N_{AIR} = 0‰$), which is similar to the $\delta^{15}N$ values of synthetic nitrogen fertilizers. To complicate further matters, conventional cropping systems may use animal manures as the main nitrogen source, which would diminish any differences in $\delta^{15}N$ values between organic and conventional crops (Bateman et al. 2005). In contrast, greater differences between the $\delta^{15}N$ values of organic and conventional growing regimes are likely in crops with high nitrogen requirements, crops grown under controlled greenhouse conditions, or crops grown hydroponically as compared to field-grown crops with relatively low nitrogen requirements or long growth cycles (Bateman et al. 2007; Rogers 2008).

In order to achieve more effective discrimination between organic and conventional cultivation, a number of studies have analyzed multiple stable isotopes (i.e., $\delta^{15}N$ with δ^2H, $\delta^{13}C$, $\delta^{18}O$, and/or $\delta^{34}S$), often combining these measurements with elemental and/or physicochemical analysis and chemometrics. Stable carbon isotope composition is considered a reliable indicator of C_3/C_4 photosynthetic pathway, environment, and climate (see Chapter 3). Agricultural management practices have also been reported to affect the carbon isotope composition of a crop—for example, crops fertilized with higher amounts of N fertilizer had enriched $\delta^{13}C$ values compared to crops that received lower or no N fertilization (Jenkinson et al. 1995). Fertilization in conventional agricultural systems often involves higher N supplies, which may result in different $\delta^{13}C$ compositions of crops from conventional and organic systems. In addition, different types of fertilizer (organic/synthetic) have been proposed to alter soil microbial activity and soil respiration rates, which in turn would affect $\delta^{13}C$ values of soil CO_2 that would be available to crops (Georgi et al. 2005). Some studies have suggested that carbon isotope composition could also be a good indicator of greenhouse cultivation since greenhouses are often heated by combustion of natural gas (methane or propane), which has very depleted $\delta^{13}C$ values, as does the CO_2 produced by combustion. This effect has been shown to produce crops with depleted $\delta^{13}C$ values, often lower than -35% (Schmidt et al. 2005; Rogers 2008). The size, density, and growth rates of crops, which often differ between organic and conventional cultivation systems, affect water uptake and evapotranspiration of crops, which in turn would have impact on the δ^2H and $\delta^{18}O$ values of leaf water (Barbour 2007). Differences in the $\delta^{34}S$ compositions of organic and conventional fertilizers can be reflected in the $\delta^{34}S$ values of crops. Organic fertilizers, which are often marine-derived, can have significantly enriched $\delta^{34}S$ compositions compared to synthetic sulfate fertilizers derived from sulfuric acid (Schmidt et al. 2005; Camin et al. 2011).

Georgi et al. (2005) analyzed bulk C, N, and S isotope compositions of cabbage, onion, lettuce, and *Chinese cabbage* together with the O isotopic composition of leaf water. Organic vegetables were found to have lower $\delta^{13}C$ values and to be significantly ^{15}N-enriched compared to conventional counterparts. The difference in the carbon isotope compositions was attributed to the higher microbial activity in organic fields, which caused higher soil respiration with lower $\delta^{13}C$ values. The authors hypothesized that higher N availability in conventional crops could also increase the rate of photosynthesis, which would lower discrimination of the enzyme Rubisco against $^{13}CO_2$. Schmidt et al. (2005) analyzed C, N, and S isotopes to discriminate organic and conventional lettuce, onions, cabbage, *Chinese cabbage*, and wheat. Rapisarda et al. (2010) performed multi-element H, C, O, N, and S isotope analyses to study the differences in the isotope compositions between organic and conventional oranges. Both of these studies, however, reported that only $\delta^{15}N$ composition provided discrimination between organic and conventional samples.

Camin et al. (2011) combined multi-element H, C, O, N, and S isotope analyses with a wide range of chemical/physical characteristics (e.g., pH, total soluble solids, colorimetric characteristics, antioxidant activity, the content of total nitrogen, organic acids, sugars, anthocyanins, and polyphenols) in order to find suitable markers for the authentication of organic oranges, clementines, strawberries, and peaches. The combination of nitrogen isotope values with several other parameters (i.e., the content of ascorbic acid and total soluble solids) was shown to be the most successful for distinguishing between organic and conventional fruits. Other isotopic parameters were shown to be less effective for discrimination, being more affected by cultivar, year, and site of production. Carbon isotope composition was found to be statistically significant for discriminating between organic and conventional peaches and strawberries, but not for oranges or clementines. The authors found large variations in $\delta^{13}C$ values between plant species, cultivars, and cultivation sites, which negated carbon isotope composition as a reliable marker for organic authentication. The effects of production system on the hydrogen and oxygen isotope ratios of crops were also found to be inconsistent. δ^2H composition was able to differentiate between certain varieties of organic and conventional strawberries, while $\delta^{18}O$ composition of juice water was found to be significant for distinguishing the production origin of some peach varieties. The authors explained these differences on the basis of the different microclimates of the production area and different cultivation density, which affect evapotranspiration resulting in isotopic differences.

A study using multiple isotopes revealed that δ^2H composition was a valuable marker for the authentication of certain organic cereals (Laursen et al. 2013). This was the only parameter that allowed complete discrimination of organic and conventional wheat and barley, but was less successful for other crop types (e.g., potatoes, faba beans). As in previous studies, $\delta^{15}N$ values were found to be indicative of the use of animal manure in non-leguminous crops: wheat, barley, and potato. This study was the first to apply MC-ICP-MS analysis of Mg isotopes ($\delta^{25}Mg$ and $\delta^{26}Mg$) for organic produce authentication based on the hypothesis that higher N uptake would result in higher chlorophyll content in conventional crops. This could result in lower $\delta^{25}Mg$ and $\delta^{26}Mg$ values compared to organic crops due to preferential incorporation of the lighter Mg isotope (^{24}Mg) into chlorophyll molecules (Black et al. 2008). The result of the study confirmed the hypothesis for conventional and organic wheat cultivated on sandy soils, but the method was unsuccessful for crops grown on loamy soils.

Most studies of organic crop authentication have focused on isotope analysis of bulk plant tissues, which reflect all of the compounds present in the bulk substance. If the isotope composition of only certain compounds differ significantly between organic and conventional crops, these differences may not be obvious in the bulk isotope composition. In contrast, compound-specific isotope analysis (CSIA) allows the measurement of stable isotope signatures of specific compounds potentially providing more in-depth information of the physiological and metabolic profile of a plant. Several emerging studies have explored CSIA approaches to identify potential markers for discrimination between organic and conventional crops (Table 12.2). Rapisarda et al. (2005) promoted the nitrogen isotope analysis of amino acids ($\delta^{15}N_{AA}$) and pulp proteins ($\delta^{15}N_{PULP}$) from juice to discriminate between organic and conventional oranges. This compound-specific approach was combined with the analysis of elemental composition, total nitrogen, and synephrine concentration, together with a range of physicochemical markers. The authors reported that $\delta^{15}N_{AA}$ and $\delta^{15}N_{PULP}$ values were significantly higher in organic fruits. Significant differences were also found in other parameters linked to the nitrogen metabolism such as total nitrogen and synephrine contents. Linear discriminant analysis (LDA) identified $\delta^{15}N_{PULP}$, synephrine, and vitamin C contents as the most significant markers to differentiate between organic and conventional oranges. A further study (Rapisarda et al. 2010) combined multi-element and isotope analysis of orange pulp (H, C, N, and S), oxygen isotope analyses of juice water, analysis of physicochemical parameters, and compound-specific nitrogen and carbon isotope analysis of amino acids ($\delta^{13}C_{AA}$ and $\delta^{15}N_{AA}$). Oranges from organically managed plots had statistically higher $\delta^{15}N$ (bulk) and $\delta^{15}N_{AA}$ values than conventional samples, confirming the results of the previous study.

Bizjak Bat et al. (2012) applied a combination of multi-element analysis, selected chemical and physical parameters (mass, antioxidant activity, content of ascorbic acid, and total phenols), bulk isotope analysis of pulp ($\delta^{13}C$, $\delta^{15}N$) and fruit juice (δ^2H, $\delta^{18}O$) as well as compound-specific carbon isotope analysis of sugars and proteins ($\delta^{13}C_{SUG}$ and $\delta^{13}C_{PROT}$, respectively) and nitrogen isotope analysis of proteins ($\delta^{15}N_{PROT}$) to differentiate the botanical, geographical, and production origins of Slovenian apples. The greatest differences in the mean values between organic and conventional apples were observed in $\delta^{13}C_{PULP}$, ascorbic acid content, and S and Ca mass fractions. The difference was, however, statistically significant only for the mean amount of ascorbic acid. These data were subjected to LDA (see Chapter 5), which revealed that $\delta^{13}C_{PROT}$, $\delta^{15}N_{PROT}$, and antioxidant activity were the most important variables for differentiating organic and conventional apples. In the case of $\delta^{13}C_{PROT}$, the authors concluded that factors such as variety, year, or site of production may influence the values more so than agricultural practice. Paolini et al. (2015) tested carbon and nitrogen isotope analysis of amino acids for potential discrimination between organically and conventionally grown wheat. The results of this study demonstrated that the combination of $\delta^{13}C$ and $\delta^{15}N$ signatures of 10 amino acids (in particular, $\delta^{13}C$ values of glutamic acid and glutamine), could improve the discrimination between conventional and organic wheat in comparison with bulk isotope analysis. The authors concluded that compound-specific stable isotope analysis of amino acids had the potential to support and improve control procedures in the organic sector.

Mihailova et al. (2014) described a compound-specific approach to oxygen isotope analysis of plant-derived nitrate ($\delta^{18}O_{NO3}$) for the discrimination between organic and conventional crops, based on the *denitrifier method,* used for stable isotope analysis of nitrate in fresh water and seawater (Sigman et al. 2001; Casciotti et al. 2002). The method involves isotope ratio analysis of nitrous oxide generated from sample nitrate by cultured denitrifying bacteria. Discrimination between organic and conventional samples using this approach was based on the differences in the nitrate sources and different $\delta^{18}O_{NO3}$ compositions of nitrate in organic and conventional cultivation systems. The oxygen isotope composition of synthetic nitrate fertilizers reflects the oxygen isotope composition of air oxygen from which they are produced, and is generally in the range between +17‰ and +25‰ (Amberger and Schmidt 1987). In contrast, nitrate produced in soil during microbial nitrification of organic fertilizers derives oxygen atoms from two sources, air and soil water (Anderson and Hooper 1983; Kumar et al. 1983), and has lower $\delta^{18}O$ values in the range of −10 to

+10‰ (Kendall 1998; Kendall et al. 2007). Stable isotope analysis of plant-derived nitrate has been reported significantly to improve the discrimination between organic and conventional potatoes from the field trials when compared to bulk isotope analysis (Laursen et al. 2013). Mihailova et al. (2014) evaluated the capability of the oxygen isotope analysis of plant-derived nitrate for the discrimination between conventional and organic retail potato samples from the UK, Jersey, Israel, and Egypt as well as between conventional and organic retail tomato samples from the UK. Conventional potatoes and tomatoes showed significantly higher $\delta^{18}O_{NO3}$ values compared to organic counterparts. LDA of retail potato samples using $\delta^{15}N_{NO3}$ and $\delta^{18}O_{NO3}$ as predictor variables resulted in approximately 85% correct classification. The authors concluded that oxygen isotope analysis of plant-derived nitrate has a potential to differentiate between organically and conventionally grown crops.

12.4 Discrimination between organic and conventional products

12.4.1 Meats

Confirming the organic or conventional production origin of flesh foods and animal products is perhaps more challenging than confirming the production of vegetable and fruit products. Few aspects of organic livestock production, necessary to comply with the regulations, are likely to produce a robust analytical "marker" (Gewin 2004). These would include day-to-day animal husbandry, humane treatment, and welfare standards, such as access to space, free-grazing, and foraging on land where exposure to synthetic pesticides, herbicides, and fertilizers is minimized. Similarly, the use of veterinary drugs such as antibiotics should be severely restricted in organic meat production and only used in the most essential cases for treatment, to avoid the spread of disease, or prevent animal suffering (Commission Regulation EC No. 889/2008). Veterinary residues are generally not a reliable marker for substitution of organic meat with conventionally-produced meat due to low concentrations in lean muscle tissue. An exception is a method using fluorescence microscopy (Kelly et al. 2006) that was able to observe permanent serial or prophylactic dosing of tetracycline antibiotic residues laid down in chicken and porcine bone cross-sections. Nevertheless, the absence of tetracycline in bone would not provide conclusive proof of organic rearing, because of the plethora of veterinary drugs that do not adhere to bone or fluoresce. The fluorescence microscope method also requires the meat sample to be obtained "on-the-bone."

Stable isotope analysis of meats might reasonably be expected to reveal differences in feeding practice (see Chapter 6). Under normal circumstances organic animals should be grazed on pasture (grasses and other crops) under organic management practices. Like the production of organic crops (discussed above), the use of synthetic fertilizers/herbicides is precluded and nitrogen supplies are usually maintained through the use of manures, composts, crop rotation (that includes the use of leguminous N-fixing plants), and the application of other plant or animal-based fertilizers. Unfortunately (as noted previously), the natural and agricultural variation of nitrogen isotope values in grazing and fodder crops means that any generalized rules for dietary effects on the stable nitrogen isotope composition of organic and conventional feed are confounded. Characterizing the transfer of nitrogen isotope "signals" from crops to animal tissues through diet and metabolism is similarly challenging, especially when compared to conventional "non-intensive" or "low-input" systems—e.g., "certified grass-fed" or "free range."

Stable carbon isotope ratios are mostly affected by the type of photosynthetic carbon fixation in plants (see Chapter 3) but can also provide additional information about water use efficiency and water stress in plants, both of which can vary substantially in conventional and organic agricultural systems. To differentiate between the diets of conventionally and organically reared livestock, the observed differences in carbon isotope values in animal tissues derived from maize/corn (C_4) feeding have been the most widely exploited (Bahar et al. 2005). However, the long-term reliability and stability of interpretation using a database of carbon isotope values must be carefully monitored as discussed below. Notwithstanding these challenges, a significant number of research papers have reported varying degrees of success in distinguishing organic and conventional animal rearing and meat production using isotopic analyses.

The use of carbon stable isotope analysis has been reported to distinguish between beef produced in conventional and organic systems in northern and southern regions of Germany (Boner and Förstel 2004). The authors argued that because of limits imposed on the proportion of fodder in the organic cattle diet, a de-fatted meat $\delta^{13}C$ value above −20‰ was not consistent with organic farming practices. It should be noted that whilst free-grazing and foraging is advocated in organic beef production, it does not exclude the use of maize fodder cultivated under organic systems, especially during winter months. Furthermore, intensive, conventional livestock farming employs a wide variety of ingredients including C_3 plant-derived materials such as wheat, soy, and brewing waste such as "spent barley." This also includes other ingredients that might not be part of a cow's natural diet, such as fishmeal. Such a diverse range of carbohydrate and protein sources means that great care needs to be taken when basing discrimination between organic and conventional meat production on differences in carbon isotope signatures of C_3 and C_4 plant materials.

The seasonality of concentrate, silage, and fodder inputs must also be considered when assessing the production authenticity of beef. Bahar et al. (2008) measured a total of 242 retail organic and conventional Irish beef samples produced over a one-year period and observed relatively little variation in $\delta^{13}C$ and $\delta^{15}N$ values for organic beef. In contrast, conventional beef showed pronounced variation with significant enrichment in $\delta^{13}C$ signatures between December and June (>2‰), but with little variation in $\delta^{15}N$ values. It should be noted that tail hair has been shown to record seasonal changes in the diet of beef cattle for periods of up to one year prior to slaughter (Osorio et al. 2011) and provide a reliable alternative to meat as a dietary record (Yanagi et al. 2012).

Although nitrogen isotope signatures might be expected to vary between conventional and organic beef production, studies have shown that this is unpredictable. Foraging of leguminous nitrogen-fixing cover crops such as clover might impart a low $\delta^{15}N$ value characteristic of synthetic nitrogen fertilizer (Schmidt et al. 2005). Conversely, the feeding of fishmeal in conventional livestock production could produce a *trophic shift* (see Chapter 4), giving rise to higher $\delta^{15}N$ values typical of manure based fertilizers.

Concerns about the recycling of animal tissues into feed [mainly originating from the bovine spongiform encephalopathy (BSE) crisis in the United Kingdom in the 1990s] led to a number of prohibitions on what may not be used in animal feed [Regulation (EC) No. 999/2001]. The use of mammalian meat and bone meal (MBM) was banned because of its link to the spread of BSE and the ban has subsequently been extended to virtually all types of processed animal protein (PAP). The use of nitrogen isotopes to detect the use of MBM and PAP feeding has been proposed for beef cattle (Delgado and Garcia 2001) and chickens (Carrijo et al. 2006).

A combination of carbon, nitrogen, and sulfur stable isotope ratios was used to differentiate between organic and conventional beef reared in Ireland (Schmidt et al. 2005). The authors proposed that ingestion of grains and/or seeds in feed-concentrates gave rise to the slightly enriched carbon isotope signatures observed in conventional beef or that it could be attributable to a maize ration in the feed. The Irish organic beef samples were also found to have depleted $\delta^{15}N$ values attributed to grazing on leguminous cover crops. The organic beef was found to have slightly enriched $\delta^{34}S$ values than conventional beef on average (+7.9 ± 0.6‰ *vs.* +7.2 ± 0.4‰), which may be attributable to a supplementary diet of seaweed, which is enriched in ^{34}S compared to terrestrial feed sources.

12.4.2 Milk

Variation in the carbon, nitrogen, and sulfur isotopes observed in bovine muscle tissue, attributable to differences in organic and conventional feeding regimes, may also be mirrored in animal products such as bovine milk (Kornexl et al. 1997; Camin et al. 2008). A 12-month study from Germany demonstrated that the carbon isotope ratios of milk fat (n = 35 samples) differed by approximately 1.5‰ between conventional and organic production (Molkentin and Giesemann 2007). A much larger difference (4.5‰) was observed when production dates were matched between organic and conventional milk samples. The difference in $\delta^{13}C$ values was attributable to maize fed to conventional dairy herds. The stable isotope ratios of nitrogen and sulfur did not provide a basis to discriminate the method of production of the milk. A larger German study (Molkentin 2009) over an 18-month period, with a collection of 286 milk samples, demonstrated that a combination of carbon isotope ratios and α-linolenic acid (C18:3ω3) concentrations allowed threshold values for retail organic milk to be established. The German organic retail milk exhibited $\delta^{13}C$ values below −26.5‰ and C18:3ω3 concentrations above 0.50% w/v. A strong negative correlation between C18:3ω3 and $\delta^{13}C$ signatures meant that it would be difficult to conceal mis-description of conventional milk as organic to take advantage of the higher retail price. A follow-up study (Molketin and Giesemann 2010) of 120 German milk samples, collected over an 18-month period, demonstrated the soundness of the proposed authenticity limits for $\delta^{13}C$ values and C18:3ω3 concentrations in milk fat. The authors also demonstrated correlations between $\delta^{13}C$ and $\delta^{15}N$ values in milk protein and α-linolenic acid concentrations that improved the possibility of detecting fraudulent mixtures of organic milk fat with conventional skimmed milk.

12.4.3 Eggs

Organic egg farming can be lucrative because organic eggs can have a retail price up to four times the price of caged eggs. Consumer preference for organic eggs can be attributed to food safety (e.g., no hormones or antibiotics fed to chickens, no pesticide residues in organic poultry feeds) and to animal welfare (e.g., organic, cage-free eggs should come from hens with access to pasture). Two types of fraud have been identified in the organic egg industry. Farmers may simply package eggs from caged, barn, or free-range farming systems under an organic label. Alternatively, farmers may substitute some or all of the organic feed with conventional feed to lower costs while still offering the hen the benefits of a cage-free lifestyle. Apart from the lack of hormones, antibiotics, and other chemical supplements given to conventional laying hens, organic hens are primarily differentiated by their free ranging status and consumption of organic feeds.

Access to forage is a very important aspect of both free-range and organic egg production, acknowledging the ability of the hen to forage at different stages of production. Stocking densities can also have a significant effect on dietary inputs. For example, intensive, free-range, and organic chicken meat production regulations stipulate significantly different stocking levels in the UK; stocking densities are 17, 13, and 10 birds/m² for intensive, free-range, and organic (free-range), respectively (Miele 2011). The free-range and organic stocking levels are still relatively dense and may result in a complete lack of foraging at later stages of the bird's life when body mass is relatively high. In a study of 25 UK commercial flocks, only 30% of the laying hens typically used the "free" range available to them (Whay et al. 2007). Regardless of consumer expectations of what free-range and organic labelling promise, if such ranging or foraging figures are common amongst modern free-range and organic egg production, developing generalized rules for the trophic effects on diet to discriminate production systems will prove challenging.

The use of stable isotope analysis to differentiate between caged, barn, free-range, and organic systems has been demonstrated for 18 brands of retail chicken eggs produced in New Zealand (Rogers 2009). This study showed that free-range and organic eggs generally exhibited relatively enriched δ^{15}N values (up to 4‰) compared to caged and barn-laid eggs; this was attributed to a trophic shift from foraged dietary inputs such as insects. The study identified one sample of "free-range" eggs and two samples of "organic" eggs that had δ^{15}N values typical of caged or barn-laid eggs. The author suggested that either the eggs were mislabeled or there was insufficient insect forage available to the hens to produce the trophic shift in the egg δ^{15}N values of typical organic and free-range farms.

Feed manufacturers can incorporate up to 15% animal protein into New Zealand organic poultry feeds, while in EU countries animal products (muscle tissue, blood, bone, offal) are strictly prohibited in animal feeds. The use of animal protein in poultry feeds in some countries will complicate any global baseline to identify organic eggs, although stable isotope analysis of eggs from the Netherlands and New Zealand provides ranges for different feed types (plant, and animal protein), as shown in Figure 12.1. Generally organic eggs separate isotopically from caged or barn-raised eggs based on δ^{15}N values, although the separation of free-range eggs is less clear due to the confounding use of pastures for both free-range and organic hens, which may provide similar foraging opportunities and isotopic overlap of diet.

12.4.4 Fish

Although the substitution of fish species is one of the most common frauds affecting consumers (Jacquet and Pauly 2008), the control of aquaculture production origin (wild-caught *vs.* farmed or organically farmed) is another major consideration for consumers. Stable isotope analysis combined with various techniques such as fatty acid profiling and multi-element analysis has been extensively reported as a means to identify dietary differences in farmed *vs.* wild-caught fish—for example salmon (Dempson and Power 2004; Thomas et al. 2008; Schröder and De Leaniz 2011), sea-bass

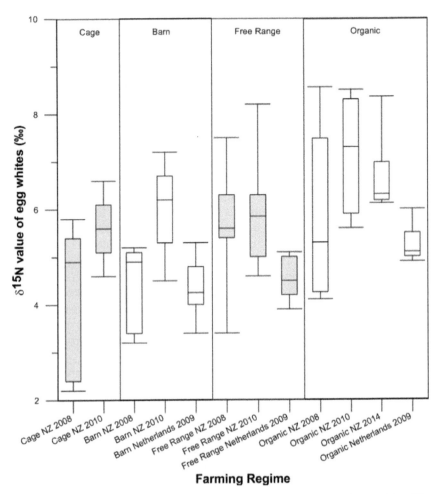

Figure 12.1 Box and whisker plot of $\delta^{15}N$ values of New Zealand and Dutch egg whites. The end whiskers represent the minimum and maximum range of $\delta^{15}N$ values, the box indicates the 1st and 3rd quartiles; the line inside the box indicates the median value. Data from K. Rogers.

(Bell et al. 2007), gilthead sea bream (Rojas et al. 2007), and blue-fin tuna (Vizzini et al. 2010) (see also Chapter 6).

In comparison, research specifically aimed to identify the organic production of fish is relatively limited. Comparable to meat production, organic certification standards have been developed to regulate the source of fish feed that should contain negligible residues of pesticides and herbicides. Fishmeal should also be derived from wild-caught fish within a similar geographical location and be the by-product of fish caught for human consumption (e.g., filleting waste). The addition of antibiotics and supplemental amino acids and pigments, such as astaxanthin (to color salmon muscle tissue) are proscribed from an organic diet. Stocking densities are also lower for organic salmon production.

Carbon and nitrogen stable isotopes ratios were determined in raw fillets of wild, conventionally, and organically aquacultured Atlantic salmon with the principal aim of developing a method for the identification of "organic" fish (Molkentin et al. 2007). Relatively high $\delta^{15}N$ values allowed organically farmed salmon to be differentiated from wild salmon, but not from conventionally farmed salmon. The authors acknowledged that to be generally applicable, a method would need to allow for variations in feed composition. A follow-up study of 130 German market trout and salmon fillets and wild-caught salmon measured the $\delta^{13}C$ and $\delta^{15}N$ values of de-fatted dry matter and permitted conventional and organic salmon/trout to be distinguished (Molkentin et al. 2015). An additional analysis of the $\delta^{13}C$ composition of fish lipid was required for the correct identification of wild *vs.* organic salmon. The authors reported that although isotope analysis provided a means to separate the three production methods, fatty acid analysis could also resolve the three production types based on the linoleic acid concentration in the fish lipids, which was lowest in wild-caught and highest in conventional salmon.

12.5 Conclusion

A large number of studies of organic *vs.* conventional crops have shown that stable isotope techniques are challenged by the diversity of fertilizer practices of both organic and conventional plant-production systems, by the intricacy of nitrogen-cycling, and by isotope fractionations between different nitrogen-pools. Emerging compound-specific isotope techniques, in combination with multi-marker strategies and multivariate statistical tests, are more promising tools for the differentiation between organic and conventional plant-based food production (Camin et al. 2011; Laursen et al. 2013; Capuano et al. 2013; Laursen et al. 2014). Large sample sets encompassing real-life variations and collaborative validation of the methods are pre-requisites for any techniques to be useful in routine organic food control, in conjunction with certification and inspection systems.

It would appear that for some countries and some commodities, such as German organic milk production, feed inputs to dairy cattle are sufficiently well regulated to establish carbon and nitrogen stable isotope (and supporting chemical concentration) values that can act as authenticity thresholds. The diversity of globalized organic food production and international trade sources for many organic products will mean that concerted and significant effort is needed to maintain robust databases that can be routinely used to monitor the authenticity of organic crops, meat, fish, and animal products. In an analogous way to organic crop cultivation, one of the most promising ways forward may be the measurement of compound-specific nitrogen stable isotope ratios of individual amino-acids, which exhibit mutually exclusive basal and dietary input trophic shifts (Chikaraishi et al. 2009).

References

Afrisco Standards for Organic Production. Version 9. Issue: July, 2011. Afrisco Certified Organic, South Africa, pp. 68.

Amberger, A. and H. L. Schmidt. 1987. Natürliche Isotopengehalte von Nitrat als Indikatoren für dessen Herkunft. *Geochimica et Cosmochimica Acta* 51: 2699–2705.

Anderson, K. K. and A. B. Hooper. 1983. O_2 and H_2O are each the source of one O in NO_2^- produced from NH_3 by *Nitrosomonas*: ^{15}N-NMR evidence. *FEBS Letters* 164: 236–240.

Bahar, B., F. J. Monahan, A. P. Moloney, P. O'Kiely, C. M. Scrimgeour and O. Schmidt. 2005. Alteration of the carbon and nitrogen stable isotope composition of beef by substitution of grass silage with maize silage. *Rapid Communications in Mass Spectrometry* 19: 1937–1942.

Bahar, B., O. Schmidt, A. P. Moloney, C. M. Scrimgeour, I. S. Begley and F. J. Monahan. 2008. Seasonal variation in the C, N and S stable isotope composition of retail organic and conventional Irish beef. *Food Chemistry* 106: 1299–1305.

Barbour, M. M. 2007. Stable oxygen isotope composition of plant tissue: A review. *Functional Plant Biology* 34: 83–94.

Bateman, A. S. and S. D. Kelly. 2007. Fertilizer nitrogen isotope signatures. *Isotopes in Environmental and Health Studies* 43: 237–247.

Bateman, A. S., S. D. Kelly and T. D. Jickells. 2005. Nitrogen isotope relationships between crops and fertilizer: Implications for using nitrogen isotope analysis as an indicator of agricultural regime. *Journal of Agricultural and Food Chemistry* 53: 5760–5765.

Bateman, A. S., S. D. Kelly and M. Woolfe. 2007. Nitrogen isotope composition of organically and conventionally grown crops. *Journal of Agricultural and Food Chemistry* 55: 2664–2670.

Beck, A. 2012. Lessons from fraud cases in organic markets. European Organic Regulations (EC) No. 834/2007, 889/2008 and 1235/2008—An evaluation of the first three years, looking for further development, 23–24.

Bell, J. G., T. Preston, R. J. Henderson, F. Strachan, J. E. Bron, K. Cooper and D. J. Morrison. 2007. Discrimination of wild and cultured European sea bass (*Dicentrarchus labrax*) using chemical and isotopic analyses. *Journal of Agricultural and Food Chemistry* 55: 5934–5941.

Bizjak Bat, K. B., R. Vidrih, M. Nečemer, B. M. Vodopivec, I. Mulič, P. Kump and N. Ogrinc. 2012. Characterization of Slovenian apples with respect to their botanical and geographical origin and agricultural production practice. *Food Technology and Biotechnology* 50: 107–116.

Black, J. R., E. Epstein, W. D. Rains, Q. Z. Yin and W. H. Casey. 2008. Magnesium-isotope fractionation during plant growth. *Environmental Science and Technology* 42: 7831–7836.

Boner, M. and H. Förstel. 2004. Stable isotope variation as a tool to trace the authenticity of beef. *Analytical and Bioanalytical Chemistry* 378: 301–310.

Bourn, D. and J. Prescott. 2002. A comparison of the nutritional value, sensory qualities, and food safety of organically and conventionally produced foods. *Critical Reviews in Food Science and Nutrition* 42: 1–34.

Brown, C. and M. Sperow. 2005. Examining the cost of an all-organic diet. *Journal of Food Distribution Research* 36: 20–26.

Canada Organic Products Regulations, SOR/2009-176. Enabling Act: Canada Agricultural Products Act: R.S.C. 1985, c. 20 (4th Supp.). Current to May 25, 2015. Last amended on September 30, 2013.

Canada Agricultural Products Act: R.S.C. 1985, c. 20 (4th Supp.). Government of Canada. Current to May 25, 2015. Last amended on February 26, 2015.

CAN/CGSB-32.310-2006, Organic Production Systems General Principles and Management Standards. Government of Canada. Canadian General Standards Board. Supersedes CAN/CGSB-32.310-99. Last amended June 2011.

CAN/CGSB-32.311-2006, Organic Production Systems Permitted Substances Lists. Government of Canada. Canadian General Standards Board. Supersedes part of CAN/CGSB-32.310-99. Last amended June 2011.

Camin, F., A. Moschella, F. Miselli, B. Parisi, G. Versini, P. Ranalli and P. Bagnaresi. 2007. Evaluation of markers for the traceability of potato tubers grown in an organic *vs.* conventional regime. *Journal of the Science of Food and Agriculture* 87: 1330–1336.

Camin, F., M. Perini, G. Colombari, L. Bontempo and G. Versini. 2008. Influence of dietary composition on the carbon, nitrogen, oxygen and hydrogen stable isotope ratios of milk. *Rapid Communications in Mass Spectrometry* 22: 1690–1696.

Camin, F., M. Perini, L. Bontempo, S. Fabroni, W. Faedi, S. Magnani, G. Baruzzi, M. Bonoli, M. R. Tabilio, S. Musmeci, A. Rossmann, S. D. Kelly and P. Rapisarda. 2011. Potential isotopic and chemical markers for characterising organic fruits. *Food Chemistry* 125: 1072–1082.

Capuano, E., R. Boerrigter-Eenling, G. van der Veer and S. M. van Ruth. 2013. Analytical authentication of organic products: An overview of markers. *Journal of the Science of Food and Agriculture* 93: 12–28.

Carrijo, A. S., A. C. Pezzato, C. Ducatti, J. R. Sartori, L. I. Trinca and E. T. Silva. 2006. Traceability of bovine meat and bone meal in poultry by stable isotope analysis. *Revista Brasileira de Ciência Avícola* 8: 63–68.

Casciotti, K. L., D. M. Sigman, M. G. Hastings, J. K. Böhlke and A. Hilkert. 2002. Measurement of the oxygen isotopic composition of nitrate in seawater and freshwater using the denitrifier method. *Analytical Chemistry* 74: 4905–4912.

Chikaraishi, Y., N. O. Ogawa, Y. Kashiyama, Y. Takano, H. Suga, A. Tomitani, H. Miyashita, H. Kitazato and N. Ohkouchi. 2009. Determination of aquatic food-web structure based on compound-specific nitrogen isotopic composition of amino acids. *Limnology and Oceanography: Methods* 7: 740–750.

Choi, W., M. Arshad, S. Chang and T. Kim. 2006. Grain ^{15}N of crops applied with organic and chemical fertilizers in a four-year rotation. *Plant and Soil* 284: 165–174.

Choi, W. J., S. M. Lee, H. M. Ro, K. C. Kim and S. H. Yoo. 2002. Natural ^{15}N abundances of maize and soil amended with urea and composted pig manure. *Plant and Soil* 245: 223–232.

Choi, W. J., H. M. Ro and S. M. Lee. 2003. Natural ^{15}N abundances of inorganic nitrogen in soil treated with fertilizer and compost under changing soil moisture regimes. *Soil Biology and Biochemistry* 35: 1289–1298.

Chung, I. -M., J. -K. Kim, M. Prabakaran, J. -H. Yang and S. -H. Kim. 2015. Authenticity of Rice (*Oryza sativa* L.) geographical origin based on analysis of C, N, O and S stable isotope ratios: A preliminary case report in Korea, China, and Philippine. *Journal of the Science of Food and Agriculture* 96: 2433–2439.

Codex Alimentarius Commission. 2013. Guidelines for the production, processing, labelling and marketing of organically produced foods—CAC/GL 32-1999/Rev. 1 - 2001. pp. 34.

Commission Regulation (EC) No. 889/2008 of 5 September 2008. Laying down detailed rules for the implementation of Council Regulation (EC) No. 834/2007 on organic production and labelling of organic products with regard to organic production, labelling and control.

Commission Regulation (EC) No. 1235/2008 of 8 December 2008. Laying down detailed rules for the implementation of Council Regulation (EC) No. 834/2007 as regards the arrangements for imports of organic products from third countries.

Commission Regulation (EC) No 889/2008 of 5 September 2008. Laying down detailed rules for the implementation of Council Regulation (EC) No. 834/2007 on organic production and labelling of organic products with regard to organic production, labelling and control.

Commission Implementing Regulation (EU) 2015/131 of 23 January 2015 amending Regulation (EC) No. 1235/2008 laying down detailed rules for implementation of Council Regulation

(EC) No. 834/2007 as regards the arrangements for imports of organic products from third countries.

Council Regulation (EC) No. 834/2007 of 28 June 2007 on organic production and labelling of organic products and repealing Regulation (EEC) No. 2092/91.

Dangour, A. D., S. K. Dodhia, A. Hayter, E. Allen, K. Lock and R. Uauy. 2009. Nutritional quality of organic foods: A systematic review. *The American Journal of Clinical Nutrition* 90: 680–685.

del Amor, F. M., J. Navarro and P. M. Aparicio. 2008. Isotopic discrimination as a tool for organic farming certification in sweet pepper. *Journal of Environmental Quality* 37: 182–185.

Delgado, A. and N. Garcia. 2001. $\delta^{15}N$ and $\delta^{13}C$ analysis to identify cattle fed on feed containing animal proteins. A safety/quality index in meat, milk and cheese. pp. 1–11. In: *Proceedings of the 6th International Symposium on Food Authenticity and Safety*, Nantes, Eurofins.

Dempson, J. B. and M. Power. 2004. Use of stable isotopes to distinguish farmed from wild Atlantic salmon, *Salmo salar*. *Ecology of Freshwater Fish* 13: 176–184.

East African Standard EAS 456:2007. East African organic products standard. East African Community, first edition 2007, 39.

EWFC. 2016. http://www.ewfc.org/en/news/news-archive/31-organic-products-scandal-italy/53-organic-food-and-feed-fraud-in-italy-the-latest-facts/ (accessed July 2016).

Federal Bureau of Investigation. 2012a. Former President of Organic Fertilizer Company Pleads Guilty to Fraud in Connection with Selling Synthetic Fertilizer to Organic Farms. https://archives.fbi.gov/archives/sanfrancisco/press-releases/2012/former-president-of-organic-fertilizer-company-pleads-guilty-to-fraud-in-connection-with-selling-synthetic-fertilizer-to-organic-farms (accessed August 2016).

Federal Bureau of Investigation. 2012b. Owner of Kern County Fertilizer Business Sentenced for Organic Fertilizer Fraud. https://archives.fbi.gov/archives/sacramento/press-releases/2012/owner-of-kern-county-fertilizer-business-sentenced-for-organic-fertilizer-fraud (accessed August 2016).

Ferrante, A. 2012. Lesson learnt from the fraud case – "Gatto con gli stivali." European Organic Regulations (EC) No. 834/2007, 889/2008 and 1235/2008—An evaluation of the first three years, looking for further development, 26–27.

Flores, P., J. Fenoll and P. Hellín. 2007. The feasibility of using $\delta^{15}N$ and $\delta^{13}C$ values for discriminating between conventionally and organically fertilized pepper (*Capsicum annuum* L.). *Journal of Agricultural and Food Chemistry* 55: 5740–5745.

Flores, P., A. López, J. Fenoll, P. Hellín and S. Kelly. 2013. Classification of organic and conventional sweet peppers and lettuce using a combination of isotopic and bio-markers with multivariate analysis. *Journal of Food Composition and Analysis* 31: 217–225.

Flores, P., P. J. Murray, P. Hellín and J. Fenoll. 2011. Influence of N doses and form on ^{15}N natural abundance of pepper plants: Considerations for using $\delta^{15}N$ values as indicator of N source. *Journal of the Science of Food and Agriculture* 91: 2255–2258.

Food Safety News. 2011. USDA Caution on Organic Fraud from China. September 3, 2011. http://www.foodsafetynews.com/2011/09/usda-warns-of-chinese-organic-fraud/#.V7wFO010yM9 (accessed August 2106).

Freyer, H. and A. Aly. 1974. Nitrogen-15 variations in fertilizer nitrogen. *Journal of Environmental Quality* 3: 405–406.

Georgi, M., S. Voerkelius, A. Rossmann, J. Graßmann and W. H. Schnitzler. 2005. Multielement isotope ratios of vegetables from integrated and organic production. *Plant and Soil* 275: 93–100.

Gewin, V. 2004. Organic FAQs. *Nature* 428: 796–798.

Hoefkens, C., I. Vandekinderen, B. De Meulenaer, F. Devlieghere, K. Baert, I. Sioen, S. De Henauw, W. Verbeke and J. Van Camp. 2009. A literature-based comparison of nutrient and contaminant contents between organic and conventional vegetables and potatoes. *British Food Journal* 111: 1078–1097.

Huber, B. 2012. How to prevent fraud in the organic sector? European Organic Regulations (EC) No. 834/2007, 889/2008 and 1235/2008—An evaluation of the first three years, looking for further development, 24–25.

Huber, B., O. Schmid and C. Mannigel. 2015. Standards and regulations: Overview. pp. 125–133. *In*: H. Willer and K. Lernoud (eds.). *The World of Organic Agriculture. Statistics and Emerging Trends 2015*. FiBL-IFOAM Report. Research Institute of Organic Agriculture (FiBL), Frick, and IFOAM—Organics International, Bonn.

IFOAM OSEA Report. 2008. Development of a regional organic standard in East Africa 2005–2007. IFOAM, Bonn, Germany, pp. 28.

IFOAM. 2014. International Federation of Organic Agriculture Movements (IFOAM). The IFOAM Norms for Organic Production and Processing, IFOAM, Germany, pp. 132.

Independent. 2016. http://www.independent.com.mt/articles/2016-04-09/local-news/Malta-placed-at-centre-of-huge-organic-products-fraud-6736156074 (accessed July 2016).

Jacquet, J. L. and D. Pauly. 2008. Trade secrets: Renaming and mislabeling of seafood. *Marine Policy* 32: 309–318.

Jaenicke, E. C. and A. C. Carlson. 2015. Estimating and investigating organic premiums for retail-level food products. *Agribusiness* 31: 453–471.

Janssen, M. and U. Hamm. 2014. Governmental and private certification labels for organic food: Consumer attitudes and preferences in Germany. *Food Policy* 49: 437–448.

Jenkinson, D. S., K. Coleman and D. D. Harkness. 1995. The influence of fertilizer nitrogen and season on the carbon-13 abundance of wheat straw. *Plant and Soil* 171: 365–367.

Kelly, S. D. and A. S. Bateman. 2010. Comparison of mineral concentrations in commercially grown organic and conventional crops—Tomatoes (*Lycopersicon esculentum*) and lettuces (*Lactuca sativa*). *Food Chemistry* 119: 738–745.

Kelly, M., J. A. Tarbin, H. Ashwin and M. Sharman. 2006. Verification of compliance with organic meat production standards by detection of permitted and nonpermitted uses of veterinary medicines (tetracycline antibiotics). *Journal of Agricultural and Food Chemistry* 54: 1523–1529.

Kendall, C. 1998. Tracing nitrogen sources and cycles in catchments. pp. 519–576. *In*: C. Kendall and J. J. McDonnell (eds.). *Isotope Tracers in Catchment Hydrology*. Amsterdam: Elsevier.

Kendall, C., E. M. Elliott and S. D. Wankel. 2007. Tracing anthropogenic inputs of nitrogen to ecosystems. pp. 375–449. *In*: R. Michener and K. Lajtha (eds.). *Stable Isotopes in Ecology and Environmental Science*. Oxford: Blackwell.

Kornexl, B. E., T. Werner, A. Roßmann and H. L. Schmidt. 1997. Measurement of stable isotope abundances in milk and milk ingredients—a possible tool for origin assignment and quality control. *Zeitschrift für Lebensmitteluntersuchung und-Forschung A* 205: 19–24.

Kumar, S., D. J. D. Nicholas and E. H. Williams. 1983. Definitive ^{15}N NMR evidence that water serves as a source of 'O' during nitrite oxidation by *Nitrobacter agilis*. *FEBS Letters* 152: 71–74.

Laursen, K. H., A. Mihailova, S. D. Kelly, V. N. Epov, S. Berail, J. K. Schjoerring, O. F. X. Donard, E. H. Larsen, N. Pedentchouk, A. D. Marca-Bell, U. Halekoh, J. E. Olesen and S. Husted. 2013. Is it really organic?—Multi-isotopic analysis as a tool to discriminate between organic and conventional plants. *Food Chemistry* 141: 2812–2820.

Laursen, K. H., J. K. Schjoerring, S. D. Kelly and S. Husted. 2014. Authentication of organically grown plants—advantages and limitations of atomic spectroscopy for multi-element and stable isotope analysis. *TrAC Trends in Analytical Chemistry* 59: 73–82.

Lim, S. -S., W. -J. Choi, J. -H. Kwak, J. -W. Jung, S. Chang, H. -Y. Kim, K. -S. Yoon and S. -M. Choi. 2007. Nitrogen and carbon isotope responses of Chinese cabbage and chrysanthemum to the application of liquid pig manure. *Plant and Soil* 295: 67–77.

McCluskey, J. J. 2000. A game theoretic approach to organic foods: An analysis of asymmetric information and policy. *Agricultural and Resource Economics Review* 29: 1–9.

Miele, M. 2011. The taste of happiness: Free-range chicken. *Environment and Planning A* 43: 2076–2090.

Mihailova, A., N. Pedentchouk and S. D. Kelly. 2014. Stable isotope analysis of plant-derived nitrate—Novel method for discrimination between organically and conventionally grown vegetables. *Food Chemistry* 154: 238–245.

Molkentin, J. and A. Giesemann. 2007. Differentiation of organically and conventionally produced milk by stable isotope and fatty acid analysis. *Analytical and Bioanalytical Chemistry* 388: 297–305.

Molkentin, J., H. Meisel, I. Lehmann and H. Rehbein. 2007. Identification of organically farmed Atlantic salmon by analysis of stable isotopes and fatty acids. *European Food Research and Technology* 224: 535–543.

Molkentin, J. 2009. Authentication of organic milk using $\delta^{13}C$ and the α-linolenic acid content of milk fat. *Journal of Agricultural and Food Chemistry* 57: 785–790.

Molketin, J. and A. Giesemann. 2010. Follow-up of stable isotope analysis of organic *vs.* conventional milk. *Analytical and Bioanalytical Chemistry* 398: 1493–1500.

Molkentin, J., I. Lehmann, U. Ostermeyer and H. Rehbein. 2015. Traceability of organic fish–Authenticating the production origin of salmonids by chemical and isotopic analyses. *Food Control* 53: 55–66.

Nakano, A., Y. Uehara and A. Yamauchi. 2003. Effect of organic and inorganic fertigation on yields, $\delta^{15}N$ values, and $\delta^{13}C$ values of tomato (*Lycopersicon esculentum* Mill. cv. Saturn). *Plant and Soil* 255: 343–349.

Oberholtzer, L., C. Dimitri and C. Greene. 2005. Price premiums hold on as U.S. organic produce market expands. VGS-308-01, Economic Research Service/USDA, 1–22.

Osorio, M. T., A. P. Moloney, O. Schmidt and F. J. Monahan. 2011. Beef authentication and retrospective dietary verification using stable isotope ratio analysis of bovine muscle and tail hair. *Journal of Agricultural and Food Chemistry* 59: 3295–3305.

Paolini, M., L. Ziller, K. H. Laursen, S. Husted and F. Camin. 2015. Compound-specific $\delta^{15}N$ and $\delta^{13}C$ analyses of amino acids for potential discrimination between organically and conventionally grown wheat. *Journal of Agricultural and Food Chemistry* 63: 5841–5850.

Pidd, H. 2010. Fraudster who conned supermarkets with free-range egg scam jailed. Caged battery hens. *The Guardian*. Thursday 11 March 2010 20.58 GMT. https://www.theguardian.com/uk/2010/mar/11/egg-fraudster-supermarkets-free-range. (Accessed August 2106).

Piva, F. 2012. The fraud case in organic sector discovered in Italy in 2011. European Organic Regulations (EC) No. 834/2007, 889/2008 and 1235/2008—An evaluation of the first three years, looking for further development, 26.

Pretty, J. N., A. S. Ball, T. Lang and J. I. L. Morrison. 2005. Farm costs and food miles: An assessment of the full cost of the UK weekly food basket. *Food Policy* 30: 1–19.

Rapisarda, P., M. L. Calabretta, G. Romano and F. Intrigliolo. 2005. Nitrogen metabolism components as a tool to discriminate between organic and conventional citrus fruits. *Journal of Agricultural and Food Chemistry* 53: 2664–2669.

Rapisarda, P., F. Camin, S. Fabroni, M. Perini, B. Torrisi and F. Intrigliolo. 2010. Influence of different organic fertilizers on quality parameters and the $\delta^{15}N$, $\delta^{13}C$, $\delta^{2}H$, $\delta^{34}S$, and $\delta^{18}O$ values of orange fruit (*Citrus sinensis* L. osbeck). *Journal of Agricultural and Food Chemistry* 58: 3502–3506.

Regulation (EC) No. 999/2001 of the European Parliament and of the Council of 22 May 2001. Laying down rules for the prevention, control and eradication of certain transmissible spongiform encephalopathies. *Official Journal of the European Communities*, L 147/1, 31.5.2001, EN.

Rogers, K. M. 2008. Nitrogen isotopes as a screening tool to determine the growing regimen of some organic and nonorganic supermarket produce from New Zealand. *Journal of Agricultural and Food Chemistry* 56: 4078–4083.

Rogers, K. M. 2009. Stable isotopes as a tool to differentiate eggs laid by caged, barn, free range, and organic hens. *Journal of Agricultural and Food Chemistry* 57: 4236–4242.

Rojas, J. M. M., F. Serra, I. Giani, V. M. Moretti, V., F. Reniero and C. Guillou. 2007. The use of stable isotope ratio analyses to discriminate wild and farmed gilthead sea bream (*Sparus aurata*). *Rapid Communications in Mass Spectrometry* 21: 207–211.

Sahota, A. 2009. The global market for organic food & drink. *In:* H. Willer and L. Kilcher (eds.). *The world of organic agriculture: Statistics and emerging trends 2009* ISBN 978-3-940946-12-6.

Schmidt, H. -l., A. Roßmann, S. Voerkelius, W. H. Schnitzler, M. Georgi, J. Graßmann, G. Zimmermann and R. Winkler. 2005. Isotope characteristics of vegetables and wheat from conventional and organic production. *Isotopes in Environmental and Health Studies* 41: 223–228.

Schmidt, O., J. M. Quilter, B. Bahar, A. P. Moloney, C. M. Scrimgeour, I. S. Begley and F. J. Monahan. 2005. Inferring the origin and dietary history of beef from C, N and S stable isotope ratio analysis. *Food Chemistry* 91: 545–549.

Schröder, V. and C. G. De Leaniz. 2011. Discrimination between farmed and free-living invasive salmonids in Chilean Patagonia using stable isotope analysis. *Biological Invasions* 13: 203–213.

Sigman, D. M., K. L. Casciotti, M. Andreani, C. Barford, M. Galanter and J. K. Böhlke. 2001. A Bacterial method for the nitrogen isotopic analysis of nitrate in seawater and freshwater. *Analytical Chemistry* 73: 4145–4153.

Šturm, M., N. Kacjan-Maršić and S. Lojen. 2011. Can δ^{15}N in lettuce tissues reveal the use of synthetic nitrogen fertiliser in organic production? *Journal of the Science of Food and Agriculture* 91: 262–267.

Šturm, M. and S. Lojen. 2011. Nitrogen isotopic signature of vegetables from the Slovenian market and its suitability as an indicator of organic production. *Isotopes in Environmental and Health Studies* 47: 214–220.

Suzuki, Y., R. Nakashita, F. Akamatsu and T. Korenaga. 2009. Multiple stable isotope analyses for verifying geographical origin and agricultural practice of Japanese rice samples. *Bunseki Kagaku* 58: 1053–1058.

Thomas, F., E. Jamin, K. Wietzerbin, R. Guérin, M. Lees, E. Morvan and C. Guillou. 2008. Determination of origin of Atlantic salmon (*Salmo salar*): The use of multiprobe and multielement isotopic analyses in combination with fatty acid composition to assess wild or farmed origin. *Journal of Agricultural and Food Chemistry* 56: 989–997.

United States Department of Agriculture. 2012. Agricultural Marketing Service. 2010–2011 Pilot Study: Pesticide Residue Testing of Organic Produce. USDA National Organic Program, USDA Science and Technology Programs, November 2012.

US Code of Federal Regulations, 7CRF—Part 205. Agricultural Marketing Service, US Department of Agriculture. National Organic Program. December 21, 2000. Last amended May 6, 2015.

USDA. 2010. US Department of Agriculture 2010 Organic Assessment of China, National Organic Program, USDA, 1–10.

USDA. 2015. Unites States Department of Agriculture, National Organic Program International Agreements: www.ams.usda.gov/NOPInternationalAgreements. Source accessed in July, 2015.

Verenitch, S. and A. Mazumder. 2015. Isotopic characterization as a screening tool in authentication of organic produce commercially available in western North America. *Isotopes in Environmental and Health Studies* 51: 332–343.

Vitòria, L., N. Otero, A. Soler and À. Canals. 2004. Fertilizer characterization: Isotopic data (N, S, O, C, and Sr). *Environmental Science & Technology* 38: 3254–3262.

Vizzini, S., C. Tramati and A. Mazzola. 2010. Comparison of stable isotope composition and inorganic and organic contaminant levels in wild and farmed bluefin tuna, *Thunnus thynnus*, in the Mediterranean Sea. *Chemosphere* 78: 1236–1243.

Whay, H. R., D. C. J. Main, L. E. Green, G. Heaven, H. Howell, M. Morgan, A. Pearson and A. J. F. Webster. 2007. Assessment of the behaviour and welfare of laying hens on free-range units. *Veterinary Record* 161: 119–128.

Willer, H. and L. Kilcher. 2010. The world of organic agriculture—statistics and emerging trends 2010. Research Institute of Organic Agriculture (FiBL), Frick, and International Federation of Organic Agriculture Movements (IFOAM), Bonn.

Winter, C. K. and S. F. Davis. 2006. Organic foods. *Journal of Food Science* 71: 117–124.

Woese, K., D. Lange, C. Boess and K. W. Bögl. 1997. A comparison of organically and conventionally grown foods-results of a review of the relevant literature. *Journal of the Science of Food and Agriculture* 74: 281–293.

Yanagi, Y., H. Hirooka, K. Oishi, Y. Choumei, H. Hata, M. Arai, M. Kitagawa, T. Gotoh, S. Inada and H. Kumagai. 2012. Stable carbon and nitrogen isotope analysis as a tool for inferring beef cattle feeding systems in Japan. *Food Chemistry* 134: 502–506.

Yu, X., Y. Binjian and G. Zhifeng. 2014. Can willingness-to-pay values be manipulated? Evidence from an organic food experiment in China. *Agricultural Economics* 45: 119–127.

Yuan, Y., M. Zhao, Z. Zhang, T. Chen, G. Yang and Q. Wang. 2012. Effect of different fertilizers on nitrogen isotope composition and nitrate content of *Brassica campestris*. *Journal of Agricultural and Food Chemistry* 60: 1456–1460.

Yun, S. -I., S. -S. Lim, G. -S. Lee, S. -M. Lee, H. -Y. Kim, H. -M. Ro and W. -J. Choi. 2011. Natural ^{15}N abundance of paddy rice (*Oryza sativa* L.) grown with synthetic fertilizer, livestock manure compost, and hairy vetch. *Biology and Fertility of Soils* 47: 607–617.

Yun, S. I., H. M. Ro, W. J. Choi and S. X. Chang. 2006. Interactive effects of N fertilizer source and timing of fertilization leave specific N isotopic signatures in Chinese cabbage and soil. *Soil Biology and Biochemistry* 38: 1682–1689.

Zhou, W., C. Hu, J. Li, P. Christie and X. Ju. 2012. Natural ^{15}N abundance of tomato and soil amended with urea and compost. *Journal of Food, Agriculture and Environment* 10: 287–293.

CHAPTER 13

Odds and Ends, or,
All That's Left to Print

Lesley A. Chesson,[1,] Brett J. Tipple,[1] Suvankar Chakraborty,[2]*
Karyne M. Rogers[3] and James F. Carter[4]

13.1 Introduction

"I am not a glutton—I am an explorer of food."

Erma Bombeck, American humorist and author; 1927–1996

We close this volume on food forensics with all the foodstuffs not easily categorized elsewhere: bottled water, flavorings and spices, sweeteners, etc. Many of these products would not be considered part of the main nutritional food groups and include some materials that might be classed as *additives*, which we should use sparingly (EUFIC 2015), if at all. Nevertheless, many products discussed in this chapter are readily available in premium food markets and, like most foodstuffs discussed in earlier chapters, they become tempting targets of economically motivated fraud.

[1] IsoForensics, Inc., 421 Wakara Way, Suite 100, Salt Lake City, UT 84108, USA.
  Email: tippleb@isoforensics.com
[2] SIRFER Lab, Department of Biology, University of Utah, 257 S 1400 E, Salt Lake City, UT
  84112, USA.
  Email: Suvankar.Chakraborty@utah.edu
[3] Environment and Materials Division, National Isotope Center, GNS Science, TE PU AO, 30
  Gracefield Road, Seaview, PO Box 31 312, Lower Hutt 5040, NEW ZEALAND.
  Email: k.rogers@gns.cri.nz
[4] Forensic and Scientific Services, Health Support Queensland, 39 Kessels Road, Coopers
  Plains QLD 4108, AUSTRALIA.
  Email: Jim.Carter@health.qld.gov.au
* Corresponding author: lesley@isoforensics.com

13.2 Bottled water

In 2015 (for the first time) water surpassed *carbonated soft drinks* (i.e., cola, pop, or soda) in terms of volume sales of bottled beverages. This was not wholly unexpected as the consumption of bottled water has risen steadily over the past decades, with per capita consumption in the U.S. increasing from 11 gallons, or approximately 40 L, in 1994 almost threefold to 32 gallons, or 120 L, by 2014 (Landi 2015). Given this growth in the bottled water market, it is not surprising that several research groups have explored stable isotopes as a means to characterize an increasingly expensive product (Bowen et al. 2005; Redondo and Yélamos 2005; Brencic and Vreca 2006; Montgomery et al. 2006; Bong et al. 2009; Chesson et al. 2010a; Dotsika et al. 2010; Rangarajan and Ghosh 2011; Kim et al. 2012; Kim et al. 2013; Raco et al. 2013; Ribeiro et al. 2014; Peng et al. 2015).

The first study to address the isotopic signatures of bottled waters presented a global survey of 234 samples (Bowen et al. 2005), collected from all continents, except Antarctica. The authors found that the δ^2H and $\delta^{18}O$ values of most bottled waters fell on the Global Meteoric Water Line (GMWL): $\delta^2H = 8 \times \delta^{18}O + 10‰$ (Craig 1961) (see also Chapter 4). Exceptions were largely attributed to evaporative processes, which enriched ^2H and ^{18}O. Comparisons of the isotope ratios of bottled water with predictions of the water at the stated source could be used to verify label claims, as demonstrated using a group of seven samples from Argentina, which claimed the same bottling location. Contrary to the source information provided on the bottle labels, the isotope ratios of the seven samples were found to comprise two distinct groups (Bowen et al. 2005).

This initial isotopic survey of bottled waters was soon followed by additional studies. A survey of Slovenian bottled waters included sparkling (both artificially and naturally carbonated) and flavored waters along with still waters (Brencic and Vreca 2006). The relationships between the δ^2H and $\delta^{18}O$ values of these bottled waters were generally similar to the GMWL, despite the possibility of a slight isotopic shift caused by the addition of small amounts of flavorings, fruit extracts, and/or sweeteners. A survey of bottled still waters from Greece also found that H and O isotope ratios fell on the GMWL or local meteoric water lines (LMWL) (Dotsika et al. 2010). Researchers in India surveyed water bottled within the country and developed a model to predict the latitude of production based on measured δ^2H and $\delta^{18}O$ values (Rangarajan and Ghosh 2011).

Montgomery et al. (2006) used Sr isotopes to characterize bottled "mineral" waters from the British Isles. A larger survey of approximately 650 European mineral waters was published soon after and was used to generate a spatial map of $^{87}Sr/^{86}Sr$ ratios in waters (Voerkelius et al. 2010). Recently, researchers in Brazil measured the $^{87}Sr/^{86}Sr$ ratios of 9 bottled waters and found that the isotopic compositions were "dominantly controlled by both the nature and stratigraphic age of the aquifer host rocks" (Ribeiro et al. 2014).

In a survey of Korean bottled waters, H and O isotope ratios along with C isotope ratios of dissolved inorganic carbon (DIC) were measured (Bong et al. 2009). This survey included samples of water with a supposed marine source (i.e., desalinized seawater), which could be discriminated from still and sparkling waters based on measured δ^2H and $\delta^{18}O$ values. A subsequent survey of Korean bottled waters (Kim et al. 2012) confirmed that desalinized water samples had δ^2H and $\delta^{18}O$ values distinctive from those of still or sparkling waters—water with a marine source had H and O isotope ratios near 0‰ with respect to both elements.

The $\delta^{13}C$ values of DIC were found to be useful for discriminating artificial and natural sparking waters, with artificial sparkling waters having on average lower carbon isotope ratios (−26.9‰) than natural sparkling waters (−16.3‰) (Bong et al. 2009). This pattern confirmed earlier work (Redondo and Yélamos 2005) to develop a "control method" based on isotopes to characterize the CO_2 in sparkling waters as either natural or artificial. Those authors found that the $\delta^{13}C$ values of DIC from artificial sparkling waters were lower than −20‰ and observed "no natural carbonic waters with values lighter than −8‰." A later survey of Italian bottled waters—including natural sparkling waters—confirmed the observation that no $\delta^{13}C_{DIC}$ values were lower than −8‰ (Raco et al. 2013).

H and O isotopes have been used to assess the authenticity of Taiwanese bottled waters that claimed to be desalinized deep seawater (Peng et al. 2015). As previously observed for waters of marine source, samples of deep seawater origin had δ^2H and $\delta^{18}O$ values of approximately 0‰—in contrast to counterfeit samples, which had δ^2H and $\delta^{18}O$ values of approximately −50‰ and −8‰, respectively. Measurement of DIC could provide an additional screening technique useful for verifying marine origin of bottled water (Kim et al. 2012) because $\delta^{13}C_{DIC}$ values in desalinized seawater have been shown to be isotopically depleted compared to those from typical seawater, possibly due to the fractionation of carbon during a reverse osmosis (desalinization) processes. Later work found that the $^{87}Sr/^{86}Sr$ ratios of desalinized seawater were higher than those of typical seawater (Kim et al. 2013).

13.3 Carbonated soft drinks

Water is the major ingredient of carbonated soft drinks, and studies have investigated the isotopic composition of this drink component to gain an understanding of the inputs to modern human body water pools. A survey of beverages commonly consumed by residents of the USA—including bottled water, soda, and beer—found that the slopes of the regression lines relating δ^2H and $\delta^{18}O$ values of the different beverage waters were not significantly different from one another or from the GMWL (Chesson et al. 2010a). In that study, water was collected from the soda samples using a cryogenic distillation technique (West et al. 2006) and analyzed via IRMS (Chesson et al. 2010a). A further study demonstrated that it was possible to measure the H and O isotopic compositions of the soda water directly via isotope ratio infrared spectroscopy (IRIS), without the need for distillation prior to analysis (Chesson et al. 2010b).

Coca-Cola® water has been collected and measured for several years as part of an annual, two-week training course offered at the University of Utah in Salt Lake City, Utah, USA. Course attendees from across the USA and other countries bring bottles and cans of Coca-Cola® from their home locations. Figure 13.1 and Table 13.1 present the extracted water δ^2H and $\delta^{18}O$ values of Coca-Cola® from the 2003 course. Surprisingly, a least absolute deviations regression between δ^2H and $\delta^{18}O$ values of these samples ($y = 9.8x + 19$) is different to the GMWL and the authors of this chapter speculate the difference may be due to the addition of highly evaporated water to the cola in the form of sugar syrup.

13.4 Caffeine

The purine alkaloid caffeine is a stimulant to the central nervous system and one of the world's most widely consumed psychoactive substances, together

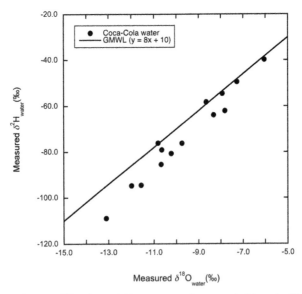

Figure 13.1 The δ^2H and $\delta^{18}O$ values of 14 water samples, extracted from bottles and cans of Coca-Cola® in 2003. The Coca-Cola® samples were purchased from around the globe. For reference, the Global Meteoric Water Line (GMWL) is shown. A least absolute deviations regression between water δ^2H and $\delta^{18}O$ values was significantly different from the GMWL.

Table 13.1 Measured hydrogen and oxygen isotope ratios of Coca-Cola® water.

City, State (or Country)	δ^2H_{water} (‰)	$\delta^{18}O_{water}$ (‰)
Los Angeles, CA	−94.4	−11.57
Los Angeles, CA	−80.6	−10.23
Boston, MA	−54.5	−7.93
Fairbanks, AK	−79.0	−10.65
Madison, WI	−62.0	−7.81
Flagstaff, AZ	−85.4	−10.68
Berkeley, CA	−76.2	−9.75
Berkeley, CA	−63.8	−8.32
Denver, CO	−108.9	−13.10
San Jose, CA	−94.6	−11.99
College Park, MD	−58.2	−8.65
Atlanta, GA	−39.8	−6.05
(France)	−49.4	−7.28
Bayreuth (Germany)	−76.1	−10.82

with alcohol and nicotine. Naturally present in beverages such as coffee or tea (see Chapter 7), caffeine is also found in many manufactured products, such as carbonated soft drinks and *energy drinks*—including caffeinated bottled waters—and even chocolate cakes. Not surprisingly, there has been significant interest in authenticating the source of caffeine used by the beverage industry. The first publication on the application of isotopes, specifically radioactive isotopes, to authenticate caffeine was authored by an employee of The Coca-Cola Export Corporation; by measuring ^{14}C, it was possible to distinguish synthetic caffeine made using "radiocarbon dead" petrochemicals from natural caffeine (Allen 1961).

The first publication of the stable isotopes of caffeine used the alkaloid drug as a proxy for morphine and measured C, H, and O isotope ratios to potentially to discriminate samples of different origin (Dunbar and Wilson 1982). Danho et al. (1992) measured ^2H of caffeine extracted from African and South American coffee beans via specific natural isotope fractionation-nuclear magnetic resonance (SNIF-NMR) and found that it was possible to distinguish the two sources. The researchers also found that the C and N isotope ratios of natural caffeine, extracted from coffee, differed from synthetic caffeine. Natural caffeine samples had $\delta^{13}C$ values between −27.7 and −25.0‰ while synthetic caffeine samples had values below −34‰; the $\delta^{15}N$ values of natural samples were all greater than 0‰ while values of synthetic caffeines were less than −20‰ (Danho et al. 1992). The significant difference in the C isotopic signatures of natural and synthetic caffeine was later confirmed using caffeine extracted from tea leaves, with natural caffeine characterized by $\delta^{13}C$ values greater than −32‰ (Weilacher et al. 1996). Similarly, natural caffeine extracted from guarana had $\delta^{13}C$ values greater than −30‰; differences between synthetic caffeine and caffeine extracted from guarana were also observed in O isotopic ratios, with synthetic samples characterized by positive $\delta^{18}O$ values and natural samples by negative $\delta^{18}O$ values (Weckerle et al. 2002a). Analysis of O isotopic ratios was found to be useful—in combination with C and H—to discriminate caffeine extracted from African and Central/South American coffee beans (Weckerle et al. 2002b). A survey of caffeine extracted from a variety of beverages (guarana, tea, coffee, mate) used ^{13}C and ^{18}O to identify products containing synthetic caffeine (Richling et al. 2003). Though no tea or mate products showed evidence of adulteration, the caffeine extracted from one coffee product and several guarana products had isotopic compositions more similar to synthetic caffeine reference materials than natural caffeine.

For isotope ratios of caffeine to become a regular adulteration-screening tool, rapid and effective analytical techniques are needed. To achieve this, researchers (Wu et al. 2012; Zhang et al. 2012) have described the application of compound-specific techniques to measure the $\delta^{13}C$ values of caffeine. Instead of measurements via EA-IRMS, the approach utilizes GC-C-IRMS (Wu et al. 2012) or high temperature reverse phase liquid chromatography (HT-RPLC-IRMS) (Zhang et al. 2012). The GC-C-IRMS method requires some sample preparation: extraction of caffeine from tea leaves with water, extraction of caffeine from the aqueous extract into chloroform, and then analysis of the chloroform extract (Wu et al. 2012). This method of extraction can be simplified by the use of Supported Liquid Extraction to isolate caffeine from brewed coffee (Carter et al. 2015) or other beverages.

13.5 Vanilla/vanillin

A common flavoring in bottled waters, carbonated soft drinks, and numerous food products, vanilla is the second-most expensive spice in the world, following saffron (Parthasarathy et al. 2008). Vanilla extract is derived from the beans of vanilla orchids, most commonly *Vanilla planifolia* or *V. tahitensis*. Due to its price, vanilla is a common target of fraud and several decades of research have been dedicated to the authentication of vanilla and *vanillin* (4-hydroxy-3-methoxybenzaldehyde), the major flavor component of vanilla bean extracts. The chemical, molecular, and physical properties of vanillin are the same regardless of its source, natural or synthetic (Toth 2012), but similar to work on caffeine, analysis of ^{14}C has proven useful to identify vanilla containing "fossil fuel derived carbon" [i.e., synthetic components (Culp and Noakes 2010; Culp and Prasad 2013)].

Early studies into the authentication of vanillin using stable carbon isotope ratios attempted to discriminate natural vanillin from synthetic (Bricout et al. 1974; Hoffman and Salb 1979; Krueger and Krueger 1983; Krueger and Krueger 1985). This proved possible because tissues of the vanilla orchid—a CAM plant—have $\delta^{13}C$ values of approximately −21‰, which are distinct from the C_3 plant materials and from petrochemicals used as the raw materials for synthetic vanillin. Recent publications of known-source samples suggest that $\delta^{13}C$ values greater than −22.6‰ characterize authentic (i.e., "natural") vanilla sources (Gassenmeier et al. 2013; Hansen et al. 2014; Bononi et al. 2015). However, the introduction of *biovanillin* to the food market may necessitate updates to this control limit. This material is produced through microbial action of "natural" precursors; thus, biovanillin is technically a natural product, but its $\delta^{13}C$ signature is reported to be as low as −36.0‰ (Bononi et al. 2015).

At least one early study of vanillin found that H isotope ratios were also useful for identifying synthetic flavoring compounds (Culp and Noakes 1992) and a further three studies have presented $\delta^{18}O$ values of vanillin (Hener et al. 1998; Bensaid et al. 2002; Werner 2003). More recent studies of vanilla authentication advocated the analysis of both $^2H/^1H$ and $^{13}C/^{12}C$ (Greule et al. 2010; Hansen et al. 2014). Hansen et al. (2014) even suggested that dual-isotope analysis of vanillin could be useful for investigating geographic source, although this has generated some debate (Greule et al. 2015a; Hansen et al. 2015).

Site-specific isotope analysis may be the most useful technique for vanilla authentication as it can detect the addition of isotopically labeled vanillin to meet a control limit (e.g., total $\delta^{13}C$ value > −22.6‰). Krueger and Krueger (1983) developed a method to measure the isotope ratio of the methyl carbon of vanillin and, two years later, the same authors published a method to measure the isotope ratio of the carbonyl carbon (Krueger and Krueger 1985). Comparison of these site-specific carbon isotope ratios with the total $\delta^{13}C$ value of vanillin could be used to identify adulterated extracts. Although useful, these site-specific isotope analyses required significant sample preparation and wet chemistry prior to isotope ratio analysis. Building on the work of Krueger and Krueger, the *Zeisel method* was applied to vanillin (Greule et al. 2010) for an easier (and faster) isolation of specific carbon atoms for isotope analysis. In this process, methoxyl groups are transformed to methyl iodide (CH_3I) by treatments with HI; methyl iodide samples can then be measured via GC-C-IRMS for C isotope ratios and via GC-TC-IRMS for H isotope ratios. SNIF-NMR techniques have also been developed to measure site-specific H isotopic compositions for vanillin authentication [e.g., (Remaud et al. 1997a; John and Jamin 2004)].

It is worth noting that most isotope investigations of vanilla authenticity have focused on the analysis of materials already available as pure compounds. While it may be far more difficult to extract flavorings from a food product than to analyze a pure compound, this scenario represents the reality faced by food forensic scientists tasked with verifying the claims on food labels. In 2009, Lamprecht and Blockberger developed a method to isolate vanillin from ice cream and yogurt for carbon isotope analysis. The procedure required protein precipitation, which was filtered; the vanillin and other components were then extracted from the filtrate. Vanillin was finally isolated by preparative HPLC for analysis via EA-IRMS (Lamprecht and Blochberger 2009). The method was tested by spiking a vanillin-free ice cream and then isolating and measuring the $\delta^{13}C$ signature of the added vanillin. More recently, Bononi et al. (2015) published a method for extracting vanillin from chocolate prior to analysis via LC-IRMS.

13.6 Essential oils

Like vanilla, *essential oils*—and the compounds they contain—are extensively used as food flavorings and are increasingly popular as alternative remedies for their perceived health benefits and curative properties. Concentrated oils extracted from plants that contain volatile aromatic compounds, essential oils are so named because they are seen to contain the "essence" of plant fragrances; common examples include lemon oil and peppermint oil. Isotope analyses have been applied to a variety of plant-derived oils and compounds contained within, mainly to discriminate natural essential oils from those of synthetic origin. In some cases, chemical differences linked to geographic origin were also investigated to verify label claims. Several publications have reviewed the analytical techniques used to authenticate essential oils (Mosandl and Juchelka 1997; Schmidt et al. 1998; Mosandl 2004; Tranchida et al. 2012; Do et al. 2015). Here, we summarize studies focused specifically are the measurement of isotope ratios of essential oils via IRMS in Table 13.2; this research is not discussed in detail in the text.

It is worth noting that several studies have combined IRMS analysis of specific oil compounds with SNIF-IRMS analysis of site-specific elements within those compounds. Hanneguelle et al. (1992) first described the "non-random distribution" of ^2H in linalool and linalyl acetate (major components of bergamot and lavender oils), with the distributions varying according to source origin—i.e., natural oils *vs.* synthetic. Remaud and colleagues (1997b) studied allyl isothiocyanate (responsible for the pungent taste of mustard) and also found distinct distributions of ^2H within natural and synthetic products. When combined with measurements of δ^2H, $\delta^{13}C$, and $\delta^{34}S$ values, principal component analysis (PCA, see Chapter 5) could separate synthetic allyl isothiocyanate from the natural product; PCA could further separate two geographic sources of the mustard oil used as the source of allyl isothiocyanate (Canada and India).

Table 13.2 Recent studies of essential oil by IRMS techniques.

Citation	Essential oil	Isotope (Analysis)
Butzenlechner et al. 1989	Bitter almond	H and C (EA-IRMS)
Hanneguelle et al. 1992	Lavender, bergamot, and coriander	H and C (DI-IRMS and EA-IRMS)
Braunsdorf et al. 1993	Lemon	C (GC-C-IRMS)
Faber et al. 1995	Peppermint	C (GC-C-IRMS)
Frank et al. 1995	Coriander	C (GC-C-IRMS)
Faulhaber et al. 1997a; Faulhaber et al. 1997b	Mandarin	C and N (GC-C-IRMS)
Remaud et al. 1997b	Mustard	C, N, and S (EA-IRMS)
Ruff et al. 2000	Bitter almond	H (GC-TC-IRMS*)
Bilke and Mosandl 2002	Lavender	H (GC-TC-IRMS)
Sewenig et al. 2003	Cinnamon	H and C (GC-TC-IRMS and GC-C-IRMS)
Jung et al. 2005	Lavender	H, C, and O (GC-TC-IRMS and GC-C-IRMS)
Nhu-Trang et al. 2006	Oregano, savory, and thyme	H (GC-TC-IRMS)
Schipilliti et al. 2010	Mandarin	C (GC-C-IRMS)
Schipilliti et al. 2011	Bergamot	C (GC-C-IRMS)
Bonaccorsi et al. 2012	Lime	C (GC-C-IRMS)
Dugo et al. 2012	Bergamot	C (GC-C-IRMS)
Schipilliti et al. 2013a	Citrus	C (GC-C-IRMS)
Schipilliti et al. 2013b	Mandarin and lemon	C (GC-C-IRMS)
Pellati et al. 2013	"Damask rose"	C (EA-IRMS and GC-C-IRMS)
Kumar et al. 2015	Lemongrass and palmarosa	C (EA-IRMS)
Schipilliti et al. 2016	*Helichrysum* ("everlasting flowers")	C (GC-C-IRMS)

* Sometimes (erroneously) referenced as GC-Py-IRMS in earlier works for "pyrolysis."

13.7 Sweeteners

13.7.1 Honey

Arguably the most publicized application of stable isotope analysis techniques to answer questions of food authenticity, honey was amongst the first foodstuffs to have an official isotope-based method accepted for its control—AOAC method 998.12 (White et al. 1998; Rossmann 2001). Honey is produced by honeybees (*Apis* spp.), with plant nectar as the major raw material. Most plant nectars are derived from C_3 plants (see Chapter 3), and so honeys can reasonably be expected to have $\delta^{13}C$ values of approximately −25‰ (Doner and White 1977), a signature that has been observed for honeys produced worldwide by various bee species [e.g., (Padovan et al. 2003; Pang et al. 2006; Kropf et al. 2010a; Kropf et al. 2010b; Schellenberg et al. 2010; Simsek et al. 2011). The extension of honey through the addition of cane sugar or corn (maize) syrup (i.e., high fructose corn syrup, HFCS)—both derived from C_4 plants and with $\delta^{13}C$ values about −10‰ (see Chapter 3)—would be expected to alter the bulk carbon isotopic composition of honey. This lead to an early recommendation that $\delta^{13}C$ values ≥ −21.5‰ were an indication of "unequivocal adulteration" of honey (White et al. 1998). However, this control limit would falsely categorize honeys produced from CAM plant nectars as adulterated; in addition, it was not possible to detect honeys extended with C_3 sweeteners, such as beet sugar or rice syrup (Tosun 2013; Guler et al. 2014).

In response to the need for a more sensitive test of honey extension with exogenous sweeteners, honey protein was proposed as an *internal standard* for adulteration (White and Winters 1989; White 1992; White et al. 1998). Based on previous work with orange juice (Bricout and Koziet 1987) (see Chapter 7), the technique was designed to measure the difference (Δ) in carbon isotope signatures of bulk honey (mostly sugars) and proteins extracted from the honey. A difference greater than 1‰ corresponded to >7% adulteration with a C_4 sweetener (White et al. 1998). To identify adulteration with C_3 sweeteners, LC-IRMS analysis of specific sugars in honey has been suggested (Cabañero et al. 2006; Elflein and Raezka 2008; Dong et al. 2016; Luo et al. 2016), using fructose, glucose, and/or sucrose as internal standards.

Recent work by Rogers et al. (2010, 2013, 2014a, b) found that applying the AOAC method to *mānuka honeys* (*Leptospermum scoparium*) from New Zealand resulted in an alarmingly high rate of detection (approximately 40%) of adulteration, although the tested samples were known to be authentic. Mānuka honey is well known for its antibacterial properties and hence commands a premium price; evidence of adulteration—false or not—would be economically detrimental to the mānuka honey market. One potential cause of the false positive adulteration test results was pollen simultaneously extracted from the honey with protein (Rogers et al. 2010). Mānuka pollen was found to be isotopically depleted relative to protein extracted from honey (Rogers et al. 2010); when present in sufficient quantities, pollen could thus increase the difference (Δ) between carbon isotope signatures of bulk honey and protein. In response, a modification of the AOAC method was proposed to include centrifugation and/or filtration to remove pollen (Rogers et al. 2013). However, the modified method "reduced the apparent C-4 sugar content not only of unadulterated honey but also of intentionally adulterated honeys (up to 15% C-4 sugar) to within acceptable limits" (Rogers et al. 2014a) and the modification was thus not recommended for routine honey screening (Frew et al. 2013; Rogers et al. 2014a).

More recently, mānuka honeys with methylglyoxal (MGO) levels above 250 mg/kg have been identified as those most likely to fail the adulteration test (Rogers et al. 2014b). MGO is an active compound found only in mānuka honey and imparts the antibacterial characteristic to the product (Adams et al. 2009). It is derived from the precursor molecule dihydroxyacetone (DHA), which is found only in the nectar of mānuka tree flowers. DHA to MGO conversion commences during the honey ripening process and is catalyzed by the reaction of naturally occurring proteins and amino acids found in nectar, pollen, and bee's digestive juices, which are mixed together during the transfer process from the tree flower to the hive by the honeybee. In honey with MGO levels exceeding approximately 250 mg/kg, the DHA to MGO reaction isotopically fractionates the bulk protein to provide more depleted carbon isotope values, consequently affecting the $\delta^{13}C_{protein}$ values measured by AOAC 998.12 method and resulting in false positive failures.

Another threat to the mānuka honey industry is mislabeling claims where non- mānuka honey is incorrectly labeled as mānuka honey, or where mānuka honey is extended with other dark honey such as bush honey, rewarewa, or kanuka honey to increase yields. To address this issue, the Ministry for Primary Industries in New Zealand is undertaking a study to investigate robust techniques to authenticate mānuka honey. Chemical profiling and DNA sequencing of pollen has been undertaken to assess the 'mānuka-ness' of honey, although issues arise where the bee may visit several different nectar-producing plants and naturally 'dilute' mānuka honey. One authentication technique growing in popularity is site-specific measurements of ^{1}H. SNIF-NMR fingerprints have been successfully used to identify adulterated honey (Spiteri et al. 2015), and more recently to identify unique compounds within mānuka nectar and honey (Spiteri et al. 2017). This technique is also effective for detecting DHA or MGO adulteration where synthetically manufactured active ingredients found in mānuka honey are fraudulently added to honey to boost its MGO content and hence its market price. Compound-specific isotope analyses (e.g., GC-IRMS and LC-IRMS) are less effective in detecting this active ingredient fraud as some synthetic DHA and MGO compounds have similar isotope ratios to naturally-derived products (Rogers et al. 2014b).

Honeybee colonies are the source of other high value products, including *royal jelly*, a substance secreted by bees to feed larvae and often marketed as a dietary supplement. Carbon isotope analysis has proven useful for screening royal jelly to identify samples from honeybee colonies that have been supplemented with cane sugars or corn syrups (Stocker et al. 2006; Daniele et al. 2011; Wytrychowski et al. 2012). In a similar manner, nitrogen isotope analysis of royal jelly may reveal instances of protein feeding—for example, the provision of yeast to honeybees (Stocker et al. 2006).

In addition to adulteration testing, origin testing of honey has become increasingly important to the market. Several studies have investigated the measurement of other isotopes of honey—including H, N, O, and S—to distinguish samples of different botanic and geographic origin (Kropf et al. 2010a; Kropf et al. 2010b; Schellenberg et al. 2010; Banerjee et al. 2015; Dinca et al. 2015). The H isotope ratios of honey protein were found to be "significantly correlated with the mean H isotopic ratios of precipitation and groundwater in the production regions" (Schellenberg et al. 2010). A year later, a study of honey with honeycomb found that "geographical variation in water δ^2H values are recorded by beeswax δ^2H values" (Chesson et al. 2011). Unfortunately, since most honey traded globally is filtered, and does not include solid beeswax, the usefulness of wax analyses for provenancing is limited. Fortunately, other honey components can potentially be used for origin testing, such as dissolved beeswax (Tipple et al. 2012), organic acids (Daniele et al. 2012), pollen (Chesson et al. 2013), and, as previously noted, protein (Schellenberg et al. 2010). Recently, elemental composition analysis (Banerjee et al. 2015; Baroni et al. 2015; Bontempo et al. 2015) and SNIF-NMR (Cotte et al. 2007; Dinca et al. 2015) has been applied to investigate the botanic and geographic originsof honey.

While the majority of honey studies have employed isotope analysis of bio-elements (e.g., H, C, N, O, and S), at least one study has measured the $^{87}Sr/^{86}Sr$ ratios of honeys from three different regions of Argentina (Baroni et al. 2015). The authors observed statistically significant differences between the strontium isotope ratios of honey from different regions. Here, we present the strontium isotope ratios ($^{87}Sr/^{86}Sr$) and elemental concentrations ([Sr]) of some authentic American honey samples collected at seven different apiaries along the east coast of the USA (Table 13.3).

We found that honeys collected from individual hives within an single apiary had relatively similar [Sr] and $^{87}Sr/^{86}Sr$ ratios. Further, we noted that honeys from apiaries in the same geographic regions may have similar $^{87}Sr/^{86}Sr$ ratios, but characteristic [Sr] (e.g., VA, ME-NH). The geochemical data from individual hives within a single apiary were consistent, suggesting that honeybees incorporate and integrate the geochemical signatures of the surrounding region. Together these data suggested that strontium concentrations and isotope ratios have potential to help determine region-of-origin of honeys.

Palynology—or the study of pollen—is still widely used to determine country- or region-of-origin of honey. Codex has prescribed levels of pollen for monofloral honey (CODEX STAN 12-1981, rev. 2001), and these are indicative of the plant availability at the origin of the honey. Pollen from indigenous species such as mānuka (New Zealand), jelly bush (Australian mānuka), *Arbutus* sp. or "strawberry tree" (Ireland, Mediterranean), "leatherwood" (*Eucryphia lucida*, Australia), and "longan" (*Dimocarpus longan*, Southern Asia) can provide a clear indicator of the honey's origin.

Table 13.3 Comparison of [Sr] and Sr isotope ratios from honeys from seven U.S. apiaries.

City	State	[Sr] (ppm)	Honey $^{87}Sr/^{86}Sr$	Average [Sr] (ppm)	[Sr] σ	Average $^{87}Sr/^{86}Sr$	$^{87}Sr/^{86}Sr$ σ	n
Newport	ME	0.051	0.71317	0.034	0.012	0.71356	0.00030	4
		0.023	0.71353					
		0.030	0.71388					
		0.032	0.71364					
Brownfield	ME	0.087	0.71411	0.065	0.030	0.71380	0.00044	2
		0.044	0.71349					
Andover	NH	0.115	0.71395	0.111	0.035	0.71361	0.00026	4
		0.151	0.71336					
		0.066	0.71348					
		0.113	0.71367					
White Plains	MD	0.092	0.71644	0.084	0.011	0.71671	0.00038	2
		0.076	0.71698					
Goodhope	WV	0.059	0.71565	0.057	0.003	0.71550	0.00021	2
		0.055	0.71535					
Arlington	VA	0.034	0.71276	0.035	0.001	0.71279	0.00005	2
		0.036	0.71283					
Stanley	VA	0.050	0.71291	0.043	0.010	0.71340	0.00069	2
		0.036	0.71389					

Nonetheless, the demand for cheap honey can entice unscrupulous packers to import and mix low price honey to extend higher value, indigenous honey. In some instances where it is advantageous not to declare the honey addition, ultrafiltration may be used to remove pollen and spores in order to disguise origin. The addition of ultrafiltered honey as an extension agent can be detected using ^{1}H-NMR fingerprinting or palynology, because of the very low pollen counts found in honey blended with filtered honey. Codex does not permit removal of pollen from honey through ultrafiltering (CODEX STAN 12-1981, rev. 2001).

13.7.2 Maple syrup

Maple syrup is made from the xylem sap of maple trees, most often the sugar maple (*Acer saccharum* Marsh). A recent study—utilizing ^{14}C measurements—found that the "sweet sap of sugar maple integrates sugars produced during several growing seasons" (Muhr et al. 2016). Like most trees, maples are C_3 plants and the $\delta^{13}C$ values of sugars from pure maple syrup should be less than −20‰, similar to the values expected for un-adulterated honey. Also akin to honey, the extension of maple syrup with C_4 sugars will affect the carbon isotopic composition and, not surprisingly, there is an official method for detecting the addition of exogenous sugars to maple syrup through the measurement of carbon isotope ratios (Carro et al. 1980; Morselli and Baggett 1984)—AOAC method 984.23. As something of a "wake-up-call," the change in $\delta^{13}C$ values of atmospheric CO_2 caused by fossil fuel combustion has been reflected in the $\delta^{13}C$ values of maple syrup collected over the past decades, with lower values observed as the atmospheric CO_2 concentration has increased (Peck and Tubman 2010). This may eventually necessitate updating the control limits set for adulteration detection of maple syrup.

Malic acid has been proposed as a suitable internal standard to detect the adulteration of maple syrups (Tremblay and Paquin 2007), analogous to the honey proteins used in AOAC method 998.12. SNIF-NMR analysis may be used alone or in conjunction with IRMS to test maple syrups for adulteration because "dueterium content at specific positions of the sugar molecules is different in maple syrup from that in beet or cane sugar" (Martin et al. 1996; Martin 2001). As a complement to SNIF-NMR analysis, one study has investigated the ^{13}C signature of ethanol from fermented maple syrup as an indicator of adulteration with cane sugar (Jamin et al. 2004). Using this technique, it was possible to calculate the minimum amount of added sugar to the original maple syrup sample. A recent publication has described the application of H, O, and C isotope ratio measurements—as well as a suite of element concentrations—for sweetener authentication (Banerjee et al. 2015). The authors found that "the addition of as little as 10% of corn syrup to pure maple syrup" was detectable using molar ratios of Na/(Na + K) and Na/

(Na + Ca). When adulteration of maple syrup with corn syrup was greater than 15%, C isotope ratio measurements were useful for authentication.

13.7.3 Other sweeteners

Glycerol ($C_3H_8O_3$) has many roles as a food ingredient or additive, including as a sweetener. Glycerol can be obtained from a variety of sources and specific dietary habits (e.g., vegan, vegetarian, etc.) may lead consumers to seek foods containing glycerol derived from plants rather than animals. Glycerol can also be added to premium drinks such as wine to improve mouth-feel, an adulteration that is illegal under most wine production regulations. Several studies have investigated the application of isotope analysis to identify glycerol source, measuring ^{13}C and/or ^{18}O (Weber et al. 1997; Zhang et al. 1998; Fronza et al. 1998; Fronza et al. 2001; Schmidt et al. 2001; Calderone et al. 2004). A method for measuring carbon isotope ratios of both glycerol and ethanol to detect adulteration has been published (Cabañero et al. 2010). Measurement of ^2H via SNIF-NMR has also been investigated as a method for glycerol authentication, with wine glycerol characterized by low deuterium content (Hermann 1999; Roßmann et al. 1998).

Further noteworthy publications on sweeteners include:

- A survey of sweeteners regularly consumed by Americans, with the goal of developing a method to quantify sugar intake by measuring the $\delta^{13}C$ values of an individual's blood (Jahren et al. 2006);
- a SNIF-NMR method to distinguish the addition of CAM *vs.* C_4 sugars to pineapple juice and tequila (Thomas et al. 2010);
- a European study, which investigated the adulteration of "sugar cane products with cheaper alternative (sugar beet)" using NMR techniques (Monakhova and Diehl 2016); and
- a survey of element concentrations combined with C, H, and O isotope ratios of several common sweeteners, including honey, maple syrup, corn syrup, and agave syrup (Banerjee et al. 2015).

13.8 Eggs

The earliest published IRMS data derived from chicken eggs focused on measurements of carbon and nitrogen isotopes of whole egg, egg white, and/or egg yolk to investigate diet [reviewed in (Rock 2012)]. This research mirrored work on flesh foods (see Chapter 6), in which early studies utilized ^{13}C and ^{15}N signatures of proteinaceous tissues to verify an animal's dietary regime. In a logical extension of this work, carbon and nitrogen isotopes were used to characterize eggs marketed in New Zealand as caged (i.e., battery), cage free, free range, or organic (Rogers 2009) (see also Chapter 12). Due to the significant price difference of these egg types, verifying the truth of label claims is very important. Several large scandals have occurred over the last 6 years in which battery farmed eggs have been labeled as free range or organic in New Zealand, Australia, Germany, and the U.K. These are only the ones that have been detected and it is suspected that many other countries have similar problems. Such mis-labeling can easily occur (intentionally or unintentionally) in factories that pack battery, free range, and organic eggs.

A follow-up study of eggs from both New Zealand and the Netherlands confirmed that a combination of C and N isotopes could discriminate organic *vs.* conventional farming regimes for chicken eggs (Rogers et al. 2015). Eggs from caged or barn-raised hens had lower $\delta^{15}N$ values consistent with their conventional feed, produced with synthetic fertilizers. In contrast, free-range hens were able to forage on pasture areas, which had been fertilized by chicken manure over time and thus had higher egg $\delta^{15}N$ values. Even though free-range hens may consume the same diet as caged or barn chickens, those hens could freely move outdoors to access pasture, affecting the nitrogen isotope ratios of eggs.

It is difficult to make global isotopic comparisons of eggs as many countries have different animal feed regulations, with some allowing fish meal or animal protein to boost the nutrient content of chicken feed. For example, in New Zealand up to 15% animal protein may be included in poultry feeds whereas some countries (i.e., those of the European Union) forbid this practice. The inclusion of animal or fish products as part of a hen's diet will affect the nitrogen isotopic signatures of eggs, making it harder to distinguish organic eggs from conventional eggs. In any geographical study, it is important to integrate information from feed manufacturers and poultry farmers and to monitor the situation over several years to identify changes in feed isotopic composition.

A survey of eggs produced under different farming regimes, stored for various times prior to analysis, and pasteurized or not demonstrated that the isotopic fingerprint of egg production was isotopically preserved (e.g., in the ^{13}C, ^{15}N, and ^{34}S of egg membrane, white, and yolk; and in the ^{13}C and ^{18}O of egg shell) throughout these processes (Rock et al. 2013).

13.9 Vinegar

Vinegar—a common cooking or pickling ingredient—is a dilute acid solution produced via acetous fermentation, which transforms alcohol into acetic acid. Vinegar is often produced from the fermentation of wine (see Chapter 8), although various other substrates can be used, including fruits, malt, or rice. Two fermentation processes are used to produce vinegar: a slow method and a fast method. In the slow method, acetic acid bacteria ferment ethanol over months or years, traditionally in wooden casks. In the fast method, oxygen is added to increase the rate of fermentation, which can require only days or weeks to produce vinegar; the process typically takes place in stainless steel tanks. The cheapest vinegars are typically made by rapid fermentation of synthetic (petroleum-derived) alcohol—products that are prepared as solutions of acetic acid (regardless of its source) must be described as "imitation vinegar." Unsurprisingly, traditional (slow fermented) vinegars command premium prices due to their complex sensory characteristics (García-Parrilla et al. 1997).

Studies using isotopes to authenticate vinegar have employed IRMS (Krueger 1992) and SNIF-NMR (Vallet et al. 1988; Remaud et al. 1992; Grégrová et al. 2012; Hsieh et al. 2013; Grégrová et al. 2014; Perini et al. 2014; Werner and Roßmann 2015). An early study demonstrated that the compound and site-specific isotope analysis methods developed for authenticating wine could also be successfully used to assess vinegars (Thomas and Jamin 2009). Work by Grégrová et al. (2012, 2014) used both SNIF-NMR and IRMS to investigate the authenticity of vinegar used for pickling cucumbers, finding that ^{2}H/^{1}H and δ^{13}C values of acetic acid could distinguish natural and synthetic samples.

Hattori et al. (2010) published a novel technique to measure the ^{2}H and ^{13}C signatures of acetic acid in vinegar by coupling headspace solid-phase microextraction (HS-SPME) with compound specific isotope ratio analysis via GC-TC-IRMS and GC-C-IRMS. The authors observed significant differences in the isotopic compositions of acetic acid of grain *vs.* rice vinegars (Hattori et al. 2010). Expanding on this technique, carbon isotope ratios were measured for the carboxyl and methyl carbons of acetic acid, with $\Delta^{13}C_{\text{carboxyl-methyl}}$ varying depending on the vinegar substrate—C_3, C_4, or CAM plant.

Thomas and Jamin (2009) first introduced the analysis of ^{18}O signatures as a tool for vinegar authentication, using these measurement to distinguish wine vinegars from vinegars made using reconstituted, dried grapes. More recently, Camin et al. (2013) surveyed a variety of authentic and suspect vinegars from Italy using oxygen isotope ratios. Setting a control limit of −2‰ for bulk raw vinegar (i.e., not diluted with water), 46 of 52 samples had lower δ^{18}O values, suggesting that these vinegars were mislabeled. Later studies have used H and C isotopes, in addition to O isotopes, to characterize balsamic vinegars for authentication (Perini et al. 2014; Werner and Roßmann 2015).

13.10 Other food products

Over the last 10 to 15 years, there have been a variety of novel applications of stable isotope analyses to food additives, nuts, seeds, spices, etc. These are summarized in Table 13.4, as it is difficult to draw broad conclusions from such diverse studies.

Table 13.4 Miscellaneous applications of isotope analysis in food forensics.

Citation	Product	Isotope(s)	Aim of the study
Spangenberg and Dionisi 2001	Cocoa butters, cocoa butter equivalents	C	Quantification of cocoa butter extension in chocolate
Diomande et al. 2015	Cocoa beans and their components	H, C, N, O	Botanic and geographic origin of fermented cocoa beans
Perini et al. 2016	Cocoa beans	H, C, N, O, S	Climate, fertilization regime, and precipitation (geographic origin)
Serra et al. 2005	Tartaric acid	C, O	Natural *vs.* synthetic origin in wine
Moreno Rojas et al. 2007	Tartaric acid	H, C, O	Natural *vs.* synthetic origin in wine
Anderson and Smith 2006	Pistachio nuts	C, N	Geographic origin
Zur et al. 2008	Pistachio nuts	^{1}H, ^{13}C (via NMR)	Geographic origin
Zhu et al. 2014	Peanuts	Sr	Geographic origin
Semiond et al. 1996	Saffron	C	Natural *vs.* synthetic origin
Maggi et al. 2011	Saffron	H, C, N	Geographic origin
Brunner et al. 2010	Paprika	Sr	Geographic origin
Horacek et al. 2010	Ginseng	H, C, N	Geographic origin (Korea *vs.* China)
Lee et al. 2011	Ginseng	Sr	Geographic origin (Korea *vs.* China)
Kim et al. 2015	Ginseng	H, C, N, O	"Regional origin" of plants from Korea
Chung et al. 2017	Ginseng	C, N, S	Cultivation soil and fertilization regime

13.11 Isotope effects during food preparation

Most foodstuffs are processed in some way (even raw vegetables are washed) and many foods are made from combinations of multiple ingredients, which may all be prepared in different ways. These preparation steps introduce the potential for isotopic changes (or exchanges) in the final product as materials are added or lost. In response, some food scientists have studied the effect of preparation on the isotopic composition of foods marketed to consumers.

The effect of baking has been investigated in Italian breads (Brescia et al. 2007), as well as sugar cookies and yeast buns (Bostic et al. 2015), using C and N isotope ratios. Brescia et al. (2007) found that $\delta^{15}N$ values of baked breads from three of the four Italian regions examined were significantly higher than those of the flours (and dough) used to make them but no significant changes in $\delta^{13}C$ values were observed. The authors suggested that the increase in $\delta^{15}N$ values from dough to bread might be due to the degradation of amino acids during dough fermentation, whereby lighter isotopes of N were lost. In contrast, Bostic et al. (2015) reported no significant differences in either the $\delta^{13}C$ or $\delta^{15}N$ values of cookies or buns pre- and post-baking—the isotopic composition of the final product could be predicted from the mass-balance of the isotopic compositions of the ingredients. A recent study on Italian pasta measured ^2H, ^{18}O, and ^{34}S in addition to ^{13}C and ^{15}N signatures to characterize the transition from wheat to flour, and finally to finished pasta (Bontempo et al. 2016). The authors observed essentially no isotopic changes through production, with mean differences between wheat, flour, and pasta of >1, 0.1, 0.1, 0.2 and 0.3‰ for H, C, N, O, and S stable isotope ratios, respectively.

Investigations of the isotopic compositions of water in food have studied the effects of brewing (Brettell et al. 2012), boiling (Brettell et al. 2012; Chesson et al. 2014), and stewing (Brettell et al. 2012), plus the loss of water during food storage (Greule et al. 2015b). Brettell et al. (2012) found that brewing increased the oxygen isotopic composition of source water by approximately 1.3‰ in finished beer products and simply boiling water to brew herbal tea (or "wort drink") increased $\delta^{18}O$ values by approximately 0.4‰. Very much in agreement with this study, a study of brewed teas and coffees from the U.S. found an average increase of approximately 0.5‰ between source water and beverage water (Chesson et al. 2014). The largest change in food water ^{18}O signatures was found to result from stewing in open pots. When water was simply evaporated over three hours, $\delta^{18}O$ values increased by up to 26.2‰ from start to finish and with meat added to the pot, the increase in water $\delta^{18}O$ values over the same period was 12.4‰ (Brettell et al. 2012). Cooking is not the only factor that can affect the isotopic composition of food as water $\delta^{18}O$ values in apples and carrots have been shown to increase due to loss during storage (Greule et al. 2015b).

In these studies, sample preparation is critical and there are a variety of methods available for water collection, including (as described earlier) cryogenic distillation (West et al. 2006), microwave extraction (Munksgaard et al. 2014), vacuum extraction (Koeniger et al. 2011), and even analysis of liquid solutions "as is" (Chesson et al. 2010b).

13.12 Conclusion

"Caveat emptor"—Let the buyer beware

In the global food market, cases of fraud and adulteration are increasing as it is no longer possible to look the farmer in the eye as you buy his or her product. While stricter and more rigorous screening now occurs for key foodstuffs (e.g., milk, meat, etc.), minor food products are particularly vulnerable to fraud and adulteration as they often do not warrant the intense scrutiny given to high production or high value foodstuffs. Country of origin labeling is still not mandatory in many countries; in other countries, it is only loosely applied. (In some cases, up to 49% of a product may be substituted or exchanged with overseas imports without affecting its origin labeling requirements.)

Look at food labels and see how often the product is described as: "Made from local and imported ingredients." In most cases the "local ingredient" will be water.

The quality, safety, and price of foodstuffs can often be highly variable. Although quality is not always directly correlated with price, it is a frequent indicator of such. Where large price discrepancies exist, products should be purchased with caution and earmarked for authentication.

References

Adams, C. J., M. Manley-Harris and P. C. Molan. 2009. The origin of methylglyoxal in New Zealand manuka (*Leptospermum scoparium*) honey. *Carbohydrate Research* 344: 1050–1053.

Allen, A. B. 1961. Differentiation of synthetic and natural caffeine. *Agricultural and Food Chemistry* 9: 294–295.

Anderson, K. A. and B. W. Smith. 2006. Effect of season and variety on the differentiation of geographic growing origin of pistachios by stable isotope profiling. *Journal of Agricultural and Food Chemistry* 54: 1747–1752.

Anon. 2001. *Codex Standard for Honey. CODEX STAN 12-1981.*

Banerjee, S., T. Kurtis Kyser, A. Vuletich and E. Leduc. 2015. Elemental and stable isotopic study of sweeteners and edible oils: Constraints on food authentication. *Journal of Food Composition and Analysis* 42: 98–116.

Baroni, M. V., N. S. Podio, R. Badini, C. M. Inga, H. A. Ostera, M. Cagnoni, E. Gautier, P. Peral-García, J. Hoogewerff and D. A. Wunderlin. 2015. Linking soil, water and honey composition to assess the geographical origin of Argentinean honey by multielemental and isotopic analyses. *Journal of Agricultural and Food Chemistry* 63: 4638–4645.

Bensaid, F. F., K. Wietzerbin and G. J. Martin. 2002. Authentication of natural vanilla flavorings: Isotopic characterization using degradation of vanillin into guaiacol. *Journal of Agricultural and Food Chemistry* 50: 6271–6275.

Bilke, S. and A. Mosandl. 2002. Authenticity assessment of lavender oils using GC-P-IRMS: ^{2}H/^{1}H isotope ratios of linalool and linalyl acetate. *European Food Research and Technology* 214: 532–535.

Bonaccorsi, I., D. Sciarrone, L. Schipilliti, P. Dugo, L. Mondello and G. Dugo. 2012. Multidimensional enantio gas chromatography/mass spectrometry and gas chromatography–combustion-isotopic ratio mass spectrometry for the authenticity assessment of lime essential oils (*C. aurantifolia* Swingle and *C. latifolia* Tanaka). *Journal of Chromatography A* 1226: 87–95.

Bong, Y. -S., J. -S. Ryu and K. -S. Lee. 2009. Characterizing the origins of bottled water on the South Korea market using chemical and isotopic compositions. *Analytica Chimica Acta* 631: 189–195.

Bononi, M., G. Quaglia and F. Tateo. 2015. Easy extraction method to evaluate δ^{13}C vanillin by liquid chromatography–isotopic ratio mass spectrometry in chocolate bars and chocolate snack foods. *Journal of Agricultural and Food Chemistry* 63: 4777–4781.

Bontempo, L., F. Camin, M. Paolini, C. Micheloni and K. H. Laursen. 2016. Multi-isotopic signatures of organic and conventional Italian pasta along the production chain. *Journal of Mass Spectrometry* 51: 675–683.

Bontempo, L., F. Camin, L. Ziller, M. Perini, G. Nicolini and R. Larcher. 2015. Isotopic and elemental composition of selected types of Italian honey. *Measurement* 98: 283–289.

Bostic, J. N., S. J. Palafox, M. E. Rottmueller and A. H. Jahren. 2017. Effect of baking and fermentation on the stable carbon and nitrogen isotope ratios of grain-based food. *Rapid Communications in Mass Spectrometry* 29: 937–947.

Bowen, G. J., D. A. Winter, H. J. Spero, R. A. Zierenberg, M. D. Reeder, T. E. Cerling and J. R. Ehleringer. 2005. Stable hydrogen and oxygen isotope ratios of bottled waters of the world. *Rapid Communications in Mass Spectrometry* 19: 3442–3450.

Braunsdorf, R., U. Hener, S. Stein and A. Mosandl. 1993. Comprehensive cGC-IRMS analysis in the authenticity control of flavours and essential oils. Part I: Lemon oil. *Zeitschrift Für Lebensmittel-Untersuchung Und -Forschung* 197: 137–141.

Brencic, M. and P. Vreca. 2006. Identification of sources and production processes of bottled waters by stable hydrogen and oxygen isotope ratios. *Rapid Communications in Mass Spectrometry* 20: 3205–3212.

Brescia, M., D. Sacco, A. Sgaramella, A. Pasqualone, R. Simeone, G. Peri and A. Sacco. 2007. Characterisation of different typical Italian breads by means of traditional, spectroscopic and image analyses. *Food Chemistry* 104: 429–438.

Brettell, R., J. Montgomery and J. Evans. 2012. Brewing and stewing: The effect of culturally mediated behaviour on the oxygen isotope composition of ingested fluids and the implications for human provenance studies. *Journal of Analytical Atomic Spectrometry* 27: 778–785.

Bricout, J., J. C. Fontes and L. Merlivat. 1974. Detection of synthetic vanillin in vanilla extracts by isotopic analysis. *Journal of the Association of Official Analytical Chemists* 57: 713–715.

Bricout, J. and J. Koziet. 1987. Control of the authenticity of orange juice by isotopic analysis. *Journal of Agricultural and Food Chemistry* 35: 758–760.

Brunner, M., R. Katona, Z. Stefánka and T. Prohaska. 2010. Determination of the geographical origin of processed spice using multielement and isotopic pattern on the example of Szegedi paprika. *European Food Research and Technology* 231: 623–634.

Butzenlechner, M., A. Rossmann and H. L. Schmidt. 1989. Assignment of bitter almond oil to natural and synthetic sources by stable isotope ratio analysis. *Journal of Agricultural and Food Chemistry* 37: 410–412.

Cabañero, A. I., J. L. Recio and M. Rupérez. 2006. Liquid chromatography coupled to isotope ratio mass spectrometry: A new perspective on honey adulteration detection. *Journal of Agricultural and Food Chemistry* 54: 9719–9727.

Cabañero, A. I., J. L. Recio and M. Rupérez. 2010. Simultaneous stable carbon isotopic analysis of wine glycerol and ethanol by liquid chromatography coupled to isotope ratio mass spectrometry. *Journal of Agricultural and Food Chemistry* 58: 722–728.

Calderone, G., N. Naulet, C. Guillou and F. Reniero. 2004. Characterization of European wine glycerol: Stable carbon isotope approach. *Journal of Agricultural and Food Chemistry* 52: 5902–5906.

Camin, F., L. Bontempo, M. Perini, A. Tonon, O. Breas, C. Guillou, J. M. Moreno-Rojas and G. Gagliano. 2013. Control of wine vinegar authenticity through $\delta^{18}O$ analysis. *Food Control* 29: 107–111.

Carro, O., G. Hillaire-Marcel and M. Gagon. 1980. Detection of adulterated maple products by stable carbon isotope ratios. *Journal of the Association of Official Analytical Chemists* 63: 840–844.

Carter, J. F., H. S. A. Yates and U. Tinggi. 2015. Isotopic and elemental composition of roasted coffee as a guide to authenticity and origin. *Journal of Agricultural and Food Chemistry* 63: 5771–5779.

Chesson, L. A., G. J. Bowen and J. R. Ehleringer. 2010b. Analysis of the hydrogen and oxygen stable isotope ratios of beverage waters without prior water extraction using isotope ratio infrared spectroscopy. *Rapid Communications in Mass Spectrometry* 24: 3205–3213.

Chesson, L. A., B. J. Tipple, B. R. Erkkila, T. E. Cerling and J. R. Ehleringer. 2011. B-HIVE: Beeswax hydrogen isotopes as validation of environment. Part I. Bulk honey and honeycomb stable isotope analysis. *Food Chemistry* 125: 576–581.

Chesson, L. A., B. J. Tipple, B. R. Erkkila and J. R. Ehleringer. 2013. Hydrogen and oxygen stable isotope analysis of pollen collected from honey. *Grana* 52: 305–315.

Chesson, L. A., B. J. Tipple, J. D. Howa, G. J. Bowen, J. E. Barnette, T. E. Cerling and J. R. Ehleringer. 2014. Stable isotopes in forensics applications. pp. 285–317. *In*: T. Cerling (ed.). *Treatise on Geochemistry, 2nd edition Volume 14: Archaeology and Anthropology*. Elsevier.

Chesson, L. A., L. O. Valenzuela, S. P. O'Grady, T. E. Cerling and J. R. Ehleringer. 2010a. Links between purchase location and the stable isotope ratios of bottled water, soda, and beer in the USA. *Journal of Agricultural and Food Chemistry* 58: 7311–7316.

Chung, I.-M., T.-J. Lee, Y.-T. Oh, B. K. Ghimire, I.-B. Jang and S.-H. Kim. 2017. Ginseng authenticity testing by measuring carbon, nitrogen, and sulfur stable isotope compositions that differ based on cultivation land and organic fertilizer type. *Journal of Ginseng Research.* 41: 195–200.

Cotte, J. F., H. Casabianca, J. Lhéritier, C. Perrucchietti, C. Sanglar, H. Waton and M. F. Grenier-Loustalot. 2007. Study and validity of ^{13}C stable carbon isotopic ratio analysis by mass spectrometry and ^2H site-specific natural isotopic fractionation by nuclear magnetic resonance isotopic measurements to characterize and control the authenticity of honey. *Analytica Chimica Acta* 582: 125–136.

Craig, H. 1961. Isotopic variations in meteoric waters. *Science* 133: 1702–1703.

Culp, R. A. and J. E. Noakes. 1992. Determination of synthetic components in flavors by deuterium/hydrogen isotopic ratios. *Journal of Agricultural and Food Chemistry* 40: 1892–1897.

Culp, R. A. and J. E. Noakes. 2010. Two decades of flavor analysis: Trends revealed by radiocarbon (^{14}C) and stable isotope ($\delta^{13}C$ and δD) analysis. pp. 9–27. *In*: C. -T. Ho, C. Mussinan, F. Shahidi and E. Tratras Contis (eds.). *Recent Advances in Food and Flavor Chemistry*. Cambridge: Royal Society of Chemistry.

Culp, R. A. and G. V. R. Prasad. 2013. Present-day radiocarbon content of select flavoring compounds reveals vanillin production pathway. *Radiocarbon* 55: 1819–1826.

Danho, D., N. Naulet and G. J. Martin. 1992. Deuterium, carbon, and nitrogen isotopic analysis of natural and synthetic caffeines. Authentication of coffees and coffee extracts. *Analysis* 20: 179–184.

Daniele, G., D. Maitre and H. Casabianca. 2012. Identification, quantification and carbon stable isotopes determinations of organic acids in monofloral honeys. A powerful tool for botanical and authenticity control. *Rapid Communications in Mass Spectrometry* 26: 1993–1998.

Daniele, G., M. Wytrychowski, M. Batteau, S. Guibert and H. Casabianca. 2011. Stable isotope ratio measurements of royal jelly samples for controlling production procedures: Impact of sugar feeding. *Rapid Communications in Mass Spectrometry* 25: 1929–1932.

Dinca, O. -R., R. E. Ionete, R. Popescu, D. Costinel and G.-L. Radu. 2015. Geographical and botanical origin discrimination of Romanian honey using complex stable isotope data and chemometrics. *Food Analytical Methods* 8: 401–412.

Diomande, D., I. Antheaume, M. Leroux, J. Lalande, S. Balayssac, G. S. Remaud and I. Tea. 2015. Multi-element, multi-compound isotope profiling as a means to distinguish the geographical and varietal origin of fermented cocoa (*Theobroma cacao* L.) beans. *Food Chemistry* 188: 576–582.

Doner, L. W. and J. W. White. 1977. Carbon-13/carbon-12 ratio is relatively uniform among honeys. *Science* 197: 891–892.

Dong, H., D. Luo, Y. Xian, H. Luo, X. Guo, C. Li and M. Zhao. 2016. Adulteration identification of commercial honey with the C-4 sugar content of negative values by an elemental analyzer and liquid chromatography coupled to isotope ratio mass spectroscopy. *Journal of Agricultural and Food Chemistry* 64: 3258–3265.

Do, T. K. T., F. Hadji-Minaglou, S. Antoniotti and X. Fernandez. 2015. Authenticity of essential oils. *TrAC Trends in Analytical Chemistry* 66: 146–157.

Dotsika, E., D. Poutoukis, B. Raco and D. Psomiadis. 2010. Stable isotope composition of Hellenic bottled waters. *Journal of Geochemical Exploration* 107: 299–304.

Dugo, G., I. Bonaccorsi, D. Sciarrone, L. Schipilliti, M. Russo, A. Cotroneo, P. Dugo, L. Mondello and V. Raymo. 2012. Characterization of cold-pressed and processed bergamot oils by using GC-FID, GC-MS, GC-C-IRMS, enantio-GC, MDGC, HPLC and HPLC-MS-IT-TOF. *Journal of Essential Oil Research* 24: 93–117.

Dunbar, J. and A. T. Wilson. 1982. Determination of geographic origin of caffeine by stable isotope analysis. *Analytical Chemistry* 54: 590–592.

Elflein, L. and K. -P. Raezka. 2008. Improved detection of honey adulteration by measuring differences between $^{13}C/^{12}C$ stable carbon isotope ratios of protein and sugar compounds with a combination of elemental analyzer—isotope ratio mass spectrometry and liquid chromatography—isotope ratio mass spectrometry ($\delta^{13}C$ EA/LC-IRMS). *Apidologie* 39: 574–587.

EUFIC. 2015. EUFIC REVIEW 10/2009: Food-based dietary guidelines in Europe. http://www.eufic.org/article/en/expid/food-based-dietary-guidelines-in-europe/.

Faber, B., B. Krause, A. Dietrich and A. Mosandl. 1995. Gas chromatography-isotope ratio mass spectrometry in the analysis of peppermint oil and its importance in the authenticity control. *Journal of Essential Oil Research* 7: 123–131.

Faulhaber, S., U. Hener and A. Mosandl. 1997a. GC/IRMS analysis of mandarin essential oils. 1. $\delta^{13}C_{PDB}$ and $\delta^{15}N_{AIR}$ values of methyl N-methylanthranilate. *Journal of Agricultural and Food Chemistry* 45: 2579–2583.

Faulhaber, S., U. Hener and A. Mosandl. 1997b. GC/IRMS analysis of mandarin essential oils. 2. $\delta^{13}C_{PDB}$ values of characteristic flavor components. *Journal of Agricultural and Food Chemistry* 45: 4719–4725.

Frank, C., A. Dietrich, U. Kremer and A. Mosandl. 1995. GC-IRMS in the authenticity control of the essential oil of *Coriandrum sativum* L. *Journal of Agricultural and Food Chemistry* 43: 1634–1637.

Frew, R., K. McComb, L. Croudis, D. Clark and R. Van Hale. 2013. Modified sugar adulteration test applied to New Zealand honey. *Food Chemistry* 141: 4127–4131.

Fronza, G., C. Fuganti, P. Grasselli, F. Reniero, C. Guillou, O. Breas, E. Sada, A. Rossmann and A. Hermann. 1998. Determination of the ^{13}C content of glycerol samples of different origin. *Journal of Agricultural and Food Chemistry* 46: 477–480.

Fronza, G., C. Fuganti, P. Grasselli, S. Serra, F. Reniero and C. Guillou. 2001. $\delta^{113}C$- and $\delta^{18}O$-values of glycerol of food fats. *Rapid Communications in Mass Spectrometry* 15: 763–766.

García-Parrilla, M. C., G. A. González, F. J. Heredia and A. M. Troncoso. 1997. Differentiation of wine vinegars based on phenolic composition. *Journal of Agricultural and Food Chemistry* 45: 3487–3492.

Gassenmeier, K., E. Binggeli, T. Kirsch and S. Otiv. 2013. Modulation of the ^{13}C/ ^{12}C ratio of vanillin from vanilla beans during curing. *Flavour and Fragrance Journal* 28: 25–29.

Grégrová, A., H. Čížková, J. Mazáč and M. Voldřich. 2012. Authenticity and quality of spirit vinegar: Methods for detection of synthetic acetic acid addition. *Journal of Food and Nutrition Research* 51: 123–131.

Grégrová, A., E. Neradová, V. Kružík, J. Mazáč, P. Havelec and H. Čížková. 2014. Determining adulteration of canned products using SNIF-NMR and IRMS: Detection of undeclared addition of synthetic acetic acid. *European Food Research and Technology* 239: 169–174.

Greule, M., A. Mosandl, J. T. G. Hamilton and F. Keppler. 2015a. Comment on Authenticity and traceability of vanilla flavors by analysis of stable isotopes of carbon and hydrogen. *Journal of Agricultural and Food Chemistry* 63: 5305–5306.

Greule, M., A. Rossmann, H. -L. Schmidt, A. Mosandl and F. Keppler. 2015b. A stable isotope approach to assessing water loss in fruits and vegetables during storage. *Journal of Agricultural and Food Chemistry* 63: 1974–1981.

Greule, M., L. D. Tumino, T. Kronewald, U. Hener, J. Schleucher, A. Mosandl and F. Keppler. 2010. Improved rapid authentication of vanillin using δ^{13}C and δ^{2}H values. *European Food Research and Technology* 231: 933–941.

Guler, A., H. Kocaokutgen, A. V. Garipoglu, H. Onder, D. Ekinci and S. Biyik. 2014. Detection of adulterated honey produced by honeybee (*Apis mellifera* L.) colonies fed with different levels of commercial industrial sugar (C$_3$ and C$_4$ plants) syrups by the carbon isotope ratio analysis. *Food Chemistry* 155: 155–160.

Hanneguelle, S., J. N. Thibault, N. Naulet and G. J. Martin. 1992. Authentication of essential oils containing linalool and linalyl acetate by isotopic methods. *Journal of Agricultural and Food Chemistry* 40: 81–87.

Hansen, A. -M. S., A. Fromberg and H. L. Frandsen. 2014. Authenticity and traceability of vanilla flavors by analysis of stable isotopes of carbon and hydrogen. *Journal of Agricultural and Food Chemistry* 62: 10326–10331.

Hansen, A. -M. S., A. Fromberg and H. L. Frandsen. 2015. Rebuttal to Comment on Authenticity and traceability of vanilla flavors by analysis of stable isotopes of carbon and hydrogen. *Journal of Agricultural and Food Chemistry* 63: 5307–5307.

Hattori, R., K. Yamada, H. Shibata, S. Hirano, O. Tajima and N. Yoshida. 2010. Measurement of the isotope ratio of acetic acid in vinegar by HS-SPME-GC-TC/C-IRMS. *Journal of Agricultural and Food Chemistry* 58: 7115–7118.

Hener, U., W. A. Brand, A. W. Hilkert, D. Juchelka, A. Mosandl and F. Podebrad. 1998. Simultaneous on-line analysis of ^{18}O/^{16}O and ^{13}C/^{12}C ratios of organic compounds using GC-pyrolysis-IRMS. *Zeitschrift Für Lebensmittel-Untersuchung Und -Forschung* 206: 230–232.

Hermann, A. 1999. Determination of site-specific D/H isotope ratios of glycerol from different sources by ^{2}H-NMR spectroscopy. *Zeitschrift Für Lebensmitteluntersuchung Und -Forschung A* 208: 194–197.

Hoffman, P. G. and M. Salb. 1979. Isolation and stable isotope ratio analysis of vanillin. *Journal of Agricultural and Food Chemistry* 27: 352–355.

Horacek, M., J. -S. Min, S. -C. Heo and G. Soja. 2010. Discrimination between ginseng from Korea and China by light stable isotope analysis. *Analytica Chimica Acta* 682: 77–81.

Hsieh, C. -W., P. -H. Li, J. -Y. Cheng and J. -T. Ma. 2013. Using SNIF-NMR method to identify the adulteration of molasses spirit vinegar by synthetic acetic acid in rice vinegar. *Industrial Crops and Products* 50: 904–908.

Jahren, A. H., C. D. Saudek, E. H. Yeung, W. H. L. Kao, R. A. Kraft and B. Caballero. 2006. An isotopic method for quantifying sweeteners derived from corn and sugar cane. *The American Journal of Clinical Nutrition* 84: 1380–1384.

Jamin, E., F. Martin and G. G. Martin. 2004. Determination of the ^{13}C/^{12}C ratio of ethanol derived from fruit juices and maple syrup by isotope ratio mass spectrometry: Collaborative study. *Journal of AOAC International* 87: 621–631.

John, T. V. and E. Jamin. 2004. Chemical investigation and authenticity of Indian vanilla beans. *Journal of Agricultural and Food Chemistry* 52: 7644–7650.

Jung, J., S. Sewenig, U. Hener and A. Mosandl. 2005. Comprehensive authenticity assessment of lavender oils using multielement/multicomponent isotope ratio mass spectrometry analysis and enantioselective multidimensional gas chromatography-mass spectrometry. *European Food Research and Technology* 220: 232–237.

Kim, G. -E., J. -S. Ryu, W. -J. Shin, Y. -S. Bong, K. -S. Lee and M. -S. Choi. 2012. Chemical and isotopic compositions of bottled waters sold in Korea: Chemical enrichment and isotopic fractionation by desalination. *Rapid Communications in Mass Spectrometry* 26: 25–31.

Kim, G. -E., W. -J. Shin, J. -S. Ryu, M. -S. Choi and K. -S. Lee. 2013. Identification of the origin and water type of various Korean bottled waters using strontium isotopes. *Journal of Geochemical Exploration* 132: 1–5.

Kim, K., J. -H. Song, S. -C. Heo, J. -H. Lee, I. -W. Jung and J. -S. Min. 2015. Discrimination of ginseng cultivation regions using light stable isotope analysis. *Forensic Science International* 255: 43–49.

Koeniger, P., J. D. Marshall, T. Link and A. Mulch. 2011. An inexpensive, fast, and reliable method for vacuum extraction of soil and plant water for stable isotope analyses by mass spectrometry. *Rapid Communications in Mass Spectrometry* 25: 3041–3048.

Kropf, U., T. Golob, M. Nečemer, P. Kump, M. Korošec, J. Bertoncelj and N. Ogrinc. 2010a. Carbon and nitrogen natural stable isotopes in Slovene honey: Adulteration and botanical and geographical aspects. *Journal of Agricultural and Food Chemistry* 58: 12794–12803.

Kropf, U., M. Korošec, J. Bertoncelj, N. Ogrinc, M. Nečemer, P. Kump and T. Golob. 2010b. Determination of the geographical origin of Slovenian black locust, lime and chestnut honey. *Food Chemistry* 121: 839–846.

Krueger, D. A. 1992. Stable carbon isotope ratio method for detection of corn-derived acetic acid in apple cider vinegar: Collaborative study. *Journal of AOAC International* 75: 725–728.

Krueger, D. A. and H. W. Krueger. 1983. Carbon isotopes in vanillin and the detection of falsified natural vanillin. *Journal of Agricultural and Food Chemistry* 31: 1265–1268.

Krueger, D. A. and H. W. Krueger. 1985. Detection of fraudulent vanillin labeled with ^{13}C in the carbonyl carbon. *Journal of Agricultural and Food Chemistry* 33: 323–325.

Kumar, A., A. Niranjan, A. Lehri, S. K. Tewari, D. V. Amla, S. K. Raj, R. Srivastava and S. V. Shukla. 2015. Isotopic ratio mass spectrometry study for differentiation between natural and adulterated essential oils of lemongrass (*Cymbopogon flexuosus*) and palmarosa (*Cymbopogon martinii*). *Journal of Essential Oil-Bearing Plants* 18: 368–373.

Lamprecht, G. and K. Blochberger. 2009. Protocol for isolation of vanillin from ice cream and yoghurt to confirm the vanilla beans origin by ^{13}C-EA-IRMS. *Food Chemistry* 114: 1130–1134.

Landi, H. 2015. Bottled water's rising tide. *Beverage World*.

Lee, A. -R., M. Gautam, J. Kim, W. -J. Shin, M. -S. Choi, Y. -S. Bong, G. -S. Hwang and K. -S. Lee. 2011. A multianalytical approach for determining the geographical origin of ginseng using strontium isotopes, multielements, and ^{1}H NMR analysis. *Journal of Agricultural and Food Chemistry* 59: 8560–8567.

Luo, D., H. Luo, H. Dong, Y. Xian, X. Guo and Y. Wu. 2016. Hydrogen ($^{2}H/^{1}H$) combined with carbon ($^{13}C/^{12}C$) isotope ratios analysis to determine the adulteration of commercial honey. *Food Analytical Methods* 9: 255–262.

Maggi, L., M. Carmona, S. D. Kelly, N. Marigheto and G. L. Alonso. 2011. Geographical origin differentiation of saffron spice (*Crocus sativus* L. stigmas)—Preliminary investigation using chemical and multi-element (H, C, N) stable isotope analysis. *Food Chemistry* 128: 543–548.

Martin, G. G., Y. -L. Martin, N. Naulet and H. J. D. McManus. 1996. Application of ^{2}H SNIF-NMR and ^{13}C SIRA-MS analyses to maple syrup: Detection of added sugars. *Journal of Agricultural and Food Chemistry* 44: 3206–3213.

Martin, Y. F. 2001. Detection of added beet or cane sugar in maple syrup by the SNIF-NMR method: Collaborative study. *Journal of the Association of Official Analytical Chemists* 84: 1511–1521.

Monakhova, Y. B. and B. W. Diehl. 2016. Authentication of the origin of sucrose-based sugar products using quantitative natural abundance ^{13}C NMR. *Journal of the Science of Food and Agriculture* 96: 2861–2866.

Montgomery, J., J. A. Evans and G. Wildman. 2006. ^{87}Sr/^{86}Sr isotope composition of bottled British mineral waters for environmental and forensic purposes. *Applied Geochemistry* 21: 1626–1634.

Moreno Rojas, J. M., S. Cosofret, F. Reniero, C. Guillou and F. Serra. 2007. Control of oenological products: Discrimination between different botanical sources of L-tartaric acid by isotope ratio mass spectrometry. *Rapid Communications in Mass Spectrometry* 21: 2447–2450.

Morselli, M. F. and L. K. Baggett. 1984. Mass spectrometric determination of cane sugar and corn syrup in maple syrup by use of ^{13}C/^{12}C ratio: Collaborative study. *Journal of the Association of Official Analytical Chemists* 67: 22–24.

Mosandl, A. 2004. Authenticity assessment: A permanent challenge in food flavor and essential oil analysis. *Journal of Chromatographic Science* 42: 440–449.

Mosandl, A. and D. Juchelka. 1997. Advances in the authenticity assessment of citrus oils. *Journal of Essential Oil Research* 9: 5–12.

Muhr, J., C. Messier, S. Delagrange, S. Trumbore, X. Xu and H. Hartmann. 2016. How fresh is maple syrup? Sugar maple trees mobilize carbon stored several years previously during early spring time sap-ascent. *New Phytologist* 209: 1410–1416.

Munksgaard, N. C., A. W. Cheesman, C. M. Wurster, L. A. Cernusak and M. I. Bird. 2014. Microwave extraction-isotope ratio infrared spectroscopy (ME-IRIS): A novel technique for rapid extraction and in-line analysis of δ^{18}O and δ^{2}H values of water in plants, soils and insects. *Rapid Communications in Mass Spectrometry* 28: 2151–2161.

Nhu-Trang, T. -T., H. Casabianca and M. -F. Grenier-Loustalot. 2006. Deuterium/hydrogen ratio analysis of thymol, carvacrol, γ-terpinene and p-cymene in thyme, savory and oregano essential oils by gas chromatography–pyrolysis–isotope ratio mass spectrometry. *Journal of Chromatography A* 1132: 219–227.

Padovan, G. J., D. De Jong, L. P. Rodrigues and J. S. Marchini. 2003. Detection of adulteration of commercial honey samples by the ^{13}C/^{12}C isotopic ratio. *Food Chemistry* 82: 633–636.

Pang, G. -F., C. -L. Fan, Y. -Z. Cao, J. -J. Zhang, X. -M. Li, Z. -Y. Li and G. -Q. Jia. 2006. Study on distribution pattern of stable carbon isotope ratio of Chinese honeys by isotope ratio mass spectrometry. *Journal of the Science of Food and Agriculture* 86: 315–319.

Parthasarathy, V. A., B. Chempakam and T. John Zachariah (eds.). 2008. *Chemistry of Spices*. Wallingford, UK; Cambridge, MA: CABI Pub.

Peck, W. H. and S. C. Tubman. 2010. Changing carbon isotope ratio of atmospheric carbon dioxide: Implications for food authentication. *Journal of Agricultural and Food Chemistry* 58: 2364–2367.

Pellati, F., G. Orlandini, K. A. van Leeuwen, G. Anesin, D. Bertelli, M. Paolini, S. Benvenuti and F. Camin. 2013. Gas chromatography combined with mass spectrometry, flame ionization detection and elemental analyzer/isotope ratio mass spectrometry for characterizing and detecting the authenticity of commercial essential oils of *Rosa damascena* Mill. *Rapid Communications in Mass Spectrometry* 27: 591–602.

Peng, T.-R., W.-J. Liang, T.-S. Liu, Y.-W. Lin and W.-J. Zhan. 2015. Assessing the authenticity of commercial deep-sea drinking water by chemical and isotopic approaches. *Isotopes in Environmental and Health Studies* 51: 322–331.

Perini, M., L. Bontempo, L. Ziller, A. Barbero, A. Caligiani and F. Camin. 2016. Stable isotope composition of cocoa beans of different geographical origin: Stable isotope composition of cocoa beans. *Journal of Mass Spectrometry* 51: 684–689.

Perini, M., M. Paolini, M. Simoni, L. Bontempo, U. Vrhovsek, M. Sacco, F. Thomas, E. Jamin, A. Hermann and F. Camin. 2014. Stable isotope ratio analysis for verifying the authenticity of balsamic and wine vinegar. *Journal of Agricultural and Food Chemistry* 62: 8197–8203.

Raco, B., E. Dotsika, A. Cerrina Feroni, R. Battaglini and D. Poutoukis. 2013. Stable isotope composition of Italian bottled waters. *Journal of Geochemical Exploration* 124: 203–211.

Rangarajan, R. and P. Ghosh. 2011. Tracing the source of bottled water using stable isotope techniques. *Rapid Communications in Mass Spectrometry* 25: 3323–3330.

Redondo, R. and J. G. Yélamos. 2005. Determination of CO_2 origin (natural or industrial) in sparkling bottled waters by ^{13}C/^{12}C isotope ratio analysis. *Food Chemistry* 92: 507–514.

Remaud, G., C. Guillou, C. Vallet and G. J. Martin. 1992. A coupled NMR and MS isotopic method for the authentication of natural vinegars. *Fresenius' Journal of Analytical Chemistry* 342: 457–461.

Remaud, G. S., Y. -L. Martin, G. G. Martin and G. J. Martin. 1997a. Detection of sophisticated adulterations of natural vanilla flavors and extracts: Application of the SNIF-NMR method to vanillin and p-hydroxybenzaldehyde. *Journal of Agricultural and Food Chemistry* 45: 859–866.

Remaud, G. S., Y. -L. Martin, G. G. Martin, N. Naulet and G. J. Martin. 1997b. Authentication of mustard oils by combined stable isotope analysis (SNIF-NMR and IRMS). *Journal of Agricultural and Food Chemistry* 45: 1844–1848.

Ribeiro, S., M. R. Azevedo, J. F. Santos, J. Medina and A. Costa. 2014. Sr isotopic signatures of Portuguese bottled mineral waters and their relationships with the geological setting. *Comunicações Geológicas* 100: 89–98.

Richling, E., C. Höhn, B. Weckerle, F. Heckel and P. Schreier. 2003. Authentication analysis of caffeine-containing foods via elemental analysis combustion/pyrolysis isotope ratio mass spectrometry (EA-C/P-IRMS). *European Food Research and Technology* 216: 544–548.

Rock, L. 2012. The use of stable isotope techniques in egg authentication schemes: A review. *Trends in Food Science & Technology* 28: 62–68.

Rock, L., S. Rowe, A. Czerwiec and H. Richmond. 2013. Isotopic analysis of eggs: Evaluating sample collection and preparation. *Food Chemistry* 136: 1551–1556.

Rogers, K. M. 2009. Stable isotopes as a tool to differentiate eggs laid by caged, barn, free range, and organic hens. *Journal of Agricultural and Food Chemistry* 57: 4236–4242.

Rogers, K. M., J. M. Cook, D. A. Krueger and K. Beckmann. 2013. Modification of AOAC *Official Method*[SM] 998.12 to add filtration and/or centrifugation: Interlaboratory comparison exercise. *Journal of AOAC International* 96: 607–614.

Rogers, K. M., S. van Ruth, M. Alewijn, A. Philips and P. Rogers. 2015. Verification of egg farming systems from the Netherlands and New Zealand using stable isotopes. *Journal of Agricultural and Food Chemistry* 63: 8372–8380.

Rogers, K. M., M. Sim, S. Stewart, A. Phillips, J. Cooper, C. Douance, R. Pyne and P. Rogers. 2014a. Investigating C-4 sugar contamination of Manuka honey and other New Zealand honey varieties using carbon isotopes. *Journal of Agricultural and Food Chemistry* 62: 2605–2614.

Rogers, K. M., M. Grainger and M. Manley-Harris. 2014b. The unique manuka effect: Why New Zealand manuka honey fails the AOAC 998.12 C-4 sugar method. *Journal of Agricultural and Food Chemistry* 62: 2615–2622.

Rogers, K. M., K. Somerton, P. Rogers and J. Cox. 2010. Eliminating false postive C_4 sugar tests on New Zealand Manuka honey. *Rapid Communications in Mass Spectrometry* 24: 2370–2374.

Rossmann, A. 2001. Determination of stable isotope ratios in food analysis. *Food Reviews International* 17: 347–381.

Roßmann, A., H. -L. Schmidt, A. Hermann and R. Ristow. 1998. Multielement stable isotope ratio analysis of glycerol to determine its origin in wine. *Zeitschrift Für Lebensmitteluntersuchung Und -Forschung A* 207: 237–243.

Ruff, C., K. Hör, B. Weckerle, P. Schreier and T. König. 2000. $^2H/^1H$ ratio analysis of flavor compounds by on-line gas chromatography pyrolysis isotope ratio mass spectrometry (HRGC-P-IRMS): Benzaldehyde. *Journal of High Resolution Chromatography* 23: 357–359.

Schellenberg, A., S. Chmielus, C. Schlicht, F. Camin, M. Perini, L. Bontempo, K. Heinrich et al. 2010. Multielement stable isotope ratios (H, C, N, S) of honey from different European regions. *Food Chemistry* 121: 770–777.

Schipilliti, L., I. Bonaccorsi, A. Cotroneo, P. Dugo and L. Mondello. 2013a. Evaluation of gas chromatography–combustion–isotope ratio mass spectrometry (GC-C-IRMS) for the quality assessment of citrus liqueurs. *Journal of Agricultural and Food Chemistry* 61: 1661–1670.

Schipilliti, L., I. L. Bonaccorsi, S. Ragusa, A. Cotroneo and P. Dugo. 2016. *Helichrysum italicum* (Roth) G. Don fil. subsp. *italicum* oil analysis by gas chromatography—carbon isotope ratio mass spectrometry (GC-C-IRMS): A rapid method of genotype differentiation? *Journal of Essential Oil Research* 28: 193–201.

Schipilliti, L., I. Bonaccorsi, D. Sciarrone, L. Dugo, L. Mondello and G. Dugo. 2013b. Determination of petitgrain oils landmark parameters by using gas chromatography–combustion–isotope ratio mass spectrometry and enantioselective multidimensional gas chromatography. *Analytical and Bioanalytical Chemistry* 405: 679–690.

Schipilliti, L., G. Dugo, L. Santi, P. Dugo and L. Mondello. 2011. Authentication of bergamot essential oil by gas chromatography-combustion-isotope ratio mass spectrometer (GC-C-IRMS). *Journal of Essential Oil Research* 23: 20–71.

Schipilliti, L., P. Q. Tranchida, D. Sciarrone, M. Russo, P. Dugo, G. Dugo and L. Mondello. 2010. Genuineness assessment of mandarin essential oils employing gas chromatography–combustion-isotope ratio MS (GC-C-IRMS). *Journal of Separation Science* 33: 617–625.

Schmidt, H., A. Rossmann and R. Werner. 1998. Stable isotope ratio analysis in quality control of flavourings. pp. 539–594. *In*: H. Ziegler and E. Ziegler (eds.). *Flavourings*, Weinheim, Germany: Wiley.

Schmidt, H., R. Werner and A. Rossmann. 2001. ^{18}O patterns and biosynthesis of natural plant products. *Phytochemistry* 58: 9–32.

Semiond, D., S. Dautraix, M. Desage, R. Majdalani, H. Casabianca and J. L. Brazier. 1996. Identification and isotopic analysis of safranal from supercritical fluid extraction and alcoholic extracts of saffron. *Analytical Letters* 29: 1027–1039.

Serra, F., F. Reniero, C. G. Guillou, J. M. Moreno, J. M. Marinas and F. Vanhaecke. 2005. ^{13}C and ^{18}O isotopic analysis to determine the origin of L-tartaric acid. *Rapid Communications in Mass Spectrometry* 19: 1227–1230.

Sewenig, S., U. Hener and A. Mosandl. 2003. Online determination of ^{2}H/^{1}H and ^{13}C/^{12}C isotope ratios of cinnamaldehyde from different sources using gas chromatography isotope ratio mass spectrometry. *European Food Research and Technology* 217: 444–448.

Simsek, A., M. Bilsel and A. C. Goren. 2011. ^{13}C/^{12}C pattern of honey from Turkey and determination of adulteration in commercially available honey samples using EA-IRMS. *Food Chemistry* 130: 1115–1121.

Spangenberg, J. E. and F. Dionisi. 2001. Characterization of cocoa butter and cocoa butter equivalents by bulk and molecular carbon isotope analyses: Implications for vegetable fat quantification in chocolate. *Journal of Agricultural and Food Chemistry* 49: 4271–4277.

Spiteri, M., E. Jamin, F. Thomas, A. Rebours, M. Lees, K. M. Rogers and D. N. Rutledge. 2015. Fast and global authenticity screening of honey using ^{1}H-NMR profiling. *Food Chemistry* 189: 60–66.

Spiteri, M., K. M. Rogers, E. Jamin, F. Thomas, S. Guyader, M. Lees and D. N. Rutledge. 2017. Combination of ^{1}H-NMR and chemometrics to discriminate manuka honey from other floral honey types from Oceania. *Food Chemistry* 217: 766–772.

Stocker, A., A. Rossmann, A. Kettrup and E. Bengsch. 2006. Detection of royal jelly adulteration using carbon and nitrogen stable isotope ratio analysis. *Rapid Communications in Mass Spectrometry* 20: 181–184.

Thomas, F. and E. Jamin. 2009. ^{2}H NMR and ^{13}C-IRMS analyses of acetic acid from vinegar, ^{18}O-IRMS analysis of water in vinegar: International collaborative study report. *Analytica Chimica Acta* 649: 98–105.

Thomas, F., C. Randet, A. Gilbert, V. Silvestre, E. Jamin, S. Akoka, G. Remaud, N. Segebarth and C. Guillou. 2010. Improved characterization of the botanical origin of sugar by carbon-13 SNIF-NMR applied to ethanol. *Journal of Agricultural and Food Chemistry* 58: 11580–11585.

Tipple, B. J., L. A. Chesson, B. R. Erkkila, T. E. Cerling and J. R. Ehleringer. 2012. B-HIVE: Beeswax hydrogen isotopes as validation of environment, part II. Compound-specific hydrogen isotope analysis. *Food Chemistry* 134: 494–501.

Tosun, M. 2013. Detection of adulteration in honey samples added various sugar syrups with ^{13}C/^{12}C isotope ratio analysis method. *Food Chemistry* 138: 1629–1632.

Toth, S. J. 2012. Comparison and integration of analytical methods for the characterization of vanilla chemistry. Ph.D., New Brunswick, New Jersey, USA: Rutgers, The State University of New Jersey, Graduate Program in Food Science.

Tranchida, P. Q., I. Bonaccorsi, P. Dugo, L. Mondello and G. Dugo. 2012. Analysis of citrus essential oils: State of the art and future perspectives. A review. *Flavour and Fragrance Journal* 27: 98–123.

Tremblay, P. and R. Paquin. 2007. Improved detection of sugar addition to maple syrup using malic acid as internal standard and ^{13}C isotope ratio mass spectrometry (IRMS). *Journal of Agricultural and Food Chemistry* 55: 197–2003.

Vallet, C., M. Arendt and G. J. Martin. 1988. Site specific isotope fractionation of hydrogen in the oxydation of ethanol into acetic acid. Application to vinegars. *Biotechnology Techniques* 2: 83–88.

Voerkelius, S., G. D. Lorenz, S. Rummel, C. R. Quétel, G. Heiss, M. Baxter, C. Brach-Papa et al. 2010. Strontium isotope signatures of natural mineral waters, the reference to a simple geological map and its potential for authentication of food. *Food Chemistry* 118: 933–940.

Weber, D., H. Kexel and H. -L. Schmidt. 1997. ^{13}C-pattern of natural glycerol: Origin and practical importance. *Journal of Agricultural and Food Chemistry* 45: 2042–2046.

Weckerle, B., E. Richling and P. Schreier. 2002a. Authenticity assessment of guarana products (*Paullinia cupana*) by caffeine isotope analysis. *Deutsche Lebensmittel-Rundschau* 98: 122–124.

Weckerle, B., E. Richling, S. Heinrich and P. Schreier. 2002b. Origin assessment of green coffee (*Coffea arabica*) by multi-element stable isotope analysis of caffeine. *Analytical and Bioanalytical Chemistry* 374: 886–890.

Weilacher, T., G. Gleixner and H. -L. Schmidt. 1996. Carbon isotope pattern in purine alkaloids: A key to isotope discriminations in C1 compounds. *Phytochemistry* 41: 1073–1077.

Werner, R. 2003. The online ^{18}O/^{16}O analysis: Development and application. *Isotopes in Environmental and Health Studies* 39: 85–104.

Werner, R. A. and A. Roßmann. 2015. Multi element (C, H, O) stable isotope analysis for the authentication of balsamic vinegars. *Isotopes in Environmental and Health Studies* 51: 58–67.

West, A. G., S. J. Patrickson and J. R. Ehleringer. 2006. Water extraction times for plant and soil materials used in stable isotope analysis. *Rapid Communications in Mass Spectrometry* 20: 1317–1321.

White, J. W. 1992. Internal standard stable carbon isotope ratio method for determination of C-4 plant sugars in honey: Collaborative study, and evaluation of improved protein preparation procedure. *Journal of AOAC International* 75: 543–548.

White, J. W. and K. Winters. 1989. Honey protein as internal standard for stable carbon isotope ratio detection of adulteration of honey. *Journal of AOAC International* 72: 907–911.

White, J. W., K. Winters, P. Martin and A. Rossmann. 1998. Stable carbon isotopic analysis of honey: Validation of internal standard procedure for worldwide application. *Journal of AOAC International* 81: 610–619.

Wu, C., K. Yamada, O. Sumikawa, A. Matsunaga, A. Gilbert and N. Yoshida. 2012. Development of a methodology using gas chromatography-combustion-isotope ratio mass spectrometry for the determination of the carbon isotope ratio of caffeine extracted from tea leaves (*Camellia sinensis*). *Rapid Communications in Mass Spectrometry* 26: 978–982.

Wytrychowski, M., G. Daniele and H. Casabianca. 2012. Combination of sugar analysis and stable isotope ratio mass spectrometry to detect the use of artificial sugars in royal jelly production. *Analytical and Bioanalytical Chemistry* 403: 1451–1456.

Zhang, B. -L., S. Buddrus, M. Trierweiler and G. J. Martin. 1998. Characterization of glycerol from different origns by ^{2}H- and ^{13}C-NMR studies of site-specific natural isotope fractionation. *Journal of Agricultural and Food Chemistry* 46: 1374–1380.

Zhang, L., D. M. Kujawinski, E. Federherr, T. C. Schmidt and M. A. Jochmann. 2012. Caffeine in your drink: Natural or synthetic? *Analytical Chemistry* 84: 2805–2810.

Zhu, Y., A. Hioki and K. Chiba. 2014. Measurement of strontium isotope ratio in nitric acid extract of peanut testa by ICP-Q-MS after removal of Rb by extraction with pure water. *Talanta* 119: 596–600.

Zur, K., A. Heier, K. W. Blaas and C. Fauhl-Hassek. 2008. Authenticity control of pistachios based on ^{1}H- and ^{13}C-NMR spectroscopy and multivariate statistics. *European Food Research and Technology* 227: 969–977.

Index

Printed and bound by CPI Group (UK) Ltd, Croydon, CR0 4YY

01/11/2024

01782624-0010